P9-EGL-370

General Chemistry for Engineers

Paul A. DiMilla

Northeastern University

Bassim Hamadeh, CEO and Publisher
Christopher Foster, General Vice President
Michael Simpson, Vice President of Acquisitions
Jessica Knott, Managing Editor
Kevin Fahey, Cognella Marketing Manager
Jess Busch, Senior Graphic Designer
John Remington, Acquisitions Editor
Jamie Giganti, Project Editor
Brian Fahey, Licensing Associate

First published in the United States of America in 2013 by Cognella, Inc.

Printed in the United States of America

ISBN: 978-1-60927-080-3 (pbk) / 978-1-62131-494-3 (hc)

Acknowledgements

This book is the product of having guided several thousand entering freshmen in the College of Engineering at Northeastern over the past decade in the study of general chemistry as it applies to their subsequent studies and careers. In addition to benefitting from the formal and informal input from fellow instructors, I have been particularly fortunate to have had the opportunity to see many of the students who took CHEM 1151 (or its predecessor, CHM U151), General Chemistry for Engineers, blossom in their subsequent studies in upper-level courses as engineers and move on to exciting careers. Special thanks should go to the undergraduate tutors who throughout the past few years have provided invaluable input and inspiration.

Preface

Overview

This book is designed to provide the entering college student interested in pursuing subsequent studies in one of the disciplines of engineering—aeronautical, chemical, civil, computer, electrical, environmental, industrial, and mechanical—with the background to understand the design and properties of materials at the microscopic and macroscopic levels, necessary for practical problem solving through the design and implementation of technological processes that improve lives as practiced by engineers. Our focus will be on developing, deepening, and broadening the student's understanding of connections between the underlying structure of matter, the energetics of chemical transformations, and observable physical properties and responses. Our primary goal will be to enable the student to practice skills in describing and predicting the properties of chemical systems and modern materials—such as alternative fuels, polymers, semiconductors, and batteries and fuel cells—using words, pictures, graphs, numbers, and equations. In particular, we seek to have the student achieve the following overall learning outcomes:

1) Appreciate how the quantization of energy at the atomic level explains the chemical properties of different elements.
2) Describe the composition of matter on atomic and molecular scales in terms of chemical formulas and structures.
3) Develop an understanding of the nature and types of chemical bonding and its consequences for molecular shape and electronic materials.
4) Predict key physical properties of gases, liquids, and solids and relate these properties to molecular shape and interactions between constituent atoms, ions, and molecules.
5) Identify the products of chemical reactions and relate the quantities of reactants consumed and products created using balanced chemical equations.
6) Determine the energy changes that occur in chemical reactions in terms of the thermodynamic properties of reactants and products, and apply these energy changes to determine whether a given reaction is spontaneous.
7) Apply, compare, and contrast kinetic and equilibrium descriptions for chemical reactions.
8) Identify the chemical reactions occurring in batteries and fuel cells and apply quantitative descriptions in terms of thermodynamics and stoichiometry to these systems.

An important underlying intention is to inspire an appreciation of the importance and relevance of chemistry in engineering and our everyday lives.

Structure

Although experience has indicated that most students using this textbook will have had some exposure to foundational concepts in general chemistry such as the periodic table, the mole, and the ideal gas law in their high school science courses, Chapter 1 provides a review of key topics central to the study of chemistry, including units, accuracy vs. precision and significant figures, and pressure and temperature.

With regard to units the focus is on recognizing and working with units. To facilitate these skills important physical constants and special conversion factors are provided on the inside of the front cover.

The first half of this book then focuses on an "atoms-first" approach to general chemistry, starting with the structure of the atom, the characterization of molecular-level phenomena in terms of quantum chemistry, electron configurations, and then moving on to chemical bonding and how to describe the structure, shape, and polarity of molecules and polyatomic ions. The description of chemical compounds is examined, and a discussion of nomenclature for organic and inorganic compounds is introduced. Based on understanding molecular polarity, the types of intermolecular forces experienced by compounds is examined as a prelude to describing their effects on the properties of pure substances in different phases, including the properties of non-ideal gases and the structure of crystal lattices. This half of the book concludes with an examination of the properties of mixtures, focusing on solutions and their properties.

The second half of this book focuses on reactions and their consequences, including thermochemistry, kinetics, chemical equilibrium, spontaneity and directionality, and electrochemistry. This material typically is less examined in high school, and some of the concepts introduced—particularly on thermodynamics—are difficult to master in any first exposure. A very well-respected thermodynamicist once told the author that one has to teach thermodynamics to understand this subject—and he was right, to the chagrin of many students. This section of the book also serves as preparation for subsequent study of these topics by many students.

Each chapter in this book starts with a set of specific learning outcomes that the student should expect to master by completion of the chapter. Because recent experience that many—if not most—students interested in majoring in an engineering discipline have been exposed to calculus in high school and are concurrently enrolled in additional study in calculus, some previous and concurrent exposure to uni-variable differential and integral calculus is assumed. The use of calculus in general chemistry can reinforce analytical concepts students have seen and are seeing in their math courses and also introduce students to the idea that calculus can be invaluable in describing an enormous variety of problems in science and engineering. The power of today's generations of calculators and computational software to solve nonlinear problems is recognized by encouraging students to use "solver" features to address problems whenever possible, including in particular for chemical equilibrium. To facilitate mastering this capability in the classroom, Appendix A provides a primer on using Solver with TI calculators. Lastly, given the prevalence and growing influence of electronic sources and fluency of students in mining these sources, this book deviates from other presentations of general chemistry in that a minimal amount of tabulated data are provided. The major exception to this exclusion is the inclusion of a table of standard reduction potentials in Appendix B.

Contents

Chapter 3 Chemical Compounds and Chemical Formulas

Chapter 4 Chemical Bonding and Structures of Molecules and Polyatomic Ions

Chapter 5 Intermolecular Forces and the Macroscopic Properties of Pure Substances

1:Fundamental Features of Matter

Learning Objectives

By the end of this chapter you should be able to:

1) Interconvert quantities expressed in different fundamental units, such as length, mass, time, and temperature and "derived" units, such as volume and energy, using the metric/international system (SI), scientific notation, and unit prefixes.
2) Distinguish between the accuracy and the precision of a set of measurements.
3) Identify the number of significant figures in a number and express the result of calculations with the correct number of significant figures.
4) Use density to convert between mass and volume in expressing the amount of a material.
5) Understand the mole concept and convert between the number of atoms, the number of moles, and the mass of an elemental substance.
6) Describe how pressure and temperature are defined and evaluated, and interconvert among different units for pressure and for temperature.
7) Us the ideal gas law to find the pressure, volume, temperature, or number of moles (n) for a pure gas given three of these properties.
8) Predict changes in pressure, volume, or temperature of a ga from an set of initial conditions to a set of different final conditions using the ideal gas law.

1.1 Fundamental and Derived Units for Chemistry

Being able to keep track and handle units is perhaps the single most important skill you can have as an Engineer! The metric/international system (SI) is the primary system for scientific and engineering measurements. The basic SI units used in chemistry are the meter (abbreviated m) for length, the kilogram (abbreviated kg) for mass, the second (abbreviated s) for time, the ampere (abbreviated A) for electric current, the mole (abbreviated mol) for amount of substance, and the kelvin (abbreviated K) for absolute temperature. The SI system is occasionally expressed as the mks (meter-kilogram-second) system (*i.e.*, lengths are expressed in units of m, masses in kg, and time in s).

Measurements in the metric/international system frequently involve scientific notation and unit prefixes that express the magnitude of a number. In scientific notation we express relatively small or large numbers in terms of powers of ten. Table 1.1 provides a set of prefixes for magnitude for units useful for chemistry and subsequent studies in engineering.

Table 1.1. Unit Prefixes for Magnitude in Metric System

Multiple	Prefix (Symbol)
10^{-15}	femto- (f)
10^{-12}	pico- (p)
10^{-9}	nano- (n)
10^{-6}	micro- (μ)
10^{-3}	milli- (m)
10^{-2}	centi- (c)
10^{-1}	deci- (d)
10^{3}	kilo- (k)
10^{6}	mega- (M)
10^{9}	giga- (G)

Example 1.1: Expressing Volume in SI Units

Volume, *V*, is an extensive property (*i.e.*, a property that depends on the amount of matter) describing the amount of space occupied by matter. What is the volume (in m^3) of a 2-L bottle of soda?

Solutions

Apply the factor-label method using the conversion factor 1 cm^3 = 1 mL, converting liters first to milliliters and then centimeters to meters (three times!), as illustrated in Figure 1.1. Note that it is easy to keep track of units by crossing out units as we proceed!

$$V = 2\,\text{L} \times \frac{10^3\ \text{mL}}{1\ \text{L}} = 2\,\text{L} \times \frac{10^3\ \text{mL}}{1\ \text{L}} \times \frac{1\ \text{cm}^3}{1\ \text{mL}} = 2\,\text{L} \times \frac{10^3\ \text{mL}}{1\ \text{L}} \times \frac{1\ \text{cm}^3}{1\ \text{mL}} \times \left(\frac{1\ \text{m}}{10^2\ \text{cm}}\right)^3$$

$$\boxed{V = 2 \times 10^{-3}\ \text{m}^3}$$

Performing calculations in mks units is usually both easier and preferred compared with equivalent calculations using other units; however, sometimes data are encountered or measurements made using non-SI units. For example, volume is commonly expressed in units of liters (L), where 1 L = 1 dm^3 = 10^{-3} m^3 = 10^3 cm^3. Conversions sometimes needed for solving problems in chemistry include English to metric units (*e.g.*, yard, foot, or inch to m, cm, or mm for length, lb_m to kg for mass, and gallon or quart to L [liter] for volume). Important common English-to-metric conversions are provided on the inside of the front cover.

The preferred method for converting between different sets of units—and, really, for any calculation involving units for any problem in any field!—is the factor-label method.

Scientific notation is preferred for expressing numbers smaller than 1/10 or larger than 1. Further, using scientific notation allows us to quickly compare the relative magnitude (*i.e.*, size) of different numbers. For example, although we could write 2×10^{-3} m^3 as 0.002 m^3, using scientific notation makes it clearer that 2×10^{-3} m^3 is less than 2×10^{-2} m^3 and greater than 2×10^{-4} m^3 simply by comparing powers of ten, rather than comparing 0.002 m^3 with 0.02 m^3 and 0.0002 m^3.

1.2 Accuracy Versus Precision and Significant Figures

Whenever we consider a measurement we distinguish between its accuracy and its precision. Accuracy is how close a measured value is to the "true" or accepted value. In contrast, precision is the degree of reproducibility of a measurement (*i.e.*, the amount of "scatter" when the same quantity is measured repeatedly). For example, sets of measurements that are more similar in value to each other are more precise but not necessarily more accurate, as illustrated in Figure 1.1.

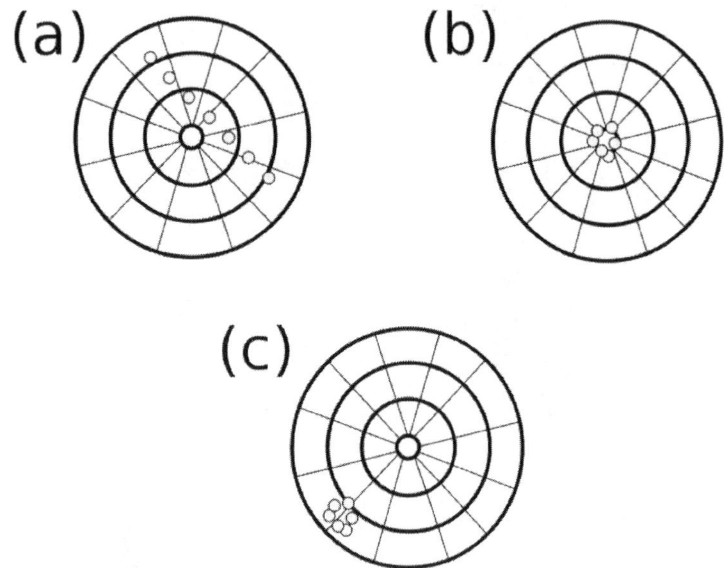

Figure 1.1. Comparison of accuracy vs. precision. (a) Neither accurate nor precise; (b) Both accurate and precise; (c) Inaccurate but precise.

The number of significant digits reported for a measurement reflects the precision of that measurement: the more precise the measurement, the greater number of significant digits. The following rules govern significant figures:

Rule 1: All nonzero digits in a number are significant. Leading zeros are not significant, and trailing zeros are only significant when a decimal point is present.

Rule 2: For addition or subtraction the result has the same number of significant figures beyond the decimal point as the quantity with the *smallest* number of such digits.

Rule 3: For multiplication or division the result contains only as many significant figures as the *least* precise quantity in the calculation (*i.e.*, as the quantity with the smallest number of significant figures).

Using scientific notation is the best strategy to correctly specify the number of appropriate number of significant figures in a measurement or calculation.

Example 1.2: Determining the Mass of a Formula Unit for a Rare Earth Magnet

Rare earth magnets are the strongest known types of permanent magnets and have important applications in modern electrodynamic devices, including wind turbine generators. One type of rare earth magnet consists of a formula unit—the smallest repeating unit in the substance (as discussed in Section 3.2)—of two atoms of neodymium (atomic symbol Nd), 14 atoms of iron (atomic symbol Fe), and one atom of boron (atomic symbol B). The mass of one atom for these elements is 144.2 amu for Nd, 55.847 amu for Fe, and 10.8 amu for B, where the atomic mass unit (amu, with1 amu = 1.6605×10^{-27} kg) can be a convenient unit for representing atomic- and molecular-scale masses.

a) Determine the number of significant figures for each mass.

b) Express each mass in scientific notation.

c) Determine the mass (in amu) for one formula unit of the magnet.

Solutions

a) The mass of an atom of Nd is expressed with four significant figures, the mass of an atom of Fe with five significant figures, and the mass of an atom of B with three significant figures.

b) The number of significant figures for each mass is seen most clearly by expressing these values in scientific notaton: $m_{Nd} = 1.442\times10^2$ amu, $m_{Fe} = 5.5847\times10^1$ amu, and $m_B = 1.08\times10^1$ amu.

c) Calculate the mass of each formula unit by writing the products of number of each atom and the mass per atom for each element on a common order-of-magnitude:

$$
\begin{aligned}
&(2 \text{ atoms Nd})\times(1.442\times10^2 \text{ amu/atom Nd}) = 2.884 \times 10^2 \text{ amu}\\
&(14 \text{ atoms Fe})\times(5.5847\times10^1 \text{ amu/atom Fe}) = 7.8186\times10^2 \text{ amu}\\
&(1 \text{ atom B})\times(1.08\times10^1 \text{ amu/atom B}) = \underline{0.108 \times 10^2 \text{ amu}}\\
&= 10.810\underline{6}\times10^2 \text{ amu} = 1.0810\times10^2 \text{ amu}
\end{aligned}
$$

where the underlined digit of 6 is not significant based on Rule 2 for significant figures. The answer should have five significant figures.

In summarizing data we often consider three types of overall metrics to describe sets of measurements. For example, consider a set of measurements of mass of a set of N standard 1 mm-thick, 2 cm-diameter pure silicon wafers used to manufacture semiconductor chips for computers, with mass of wafer i denoted as m_i. The average (also known as mean) mass, $\bar{m}$, is the sum of the masses for individual wafers divided by the number of wafers:

$$
\bar{m} \equiv \frac{m_1 + m_2 + \ldots + m_N}{N} = \frac{\sum_{i=1}^{i=N} m_i}{N} \quad [1.1]
$$

The average represents the expected value for a large number of samples. The standard deviation (represented by the Greek symbol σ) and is determined as:

$$\sigma \equiv \sqrt{\frac{\sum_{i=1}^{i=N}(m_i - \bar{m})^2}{N-1}} \quad [1.2]$$

The standard deviation provides a representation of the precision of the data. Lastly, the percent error (also known as percentage error) provides a representation of how close the average (expected) value is to the true or actual value, m_{true}, how precise the data are, and is determined as:

$$\% - error \equiv \left| \frac{\bar{m} - m_{true}}{m_{true}} \right| \times 100\,\% \quad [1.3]$$

The percent error provides an estimate for the accuracy of the data.

Example 1.3: Evaluating the Variation in Thickness Across a Silicon Wafer

Schmitz *et al.* (*Proc. of Am. Soc. Prec., Eng. Winter Topical Mtg.*, 98, 2003) reported the following set of optical measurements for the thickness of a 200 mm-diameter, nominally 750 μm-thick silicon wafer: 754.4 μm, 754.6 μm, 754.7 μm, 754.4 μm, and 754.7 μm.

a) What is the average thickness (in μm) for this wafer?

b) What is the standard deviation in thickness (in μm) for this wafer?

c) What is the percent error between the measured thickness of this wafer and the nominal thickness?

Solutions

a) Determine the average thickness, $\bar{h}$, of the $N = 5$ measurements using Equation [1.1] as:

$$\bar{h} = \frac{\sum_{i=1}^{i=N} h_i}{N} = \frac{h_1 + h_2 + h_3 + h_4 + h_5}{5}$$

$$= \frac{(754.4\ \mu m) + (754.6\ \mu m) + (754.7\ \mu m) + (754.6\ \mu m) + (754.7\ \mu m)}{5}$$

$$\bar{h} = 754.6\ \mu m$$

b) Determine the standard deviation in thickness, σ_h, using Equation [1.2] as:

$$\sigma_h = \sqrt{\frac{\sum_{i=1}^{i=N}(h_i - \bar{h})^2}{N-1}} = \sqrt{\frac{(h_1-\bar{h})^2}{4}+\frac{(h_2-\bar{h})^2}{4}+\frac{(h_3-\bar{h})^2}{4}+\frac{(h_4-\bar{h})^2}{4}+\frac{(h_5-\bar{h})^2}{4}}$$

$$= \frac{1}{2}\left[\begin{array}{l}[(754.4\ \mu m)-(754.6\ \mu m)]^2+[(754.6\ \mu m)-(754.6\ \mu m)]^2 \\ +[(754.7\ \mu m)-(754.6\ \mu m)]^2+[(754.6\ \mu m)-(754.6\ \mu m)]^2 \\ +[(754.7\ \mu m)-(754.6\ \mu m)]^2\end{array}\right]$$

$$\sigma_h = 0.1\ \mu m$$

where we have applied the rules for significant figures to specify the precision of our result.

c) Based on our results from parts a and b the measured thickness of the wafer is 754.6±0.1 μm. Using Equation [1.3] to compare with the nominal value h_{true} = 750 μm we find:

$$\%-error = \left|\frac{\bar{h}-h_{true}}{h_{true}}\right| \times 100\,\% = \left|\frac{(754.6\ \mu m)-(750\ \mu m)}{(750\ \mu m)}\right| \times 100\,\% = 0.6\,\%$$

1.3 Mass, Volume, and Density

Density is an intensive property (*i.e.*, a property that does not depend on the amount of matter) describing the ratio of the mass of a substance or object to its volume:

$$\rho \equiv \frac{m}{V} \qquad [1.4]$$

Density is a characteristic physical property of substances and sometimes can be used as part of a scheme to identify a substance. Given two out of three of the set $\{\rho, m, V\}$, the unknown third property can be determined algebraically. Whenever possible it is greatly preferred that you perform the symbol algebra before substituting numbers and units. Although you may find it more difficult initially to implement this strategy, this approach is guaranteed to make your studies and work easier as you proceed through an engineering curriculum and your careers!

Example 1.4: Expressing Volume in SI Units

The density of polyethylene terephthalate (abbreviated PET), the plastic used for many food and beverage containers, has a density of 1.4 g/cm^3. What is the volume (in cm^3) occupied by 20.0 g of PET?

Solutions

Rearrange the mathematical definition for density, Equation [1.4], using symbolic algebra to solve explicitly for the volume in terms of the given mass and density:

$$V = \frac{m}{d} = \frac{(30.0\ \text{g})}{(1.4\ \text{g/cm}^3)} = 2.1 \times 10^2\ \text{cm}^3$$

Our answer should have two significant figures based on the rules for multiplication.

1.4 The Mole Concept

In chemistry a special unit of great utility is the mole (abbreviated mol), which describes the amount of a substance in terms of the number of atoms or molecules in the substance: 1 mol of atoms, molecules, or anything is equal to 6.022×10^{23} of these objects. We summarize this conversion of units as Avogadro's number, $N_A = 6.022 \times 10^{23}\ \text{mol}^{-1}$. The symbol n is used to denote the number of moles of a substance.

The usefulness of the unit of moles arises because the molar mass of an element—the mass in grams for 1 mol of the element (denoted with the symbol M)—has a value equal to the value of the atomic mass given underneath the symbol for each element in the periodic table (provided on the inside of the front cover; the general structure of the periodic table is introduced in Section 2.2). For example, the molar mass of oxygen (atomic symbol O) is 15.9994 g/mol. Frequently, we find it useful to truncate values for molar masses to two significant digits beyond the decimal point (*i.e.*, $M_O = 16.00$ g/mol. Note that the value for the molar mass for an element also represents the mass of an atom of that element in atomic mass units (abbreviated amu), where 1 amu = 1.661×10^{-27} kg. The molar mass, however, is a more useful representation for most calculations encountered by engineers in which we care about macroscopic properties. Given the periodic table and Avogadro's number, it is more straightforward to use the factor-label method to relate numbers of atoms, moles, and mass for an element.

Example 1.5: Relating Number of Atoms, Number of Moles, and Mass for Helium

Helium (He) is an inert gas lighter than air used for advertising blimps. What is the mass (in g) of helium corresponding to 3.65×10^{25} atoms He?

Solutions

First convert the number of atoms to number of moles using Avogadro's number:

$$3.65 \times 10^{25}\ \text{atoms He} \times \frac{1\ \text{mol He}}{6.022 \times 10^{23}\ \text{atoms He}} = 6.06 \times 10^1\ \text{mol He}$$

Now convert the number of moles to the corresponding mass using a molar mass of 4.00 g/mol for He:

$$6.06 \times 10^1\ \text{mol He} \times \frac{4.00\ \text{g He}}{1\ \text{mol He}} = 2.42 \times 10^2\ \text{g He}$$

1.5 Pressure and Temperature

Pressure and temperature are two key intensive properties describing the state of a sample of matter. Pressure (P) is defined as the force exerted isotropically (*i.e.*, equally in all directions) per unit area by matter. Pressure arises at the microscopic level from the collision of particles of matter with physical boundaries. Historically, pressure has been measured with barometers and manometers based on heights of columns of relatively dense fluids, such as mercury and water. The force exerted by these fluids per unit area is equivalent to pressure. More practically for engineers, pressure is measured with analog and digital gauges based on the mechanical flexing of metals or changes in the thermal properties of a detector. For example, electronic pressure meters frequently use a Bourdon gauge in which a metallic tube changes curvature in response to changes in pressure. These gauges can be calibrated using barometers.

Values for pressure are expressed in a variety of units. The SI unit for pressure is the pascal (Pa), defined as 1 Pa ≡ 1 N/m^2 (equivalent to 1 N-m/m^3 = 1 J/m^3, where 1 joule (J) = 1 N-m). The pascal is a very small unit for pressure relative to our everyday experiences, such that we commonly represent pressure in other units, including bars (1 bar = 10^5 Pa), atmosphere (atm, based on an average atmospheric pressure at sea level of 1 atm = 1.013×10^5 Pa), pounds per square inch (psi, with 14.7 psi = 1 atm), and torr (often used in work with vacuums, with 760 torr = 1 atm). Pressure also sometimes is expressed in units of millimeters mercury (mm Hg) and feet of water (ft H_2O); these units reflect the pressure that would be exerted underneath columns of these fluids, with 1 mm Hg = 1 torr. Conversion factors relating different units for pressure are provided on the inside of the front cover.

Example 1.6 Expressing Tire Pressure in SI Units

A typical tire pressure for optimal performance of an automobile's tire is 32 psi. What is this pressure in Pa?

Solutions

Convert the pressure using the conversion factors 1 atm = 14.7 psi = 1.013×10^5 Pa:

$$32\text{ psi}\times\frac{1.013\times10^5\text{ Pa}}{14.7\text{ psi}}=2.2\times10^5\text{ Pa}$$

Temperature (T) is a property reflecting the "degree of hotness" and amount of energy stored in matter. Temperature is related at the microscopic level to the motion of particles comprising the material. For noble gases at relatively low pressures temperature is proportional to the average kinetic energy of gas particles, as we will discuss in Section 5. In chemistry values for temperature are expressed most commonly in SI units of degrees Celsius (°C)—0 °C is the temperature at which pure water freezes and 100 °C is the temperature at which pure water boils at standard pressure—and Kelvin (K, not "degrees Kelvin). The Kelvin scale represents absolute temperature and is related to the Celsius scale as:

$$T(\text{K}) = T(^\circ\text{C}) + 273.15 \qquad [1.5]$$

We define as a useful reference a standard temperature $T^\circ \equiv 0.00$ °C = 273.15 K and a standard pressure $P^\circ \equiv 10^5$ Pa = 1 bar, such that standard temperature and pressure (abbreviated "STP") is defined as 0.00 °C and 1 bar. Another reference temperature is "room temperature," commonly defined as 25.00 °C.

Example 17: Expressing Room Temperature in K

What is room temperature in K?

Solutions

Convert °C to K as 25.00 °C + 273.15 = 298.15 K. Note that the number of significant figures for temperature changes when we convert from degrees Celsius to Kelvin based on the rules for addition from Section 1.2!

1.6 The Ideal Gas Law

For gases equations of state provide relationships between pressure, volume, temperature, and the number of moles of the gas. We focus in this section on pure gases, deferring discussion of mixtures of gases to Section 6.1. Under conditions of relatively high temperature and relatively low pressure, three simple gas laws were identified experimentally during the seventeenth, eighteenth, and nineteenth centuries, respectively, and are subsequently named after their discoverers:

1. Boyle's Law (1662): At constant temperature and number of moles, pressure is inversely proportional to volume (*i.e.*, $P \alpha V^{-1}$, where the symbol "α" means proportional; see Figure 1.2a).

2. Charles' Law (1787): At constant pressure and number of moles, volume is proportional to temperature (*i.e.*, $V \alpha T$). As illustrated in Figure 1.2b, extrapolating this linear relationship, one finds that $V = 0$ corresponds to a temperature of -273.15 °C (*i.e.*, 0 K).

3. Avogaro's Law (1811): At constant pressure and temperature, volume is proportional to the number of moles (*i.e.*, $V \alpha n$; see Figure 1.2c).

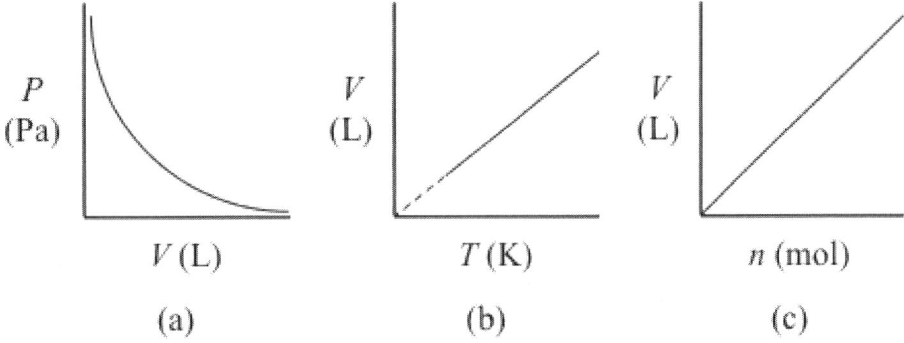

Figure 1.2 Observations leading to simple gas laws. (a) Boyle's observations; (b) Charles' observations; (c) Avogadro's observations.

Boyle's law tells us that the product $P \times V$ is a constant, Charles' law that the ratio V/T is a second constant, and Avogaro's Law that the ratio V/n is a third constant. Because all three laws are valid simultaneously, we find that the ratio $\frac{PV}{nT}$ is a constant that we define as the gas constant, $R = 8.315$ J/mol-K = 8.206×10^{-2} atm-L/mol-K. Note that the gas constant can be expressed in a variety of units. We choose the units based on the application, with the gas constant in units of atm-L/mol-K most useful when the pressure is expressed in units of atm and the volume in L. Alternatively, if the pressure is expressed in units of Pa and volume in m^3, it is more convenient to use the gas constant in units of J/mol-K. A value of $R = 8.314$ J/mol-K is perhaps the most general form for the gas constant, as this value offers more insight from the perspective of changes in energy associated with reactions.

The relationship $\frac{PV}{nT} = R$ is termed the generalized gas equation and is useful for solving some practical problems. However, we encounter this simple equation of state most frequently in the form of the ideal gas law:

$$PV = nRT \qquad [1.6]$$

Note that both the generalized gas equation and the ideal gas law require that temperature be expressed as absolute temperature (*i.e.*, in units of K, not °C).

The relationship 1 mol = 22.4 L at STP is often used for gases. What is the origin of this relationship? Consider n = 1.00 mol of an ideal gas at $P = 10^5$ Pa and T = 73.15 K. Rearrange the ideal gas law, converting pressure from Pa to atm and expressing the gas constant in units of atm-L/mol-K:

$$V = \frac{nRT}{P} = \frac{(1.00\text{ mol})(8.206\times10^{-2}\text{ atm}-\text{L/mol}-\text{K})(273.15\text{ K})}{\left(10^5\text{ Pa}\times\frac{1.00\text{ atm}}{1.013\times10^5\text{ Pa}}\right)} = 22.4\text{ L} \qquad [1.7]$$

Many practical problems can be addressed using the ideal gas law and its formulation as the generalized gas equation. For example, consider a pure gas in a container, with an initial state (denoted with the subscript *i*) of n_i moles at a pressure P_i, volume V_i, and temperature T_i and transformed to a final state (denoted with the subscript *f*) of n_f moles at a pressure P_f, volume V_f, and temperature T_f:

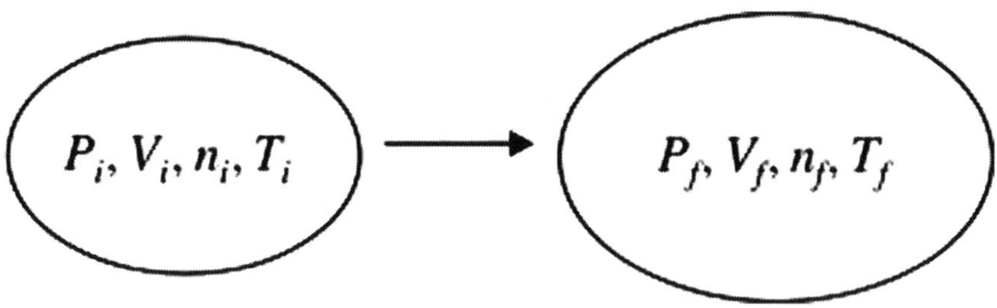

Applying the generalized gas equation we find:

$$\frac{P_iV^i}{n_iT^i} = R = \frac{P_fV^f}{n_fT^f} \qquad [1.8]$$

Note that if we are given information for the values for seven of the eight properties in the set $\{P_i, P_f, V_i, V_f, T_i, T_f, n_i, n_f\}$, we can determine the unknown eighth property by algebraic manipulation.

Example 1.8: Tire Temperature After Driving

A car tire is inflated at 22 °C to a pressure of 1.80 atm. After the car has been driven for several hours, the volume of the tire has expanded from 7.20 L to 7.80 L, and the pressure has increased to 1.90 atm. What is the tire's final temperature (in °C)?

Solutions

First, identify and group the given information, also identifying unspecified properties:

$P_i = 1.8$ atm	$P_f = 1.9$ atm
$V_i = 7.2$ atm	$V_f = 7.8$ atm
$T_i = 22.$ °C = 295. K	$T_f = ?$
$n_i = ?$	$n_f = ?$

Assume that the tire has no leaks (reasonable in the absence of further information), such that $n_i = n_f$, and apply the generalized gas equation to find:

$$\frac{P_iV_i}{T^i} = \frac{P_fV_f}{T^f}$$

Rearrange to solve for the final temperature as:

$$T_f = \frac{P_fV_fT_i}{P_iV_i} = \frac{(1.90 \text{ atm})(7.80 \text{ L})(295, \text{ K})}{(1.80 \text{ atm})(7.20 \text{ L})} = 337.\text{ K} = 64.\,^\circ\text{C}$$

Note that we did not need to use a value for the gas constant here!

2: Atomic Structure and Electron Configurations

Learning Objectives

By the end of this chapter you should be able to:

1) Outline the experiments by which the basic structure of the atom was discovered.
2) Apply the definitions of atomic number, mass number, and net charge with the periodic table of elements to find the numbers of neutrons, protons, and electrons in an isotope.
3) Compare the properties of light and electrons using the concepts of wave-particle duality, the Heisenberg uncertainty principle, and quantized energy.
4) Explain how atomic excitation and emission spectra result from the quantization of electron energies.
5) Describe solutions to the Schrödinger equation for the hydrogen atom in terms of electron orbitals and quantum numbers.
6) Assign electron configurations to atoms using the Pauli exclusion principle, the aufbau process, and Hund's rule.

2.1 Atomic Structure Based on the Rutherford Model

By the start of the nineteenth century it was recognized that the universe consisted of matter, with the amount of matter characterized by its mass and the amount of space the matter occupied by its volume, and by the end of the nineteenth century the following fundamental description was accepted:

- Based on the law of conservation of matter, the total mass of the universe is constant.
- Elements are the simplest substances of matter.
- Based on Dalton's atomic Theory, matter is composed of microscopic particles known as atoms that are indivisible and cannot be converted into atoms of another element, with all atoms for a given element identical in properties such as mass but different from atoms of all other elements.

However, around 1900, however, three important sets of experiments produced results that revolutionized this picture. In 1897 J. J. Thomson established that atoms were not the simplest particles of matter by identifying the existence of electrons (e^-), negatively-charged subatomic particles generated as cathode rays from metals subjected to high voltages or heated in a discharge tube containing gas at very low pressure. Thomson found that these particles traveled in straight lines that could be bent by electric and magnetic fields and identified their ratio of mass-to-charge using an apparatus depicted in Figure 2.1. An animation of this apparatus is available at the website http://www.aip.org/history/electron/jjappara.htm.

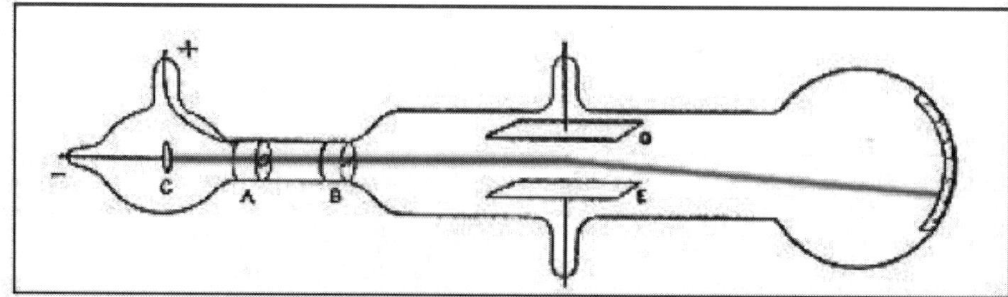

Figure 2.1. Schematic of Thomson's apparatus. Rays from a negatively-charged cathode (C) pass first through a slit in a positively-charged anode (A) and then through a slit in a grounded metal plug (B). The deflection of the rays due to an electrical voltage between a pair of aluminum plates (D and E) is measured at the end of the tube. From J. J. Thomson, "Cathode Rays," The London, Edinburgh, and Dublin Philosophical Magazine and Journal of Science, Fifth Series, October 1897, p. 296.

In 1909 Robert Millikan established that the charge of an individual electron is $q_{e^-} = -1.609 \times 10^{-19}$C, where the coulomb, C, is the SI unit for charge. In an experiment now known as the "Millikan oil drop experiment" (depicted in Figure 2.2), the motion of a series of ionized oil drops in response to an applied electric field was measured. Millikan found that each drop had a charge which was an integer multiple of q_{e^-}. Combined with the results of Thomson's experiment, Millikan's results led to the conclusion that the rest mass of an electron is $m_{e^-} = 9.11\times10^{-31}$ kg.

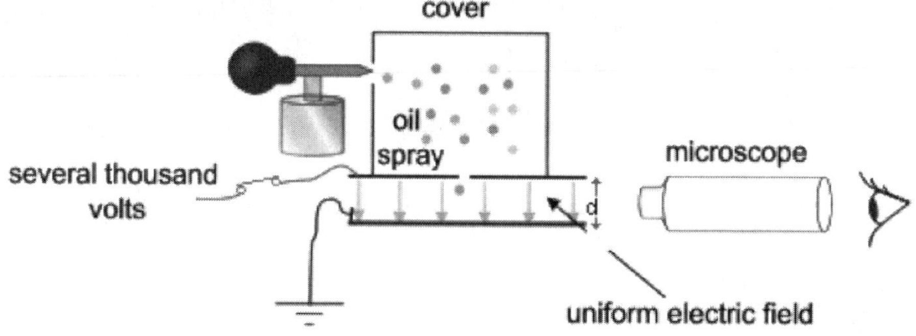

Figure 2.2. Schematic of the Millikan oil drop experiment

In 1910 Ernest Rutherford established that the majority of the mass of an atom and all of its positive charge is localized at a dense center, termed the nucleus, occupying a tiny fraction of the atom's volume. Rutherford reached these conclusions by examining the scatter of a beam of positively-charged alpha particles by gold foil. The existing "plum pudding" model of the atom, with a diffuse distribution of the mass and positive charger, predicted that all alpha particles would pass through the foil undisturbed, but Rutherford observed a subset of the incident particles were deflected (Figure 2.3). Rutherford established in subsequent experiments that the proton is the fundamental subatomic particle with a positive charge equal in magnitude but opposite in sign to the charge of an electron, and predicted the existence of electrically-neutral subatomic particles, termed neutrons, also localized to the nucleus and with masses similar to protons.

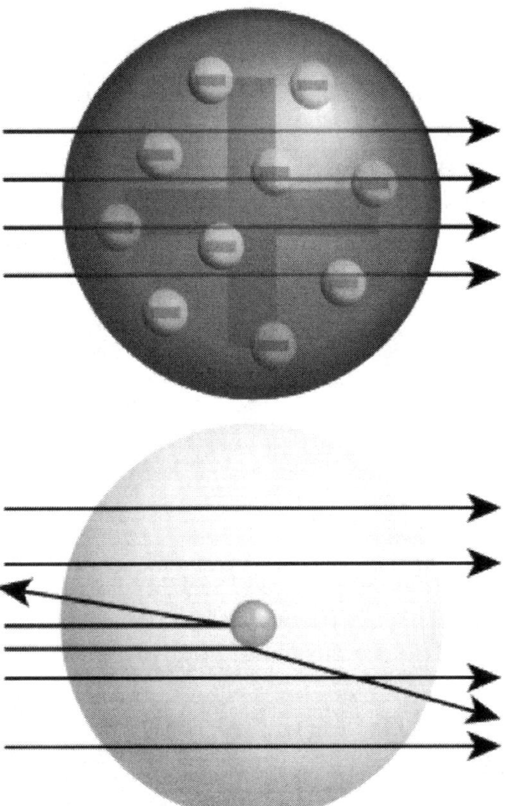

Figure 2.3. Rutherford's gold foil experiment. Top: Expected outcome that alpha particles pass through atom undisturbed based on "plum pudding" model. Bottom: Observed outcome that some alpha particles were deflected, including some back-scattered, leading to Rutherford's conclusion that the positive charge of the atom must be concentrated into a very small location.

The experiments of Thomson, Millikan, and Rutherford led to our currently-accepted model for the nuclear atom in which atoms consist of protons (electric charge +1), neutrons (charge 0), and electrons (charge −1), with electrons occupying most of the atom's volume but accounting for a tiny fraction of its mass. The number of protons in a nucleus of an atom is defined as its atomic number (Z), the sum of the number of protons and neutrons its atomic mass number (A), and the number of protons less the number of electrons its net charge.

2.2 Identifying Numbers of Subatomic Particles for an Atom Using the Periodic Table

By the beginning of the twentieth century Dmitri Mendeleev was able to organize elements in the form of a periodic table based on systematic trends in their properties. The periodic table arranges elements according to increasing atomic number in horizontal rows termed periods and vertical groups or families that have similar chemical properties. Entries in the periodic table provide the atomic symbol, atomic number, and atomic mass for each element. Nonmetals are found in the upper right-hand corner and noble gases in the right-most column of the table, respectively. Most of the other elements are metals. Metalloids lie between non-metals and metals. Over the course of this book we will explain why different elements have different properties.

Each element is represented in the periodic table based on a unique atomic symbol—denoted "E" (see Figure 2.4)—and has a unique atomic number. Ions are species formed when atoms lose or gain electrons.

Yet Dalton's hypothesis that all atoms for a given element are identical is incorrect: atoms can occur as isotopes with a different number of neutrons, such that all atoms for a given element do *not* have identical atomic mass numbers! Atomic masses given in the periodic table are averages of atomic mass numbers for the naturally-occurring isotopes for the element, weighted by the corresponding naturally-occurring abundances for each isotope. Note that for a given atom the atomic number Z in the schematic below is often not specified because it can be deduced from the element's identity and its place in the periodic table.

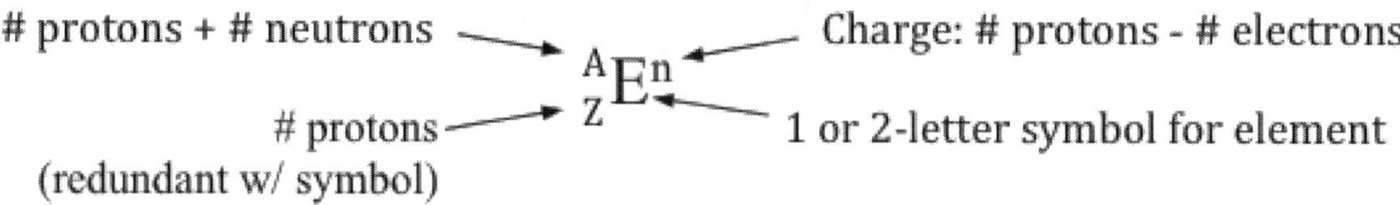

Figure 2.4. Interpreting information for an element in the periodic table

Example 2.1: Reading Symbols on the Periodic Table

Consider an isotope of oxygen (atomic symbol O) with atomic mass number 18 and a charge of -2. How many protons, neutrons, and electrons does it have?

Solutions

Based on its position in the periodic table, all isotopes of oxygen have 8 protons. Based on an atomic mass number of 18, the number of neutrons is the atomic mass number less the number of protons, 18 – 8 = 10. Based on a charge of -2, the number of electrons is the number of protons less the charge, 8 – (-2) = 10.

2.3 Light and Matter Have Quantized Properties According to Quantum Mechanics

The initial interpretation for the nuclear model for atomic structure was that negatively-charged electrons orbit a positively-charged nucleus. This model was part of an overall description for nature, termed classical physics, in which matter is a collection of atoms whose behavior was governed by Newton's laws of motion, and light is an electromagnetic wave governed by Maxwell's theory of electromagnetic fields. As a wave, light has a wavelength, λ, and frequency, ν, which are related as:

$$\lambda\nu = c \qquad [2.1]$$

where $c = 300\times10^{8}$ m/s is the speed of light. The wavelength of light corresponds to the distance between peaks on the wave. According to classical physics matter is only a collection of particles—not waves—and light is only a wave—not a particle!

As Rutherford was formulating the nuclear model, however, problems were being encountered with some of the descriptions of classical physics. For example, classical physics predicts that negatively-charged electrons should not be stably associated with a positively-charged nucleus, which fails to describe chemical bonds, cannot describe the absorption and emission of light at discrete wavelengths (occurring in phenomena such as fluorescence and phosphorescence), and cannot predict the diffraction of atoms and subatomic particles by crystals. From the perspective of items we encounter in our lives in the twenty-first century, classical physics also cannot explain the existence or operation of modern solid-state devices such as transistors, photovoltaic devices (*e.g.*, solar cells), light-emitting diodes (LEDs), lasers, plasma screens, and magnetic resonance imaging (MRI) systems (Figure 2.5).

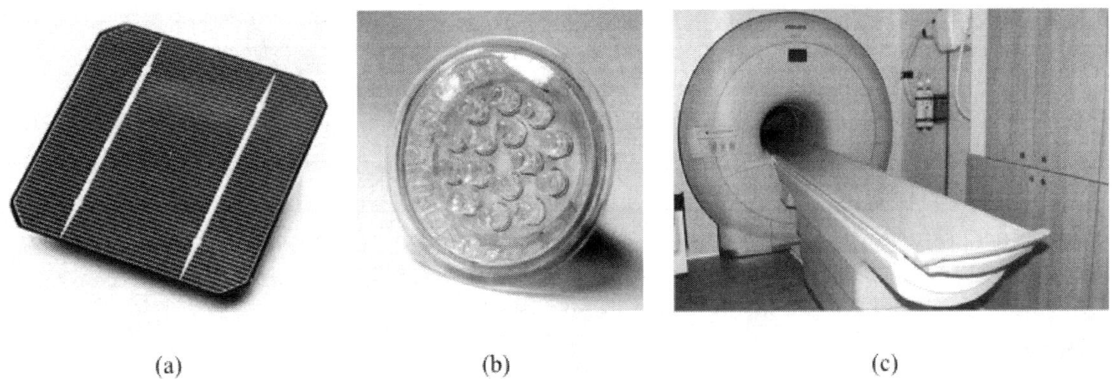

(a) (b) (c)

Figure 2.5. Examples of modern solid-state devices based on quantum mechanics. (a) Photovoltaic device (solar cell); (b) energy-efficient lamp consisting an array of LEDs; (c) MRI system.

The underlying basis for photovoltaic devices, the photoelectric effect—the emission of photoelectrons from a metal upon irradiation with light (Figure 2.6)—was a source of great anxiety to scientists at the turn of the twentieth century. Examining the photoelectric effect (*e.g.*, using the user-friendly simulation available at http://phet.colorado.edu/en/simulation/photoelectric) provides insight into the issues encountered with classical physics. Figure 2.7 demonstrates that the predictions of classical physics regarding the effect of frequency and intensity of incident light on the kinetic energy of ejected photoelectrons (KE_{e^-}) are incompatible with experimental observations. For example, classical physics predicts that the frequency of incident light does not affect whether a photoelectron is emitted nor their KE_{e^-}, but that KE_{e^-} should be proportional to the intensity of light. In contrast, experimentally we observe that there is a threshold frequency only above which photoelectrons are emitted, with KE_{e^-} increasing linearly with frequency above the threshold frequency, and the intensity of incident light does not affect their kinetic energy but rather the number of photoelectrons is proportional to the intensity of incident light.

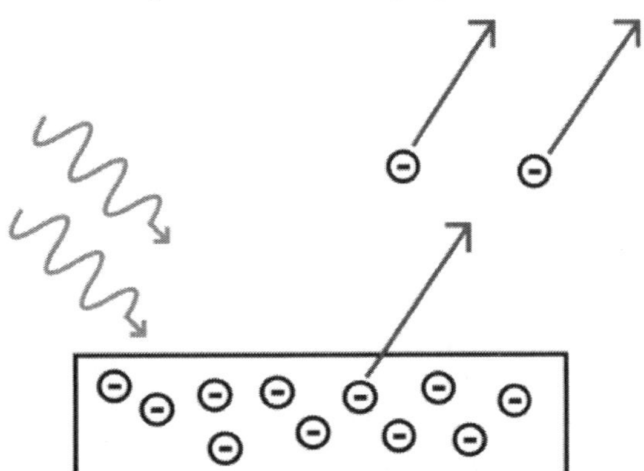

Figure 2.6. In the photoelectric effect electrons are ejected from a metal upon irradiation with light.

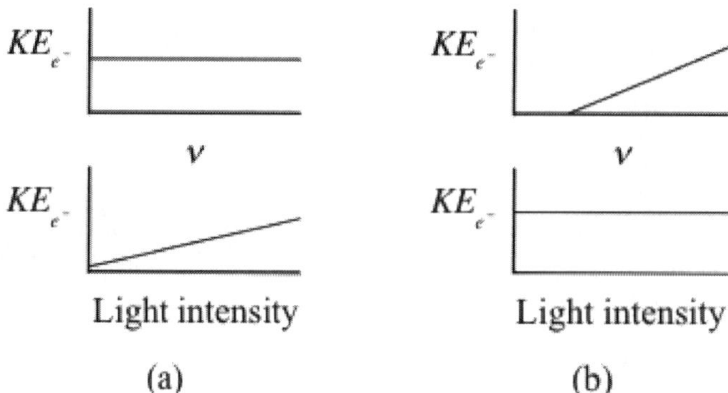

Figure 2.7. Comparison of predictions of classical physics (a) versus experimental observations (b) for the kinetic energy of ejected photoelectrons upon irradiation of a metallic surface with light

Addressing the problems encountered with classical physics, such as the mismatch between theory and experiment for the photoelectric effect), required developing an entirely new perspective in science, quantum mechanics. At its core quantum mechanics has two foundational postulates. As a first foundational postulate Max Planck proposed in 1901 that energy is quantized (*i.e*, discrete in values) at the atomic level. As a second foundational postulate Louis de Broglie proposed in 1924 wave-particle duality—all matter is both a particle and a wave—with the wavelength for a particle given by the de Broglie relation:

$$\lambda \nu = h \qquad [2.2]$$

where Planck's constant is $h \equiv 6.626\times10^{-34}$ J-s and the particle's momentum (p) and kinetic energy respectively are:

$$p = mv \qquad [2.3a]$$

$$KE = \frac{1}{2}mv^2 \qquad [2.3b]$$

Classical physics works quite well with macroscopic objects because the wavelength of objects we can see with our naked eye is too small to observe!

Albert Einstein in 1905 explained experimental observations for the photoelectric effect by invoking Planck's concept and proposing that light is both a wave and a particle, termed a photon, with energy:

$$E_{photon} = mv \qquad [2.4]$$

Einstein then identified that the relationship between KE of an ejected electron and frequency of incident light is given as:

$$KE_{e^-} = \frac{1}{2}m_{e^-}v^2 = \begin{cases} 0 \text{ for } h\nu < \Phi \\ h\nu - \Phi \text{ for } h\nu > \Phi \end{cases} \qquad [2.5]$$

where Φ is defined as the n, a characteristic property of a metal (Table 2.1). Einstein was awarded a Nobel Prize in physics in 1921 for this work, which is the basis for modern solid-state electronics and surface-sensitive spectroscopies. Work functions have units of energy and commonly are reported in molecular-scale units of electron volts (eV), where 1 eV = 1.602×10^{-19} J is the energy corresponding to an electron accelerated in an electric field of 1 V. The CRC *Handbook of Chemistry and Physics* (available on-line at http://www.hbcpnetbase.com/) provides a source of values for work functions for different elements.

Example 2.2: Photoelectric Effect and Comparison of Properties of Light and Electrons

Metallic rubidium (Rb), which has a work function of 2.09 eV, is illuminated with 200-nm light for 60.0 s using a 200 W bulb (where 1 W = 1 J/s).

a) What is the frequency (in s^{-1}) associated with this incident light?

b) What is the energy (in J/photon) of a photon of this incident light?

c) How many photons strike metallic rubidium during the period of illumination?

d) What is the kinetic energy (in J) of an electron ejected from metallic rubidium upon excitation with this light?

e) What is the speed (in m/s) of the ejected electron?

f) What is the wavelength (in nm) associated with the ejected electron?

Solutions

a) The frequency of light with a wavelength of 200 nm can be determined by rearranging Equation [2.1] as:

$$\nu = \frac{c}{\lambda} = \frac{(3.00\times10^{8}\ \text{m/s})}{\left(200\ \text{nm}\times\frac{10^{-9}\ \text{m}}{1\ \text{nm}}\right)} = 1.50\times10^{15}\ \text{s}^{-1}$$

b) The energy of a photon of light with a wavelength of 200 nm can be determined from Equation [2.4] as:

$$E_{photon} = h\nu = \frac{hc}{\lambda} = (6.626\times10^{-34}\ \text{J}-\text{s})(1.50\times10^{15}\ \text{s}^{-1}) = 9.94\times10^{-19}\ \text{J/photon}$$

c) We can determine the number of photons based on the total energy of light that illuminates the sample (corresponding to a power $P = 200.$ J/s applied for a time $t = 60.0$ s) and the energy per photon as:

$$E_{Total} = Pt = nE_{photon}$$

$$n = \frac{Pt}{E_{photon}} = \frac{(200.\ \text{J/s})(60.0\ \text{s})}{(9.94\times10^{-19}\ \text{J/photon})} = 1.21\times10^{22}\ \text{photon}$$

d) The kinetic energy of an electron ejected from rubidium upon excitation by 200-nm light can be determined from Equation [2.5] as:

$$KE_e = h\nu - \Phi = \frac{hc}{\lambda} - \Phi$$

$$= (6.626\times10^{-34}\ \text{J}-\text{s})(1.50\times10^{15}\ \text{s}^{-1}) - \left(2.09\ \text{eV}\times\frac{1.602\times10^{-19}\ \text{J}}{1\ \text{eV}}\right)$$

$$KE_e = 6.59\times10^{-19}\ \text{J}$$

e) The speed of the ejected electron can be determined by rearranging Equation [2.3b] as:

$$KE_{e^-} = \frac{1}{2} m_{e^-} v^2$$

$$v = \sqrt{\frac{2KE_{e^-}}{m_{e^-}}} = \sqrt{\frac{2(6.59\times10^{-19}\ \text{J})}{(9.11\times10^{-31}\ \text{kg})}} = 1.20\times10^{6}\ \text{m/s}$$

f) The wavelength associated with the ejected photoelectron can be determined applying the de Broglie relation, Equation [2.2]:

$$\lambda = \frac{h}{p} = \frac{h}{m_{e^-} v} = \frac{(6.626\times10^{-34}\ \text{J}-\text{s})}{(9.11\times10^{-31}\ \text{kg})(1.20\times10^{6}\ \text{m/s})} \times \frac{10^{9}\ \text{nm}}{1\ \text{m}} = 0.605\ \text{nm}$$

2.4 The Quantization of Electron Energy Can Explain Observed Atomic Excitation and Emission Spectra

Consider the interaction between matter and light further. White light is visible light with a continuous range of wavelengths between ~400 nm and ~800 nm. Refraction of white light through a prism results in a spectrum of colors. Note that not all light is visible (Figure 2.8): infrared (IR) light has wavelengths longer than 800 nm, and ultraviolet (UV) light has wavelengths shorter than 400 nm. Microwaves and X-rays are other commonly encountered forms of electromagnetic radiation (*i.e.*, light).

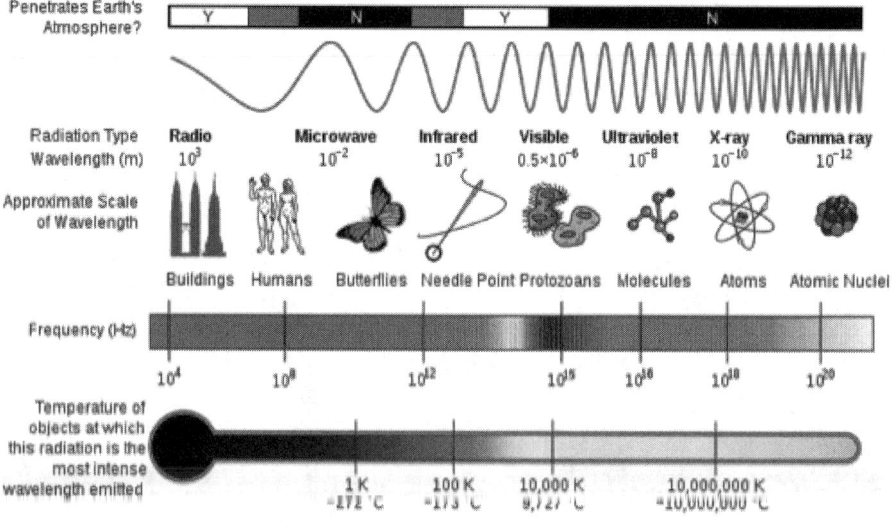

Figure 2.8. The electromagnetic spectrum and its properties.

Spectroscopy is the use of light to examine the properties of materials (Figure 2.9) One of the key features of quantum chemistry is its ability to predict the absorption and emission of light at discrete wavelengths. For example, only discrete wavelengths of light are emitted when a pure element is heated in a flame. This phenomenon is called atomic emission. Similarly, only discrete wavelengths are absorbed when a pure element is illuminated with white light. This phenomenon is called atomic absorption. Data for atomic spectra for different elements are available on-line through the National Institute of Standards and Technology at the website http://www.nist.gov/pml/data/atomspec.cfm.

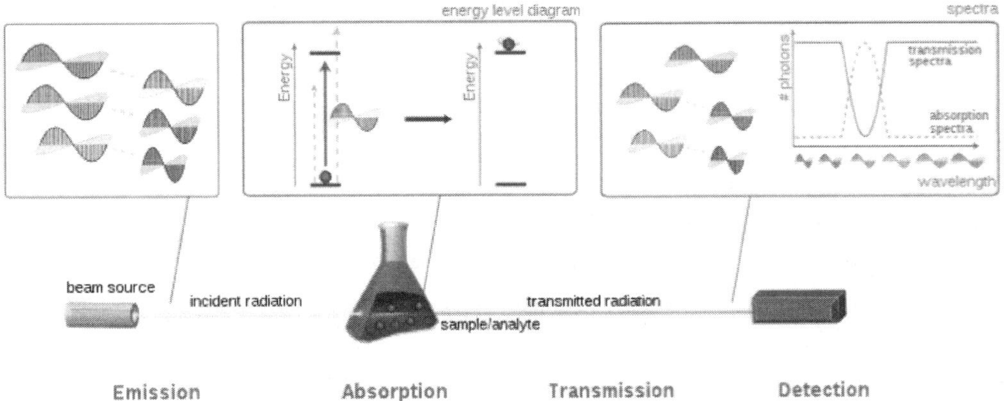

Figure 2.9. Spectroscopy is the use of electromagnetic radiation to examine the properties of matter. Only photons with energies corresponding to transitions between quantized energy levels in a sample are absorbed; unabsorbed photons are transmitted. A detector measures the wavelengths associated with the transmitted light and produces a spectrum for the sample.

At the turn of the twentieth century scientists discovered that the wavelengths of light emitted and absorbed by elemental hydrogen could be described as:

$$\frac{1}{\lambda} = R_H\left(\frac{1}{n_1^2} - \frac{1}{n_2^2}\right) \qquad [2.6]$$

where $R_H = 1.097 \times 10^5\ \text{cm}^{-1}$ is the Rydberg constant and $\{n_1, n_2\}$ are positive integers with $n_1 > n_2$. This relationship could correlate data from various observed series (Figure 2.7). Equation [2.6], however, is phenomenological: it provides no mechanistic insight but rather merely correlates the data.

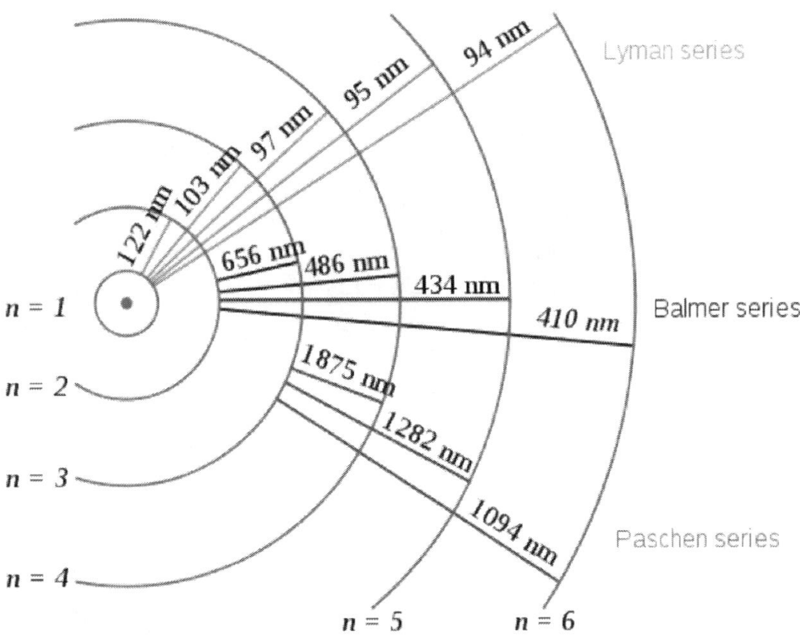

Figure 2.10. Atomic absorption series for the hydrogen atom.

In 1913 Niels Bohr proposed a mechanistic model that successfully explained the experimental observations and predicted the observed phenomenological relationship. Bohr postulated that electrons "orbit" the nucleus in stable states without radiating light. These states have discrete, quantized energies E_n, where the quantum number $n = 1, 2, 3, \ldots$ is a positive integer. Bohr also suggested that transitions between states differing in energy by $\Delta E = E_f - E_i$ occurs upon the absorption or emission of a photon of light with frequency ν and wavelength λ, where:

$$\Delta E = h\nu = \frac{hc}{\lambda} \qquad [2.7]$$

Lastly, Bohr proposed that absorption of light results in the promotion of the electron to a higher energy state and that emission of light results in the fall of the electron to a lower energy state (Figure 2.11).

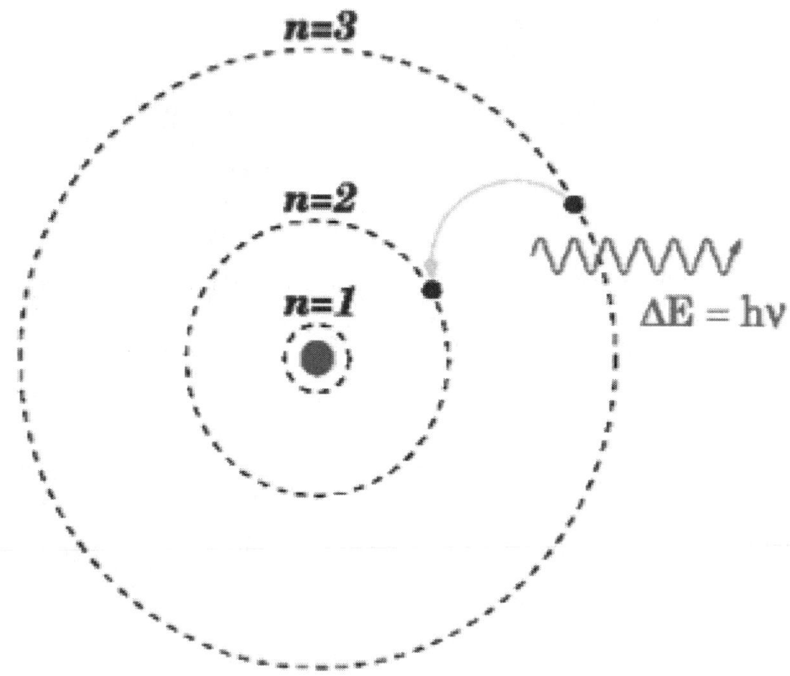

Figure 2.11. Bohr model for the hydrogen atom.

The Bohr model can describe the observed data for hydrogen if the transition $n_2 \rightarrow n_1$ represents emission and $n_1 \rightarrow n_2$ represents absorption. Note that the lowest energy state $n_i = 1$ is called the ground state and that higher energy states are called excited states; the state with the second-lowest energy is the 1st first excited state, the state with the third-lowest energy is the 2nd second excited state, *etc.* This model for hydrogen also can be extended by analogy to hydrogenic (hydrogen-like) atoms—atoms with only one electron, such as He^+ and Li^{2+}—by introducing the parameter Z for nuclear charge number (*i.e.*, $Z = 1$ for H, $Z = 2$ for He, $Z = 3$ for Li …), resulting in the Rydberg equation:

$$\frac{1}{\lambda} = Z^2 R_H\left(\frac{1}{n_1^2} - \frac{1}{n_2^2}\right) \text{ with } n_1 \in \{1, 2, \ldots\} \text{ and } n_2 \in \{n_1 + 1, n_1 + 2, \ldots\} \qquad [2.8]$$

Example 2.3: The Lyman Series for Hydrogen

Atomic spectra occur in "series" based on the value for n_1. Consider the Lyman Series for absorption of ultraviolet light, corresponding to excitation of an electron for hydrogen from the ground state (*i.e.*, for $n_1 = 1 \rightarrow n_2 = 2, 3, \ldots$).

a) What is the wavelength (in nm) of light absorbed for the transition from the ground state to the 1st first excited state?

b) What is the frequency (in s^{-1}) associated with this light?

c) What is the energy (in J/photon) of a photon of this incident light?

d) What is the speed (in m/s) of an electron moving with the same wavelength?

Solutions

a) Apply Equation (2.6) with $n_1 = 1$ and $n_2 = 2$, rearranging to solve for the wavelength and converting from units of cm to nm:

$$\frac{1}{\lambda} = R_H\left(\frac{1}{n_1^2} - \frac{1}{n_2^2}\right)$$

$$\lambda = R_H^{-1}\left(\frac{1}{n_1^2} - \frac{1}{n_2^2}\right)^{-1} = (1.097\times10^5 \text{ cm}^{-1})^{-1}\left[\frac{1}{(1)^2} - \frac{1}{(2)^2}\right]^{-1} = 1.215\times10^{-5} \text{ cm}\times\frac{10^7 \text{ nm}}{1 \text{ cm}}$$

$$\lambda = 121.5 \text{ nm}$$

b) The frequency of light with a wavelength of 121.5 nm can be determined by rearranging Equation (2.1) as:

$$\nu = \frac{c}{\lambda} = \frac{(3.00\times10^8 \text{ m/s})}{\left(121.5 \text{ nm}\times\frac{10^{-9} \text{ m}}{1 \text{ nm}}\right)} = 2.47\times10^{15} \text{ s}^{-1}$$

c) The energy of a photon of light with a wavelength of 121.5 nm can be determined using Equation (2.4) as:

$$E = h\nu = \frac{hc}{\lambda} = (6.626\times10^{-34} \text{ J} - \text{s})(2.47\times10^{15} \text{ s}^{-1}) = 1.64\times10^{-18} \text{ J}$$

Note that this energy also corresponds to the difference in energy, ΔE, between the ground state and the first excited state for an electron for hydrogen.

d) Determine the momentum associated with a particle with a wavelength of 121.5 nm by applying the de Broglie relation, Equation (2.2):

$$p = \frac{h}{\lambda} = \frac{(6.626\times10^{-34} \text{ J} - \text{s})}{\left(121.5 \text{ nm}\times\frac{1 \text{ m}}{10^9 \text{ nm}}\right)} = 5.45\times10^{-27} \text{ kg} - \text{m/s}$$

Determine the speed of an electron with this momentum by rearranging Equation [2.3a]:

$$p = m_e v$$

$$v = \frac{p}{m_e} = \frac{(5.45\times10^{-27}\ \text{kg - m/s})}{(9.11\times10^{-31}\ \text{kg})} = 5.98\times10^{3}\ \text{m/s}$$

The concept of quantization of electronic energies is the foundation for UV–vis spectroscopy with visible and ultraviolet light, as well as part of the underlying basis for how lasers work. In 1917 Einstein first proposed the concept of a laser—an acronym for light amplification by stimulated emission of radiation—by evaluating the probabilities of absorption of light by atoms in their ground state, spontaneous emission of light by atoms in an excited state, and stimulated emission of light by atoms in an excited state in response to incident photons with energies corresponding to the electronic transition from the excited state back to the ground state (Figure 2.12). Technical hurdles, including the need to achieve a population inversion in which the occupancy of excited states was increased dramatically, prevented the practical construction of a laser for about forty years.

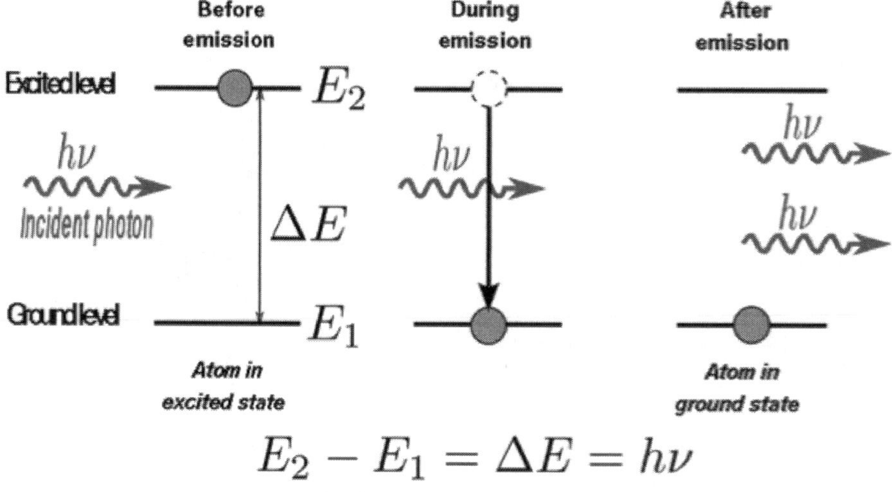

Figure 2.12. Process of stimulated emission of light as it occurs in a laser.

2.5 Electron Orbitals and Quantum Numbers for Hydrogenic Atoms

Although the Bohr model represented an important advance, significant unresolved issues remained, including an inability to extend the Bohr model to explain observations for non-hydrogenic atoms (*i.e.*, atoms with more than one electron). More profoundly, however, the Bohr model is incorrect mechanistically because it is an incomplete quantum-mechanical model: Bohr treated an electron as a classical particle orbiting the nucleus in defined, discrete orbits. In reality, we cannot state precisely where an electron is in space as a consequence of the Heisenberg uncertainty principle: the more precisely we know where a particle is (evaluated based on its uncertainty in position, Δx), the less precisely we know its momentum (*i.e.*, its motion, evaluated based on its uncertainty in momentum Δp). The Heiseberg uncertainty principle can be expressed mathematically as:

$$\Delta x \Delta p \geq \frac{h}{4\pi} \qquad [2.9]$$

Example 2.4: Imaging Using a Conventional Tube Television with Cathode Rays

Consider a classical tube television using cathode rays, for which each electron in the beam has a momentum of $5.410 \pm 0.054 \times 10^{-23}$ kg-m/s (*i.e.*, the uncertainty in momentum is 5.4×10^{-25} kg-m/s). What is the minimum uncertainty in the position (in nm) where an electron lands on the tube's phosphor?

Solutions

Apply Equation [2.9], rearranging to solve for the uncertainty in position:

$$\Delta x \geq \frac{h}{4\pi\Delta p} = \frac{(6.626 \times 10^{-34}\,\text{J - s})}{4\pi(5.4 \times 10^{-25}\,\text{kg - m/s})} = 9.8 \times 10^{-11}\,\text{m} \times \frac{10^{9}\,\text{nm}}{1\text{m}} = 9.8 \times 10^{-2}\,\text{nm}$$

This uncertainty is much smaller than the diameter of a typical electron beam, such that this uncertainty has no observable effect on the quality of a TV picture.

In the 1920s Schrödinger offered a better explanation for the electronic structure of a hydrogen atom by treating an electron fully as both a particle and a wave. The mathematical details of the resulting Schrödinger equation require an understanding of advanced calculus concepts significantly beyond the background of a typical student studying general chemistry; however, the outcomes of solutions of Schrödinger's model are within our grasp. The solutions to the Schrödinger equation describe electrons as occupying discrete orbitals, which can be related to probabilities of finding an electron in particular regions of space, with energies quantized based on a set of three quantum numbers. The principal quantum number, $n \in \{1, 2, \ldots\}$ (*i.e.*, n is a positive integer), determines the shell (alternatively called level), with pairings given in Table 2.1. The orbital angular momentum quantum number (also known as the azimuthal quantum number), $l \in \{0, 1, 2, \ldots, n\text{-}1\}$ (*i.e.*, $l < n$), determines the subshell (alternatively called sublevel), with pairings given in Table 2.2. The magnetic quantum number, $m_l \in \{-l, -l+1, -l+2, \ldots, -1, 0, 1, \ldots, l\text{-}2, l\text{-}1, l\}$ (*i.e* $|m_l| \leq l$), determines the spatial orientation of the orbital.

Table 2. Pairings Between Principal Quantum Number and Electron Shell

Principal Quantum Number (n)	Shell Name
1	K
2	L
3	M
4	N
5	O

Table 2.2 Pairings Between Orbital Angular Momentum Quantum Number and Electron Subshell

Orbital Angular Momentum Quantum Number (l)	Shell Name
0	s
1	p
2	d
3	f

In general, orbital energy increases as n increases or as l increases for a fixed n and is independent of m_l and m_s. The shape of an orbital—depicting the highest probability of finding an electron in that orbital—depends on the set of $\{n, l, m_l, m_s\}$ for the orbital (Figure 2.13). In particular, s orbitals are spherically symmetric, p and d orbitals are lobular, and f orbitals have very complex shapes.

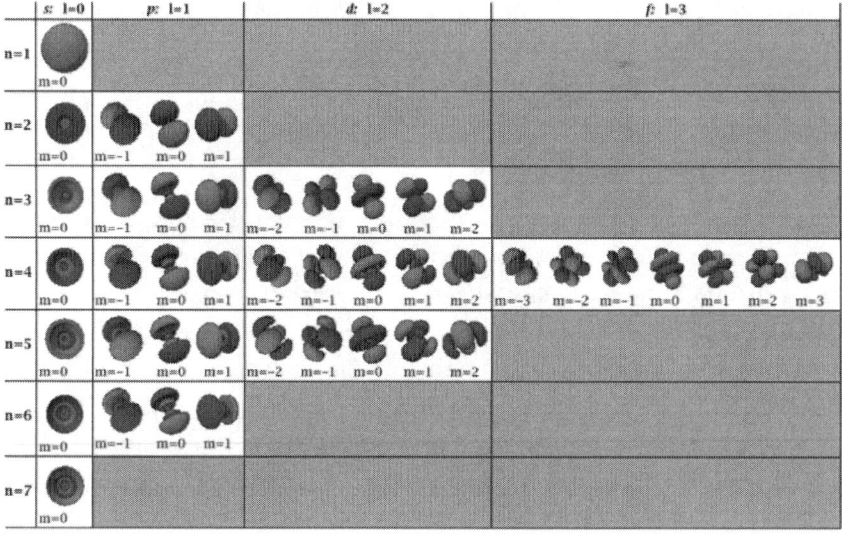

Figure 2.13. Atomic orbitals for hydrogenic atoms. (Credit: Richard Parsons)

Subsequent study of the effects of magnetic fields on observed atomic spectra led to the requirement of a fourth quantum number, the electron spin quantum number, $m_s \in \{-1/2, +1/2\}$.

Example 2.5: Forbidden and Allowed Quantum Numbers

Which of the following sets of quantum numbers are verboten and which are allowed?

Set A: $n = 2$, $l = 0$, $m_l = 0$, $m_s = -1$

Set B: $n = 2$, $l = 1$, $m_l = 1$, $m_s = +1/2$

Set C: $n = 1$, $l = 0$, $m_l = -1$, $m_s = +1/2$

Set D: $n = 1$, $l = 1$, $m_l = 0$, $m_s = -1/2$

Set E: $n = 2$, $l = 1$, $m_l = 0$, $m_s = -1/2$

Solutions

Set A is forbidden because the electron spin quantum number, m_s, cannot be any value other than -1/2 or +1/2. Set C is forbidden because the magnitude of the magnetic quantum number, m_l, cannot be less than the orbital angular momentum quantum number, l. Set D is forbidden because the principal quantum number, n, must be greater than the orbital angular momentum quantum number. Sets B and E are allowed because the rules for allowable quantum numbers are met.

2.6 Assigning Electron Configurations to Atoms

To understand and predict the chemical properties (*e.g.*, types of compounds formed) for an element, we need to identify how electrons for the element populate its orbitals (*i.e.*, shells and subshells). In particular, we must examine how to assign electron configurations, describing how electrons are distributed among the different orbitals for atoms in their ground and excited state. This process also will provide further insight into the structure of the periodic table.

There are three basic rules for assigning electron configurations:

1. The Pauli exclusion principle: No two electrons for an atom can have the same set of four quantum numbers. As a consequence of this rule, we find that only two electrons may occupy any given orbital and the two electrons must have opposing spins (*i.e.*, +1/2 and -1/2). Further, we identify that each s subshell consists of a single orbital (with $m_l = 0$) holding a maximum of two electrons, that each p subshell consists of three orbitals (with m_l = -1, 0, or 1) holding a maximum of six electrons, that each d subshell consists of five orbitals (with m_l = -2, -1, 0, 1, or 2) holding a maximum ten electrons, and that each f subshell consists of seven orbitals (with m_l = -3, -2, -1, 0, 1, 2, or 3) holding a maximum of fourteen electrons.

2. Assign electrons to atoms in their ground state in order of lowest to highest energies to minimize the total electronic energy. We observe that the quantized energy of an atomic orbital increases in order of 1s, 2s, 2p, 3s, 3p, 4s, 3d, 4p, 5s, 4d, 5p, 6s, 4f, 5d, 6p, 7s, 5f, 6d, 7p, … as depicted in Figure 2.14.

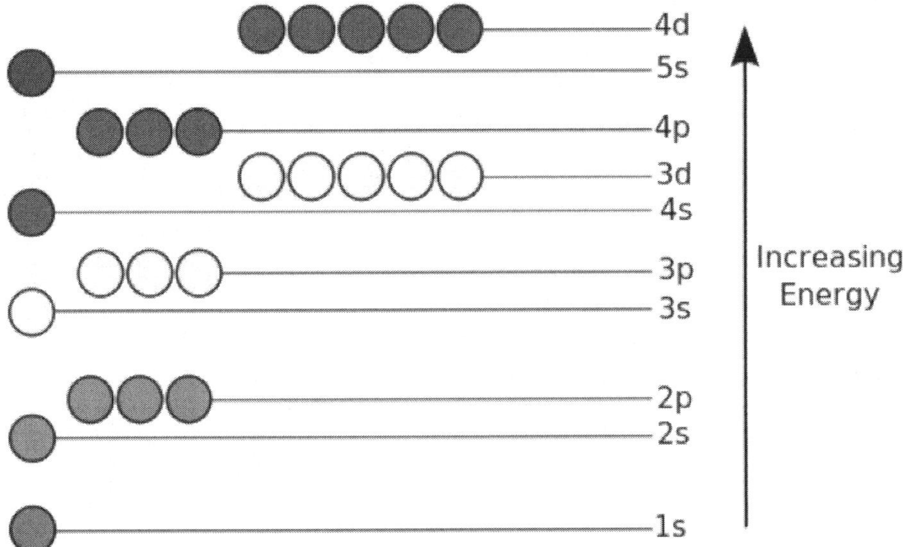

Figure 2.14. Energy level diagram for atomic orbitals.

3. Hund's Rule: When assigning electrons to a set of orbitals with the same energies, assign the electrons first as unpaired and then pair as needed.

Rather than using rote memorization for the ordering of energies for shells and subshells, consider the structure of the periodic table to understand this ordering. We apply what is known as the aufbau process (from the German for "building up") for assigning electron configurations in order of increasing atomic number, as depicted in Figure 2.15.

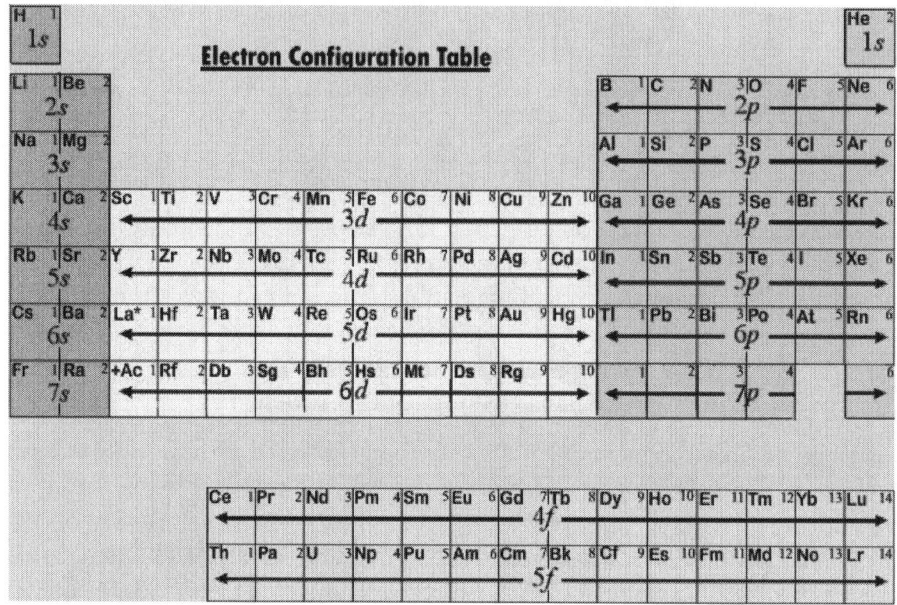

Figure 2.15. The aufbau process for assigning electron configurations.

There are three alternative representations for any given electron configuration. Consider these alternatives by examining as a specific example the electron configuration for iron (Fe), which has twenty-six electrons as a neutral atom. We focus on the ground-state configuration, the distribution of electrons for an iron atom with lowest energy.

- Alterative 1: An orbital diagram is the most graphical but also the most time-consuming to write.

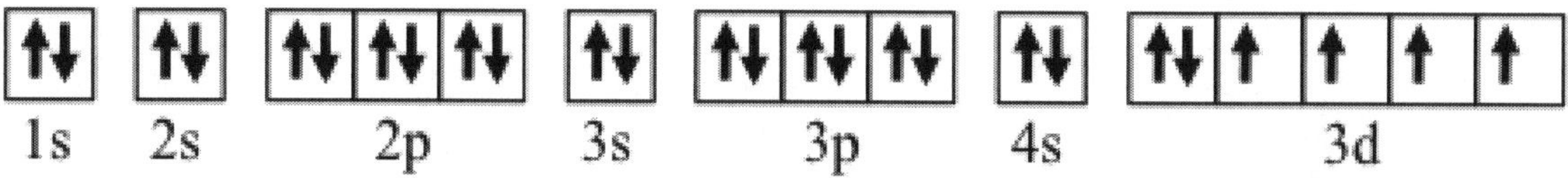

- Alterative 2: A condensed spdf notation (also known as spectroscopic notation) is more abbreviated. $1s^22s^22p^63s^23p^64s^23d^6$.
- Alterative 3: A valence shell notation is most abbreviated and is based on appending the outermost (valence) electrons to the abbreviated electron configuration for the nearest lower-total-energy noble gas core: [Ar] $4s^23d^6$.

Example 2.6: Assigning Electron Configurations

Assign the ground-state electron configuration to nitrogen (N), silicon (Si), neon (Ne), and bromine (Br) using:

a) Orbital diagrams

b) Condensed spdf notation

c) Valence shell notation

Solutions

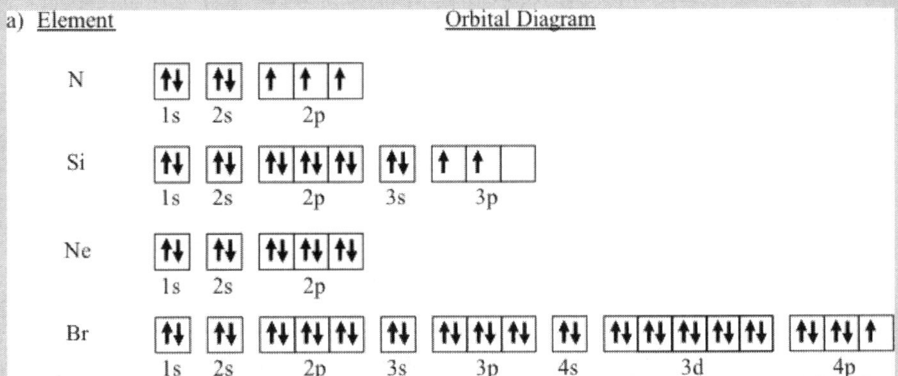

b) N: $1s^22s^22p^3$

Si: $1s^22s^22p^63s^23p^2$

Ne: $1s^22s^22p^6$

Br: $1s^22s^22p^63s^23p^64s^23d^{10}4p^5$

c) N: [He] $2s^22p^3$

Si: [Ne] $3s^23p^2$

Ne: [Ne] or [He] $2s^22p^6$

Br: [Ar] $4s^23d^{10}4p^5$

Example 2.7: Characterizing Electron Configurations

Identify the neutral atom corresponding to the following valence shell configurations and whether the configuration corresponds to the ground state or an excited state.

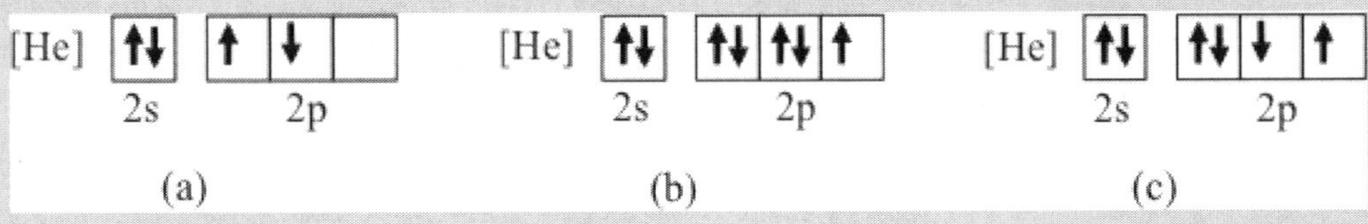

Solutions

a) The neutral atom is carbon (C). The configuration corresponds to an excited state because the two electrons for the 2p orbitals have opposing spins (in the ground state all unpaired electrons for a given orbital have the same spin).

b) The neutral atom is fluorine (F). The configuration corresponds to the ground state.

c) The neutral atom is oxygen (O). The configuration corresponds to an excited state because the two unpaired electrons for the 2p orbitals have opposing spins.

3:Chemical Compounds and Chemical Formulas

Learning Objectives

By the end of this chapter you should be able to:

1) Distinguish between physical versus chemical properties of matter and homogeneous versus heterogeneous mixtures.
2) Distinguish between an empirical formula, a molecular formula, a structural formula, and a formula unit for representing a chemical compound.
3) Determine the molar mass of a substance from its formula.
4) Calculate the mass percent composition of a substance based on its chemical formula.
5) Relate the empirical or molecular formula of a substance to its percent composition.

3.1 Fundamental Concepts for Describing the Properties of Substances

Being able to assign electron configurations for atoms allows us to identify the outermost (valence) electrons for an atom. It is these electrons that play a fundamental role in chemical bonding. Before considering how atoms form bonds, however, we need to distinguish different classes of properties of matter and how to classify matter in general.

The composition of any sample of matter tells us the relative proportions of each element in the substance. A compound is a substance that consists of more than one element and in which the different atoms are held together by chemical bonds. Physical properties, such as volume, color, density, and hardness, are any properties that are not associated with changes in composition. Physical changes correspond to processes in which one or more physical properties for the substance change without a change in composition. For example, the state (or phase) of a substance, such as whether it is a solid, liquid, or gas, is an important physical property, and phase changes occur without change in composition. In contrast, chemical properties are associated with changes in the composition of a substance. Chemical changes correspond to changes in composition and are associated with reactions.

Substances can exist either in their pure forms or in mixtures with one or more other substances. Mixtures are characterized as either homogeneous—with uniform composition and physical properties through the mixture—or heterogeneous—with composition and physical properties that vary spatially within the mixture (Figure 3.1). Solutions are homogeneous mixtures in which one or more substances are in large excess. For example, milk is a solution in which water is the substance in large excess.

Figure 3.1. Examples of heterogeneous mixtures found in nature. (a) Granite; (b) mud; (c) fog.

3.2 Types of Formulas for Representing Compounds

Formulas represent the composition of chemical compounds in the form of information on the ratios of the numbers of each element in a compound. We encounter three types of formulas. An empirical formula provides the relative number of each element in a compound (*i.e.*, the smallest combining ratios). Empirical formulas are the simplest but least representative type of formula. A molecular formula provides the absolute number of each element in a compound. Molecular formulas are more complex and representative than empirical formulas. A structural formula provides a representation of the connectivity between atoms in a compound. Structural formulas are the most complex and representative type of formula.

How we apply the different types of formulas to specific compounds depends on the type of the compound. We distinguish two different types of compounds: molecular compounds (molecules) ionic compounds. Atoms in a molecule are held together by covalent bonds, shared pairs of electrons. Non-metals form molecules. Table 3.1 provides examples of formulas or molecules. Note that two molecules, such as ethene and polyethylene (a polymer formed from ethene), may have different molecular formulas but share a common empirical formula. Also, recognize that different molecules may share a common molecular formula but have different structural formulas. We will examine structural formulas for molecules, which uniquely represent a given molecule, in Chapter 4.

Table 3.1. Examples of Formulas for Molecules

Name	Molecular Formula	Empirical Formula
water	H_2O	H_2O
hydrogen peroxide	H_2O_2	HO
ethene	C_2H_4	CH_2
polyethylene	$(CH_2)_n$	CH_2

In an ionic compound, cations (positively-charged ions) and anions (negatively-charged ions) are held together by ionic bonds formed as the result of the transfer of electrons among atoms, with the overall compound having a zero net charge. Metals and non-metals combine to form ionic compounds Ions can be either monatomic, consisting of a single ionized atom (*e.g.*, Na^+), or polyatomic, consisting of an ionized collection of atoms held together by covalent bonds (*e.g.*, SO_4^{2-}). No single, unambiguous connectivity exists between individual ions, such that the concept of a "molecular formula" does not apply to ionic compounds. Instead, we define a formula unit as the empirical formula for an ionic compound, representing the simplest electrically neutral collection of ions for the compound. Table 3.2 provides examples of formula units and compositions for ionic compounds. We will examine structural models for ionic compounds in Chapter 5.

Tab 3.2. Examples of Formula Units for Ionic Compounds

Name	Formula Unit	Ions
sodium chloride	$NaCl$	Na^+ and Cl^-
barium sulfate	$BaSO_4$	Ba^{2+} and SO_4^{2-}
calcium fluoride	CaF_2	Ca^{2+} and F^-

3.3 Applying the Mole Concept and Molar Mass to Chemical Compounds

The mole concept, introduced in Section 1.4 for elements, also applies to chemical compounds. We remind ourselves that 1 mol of any substance—be it an element or a compound—is equivalent to Avogadro's number of atoms (for an element), molecules (for a molecular compound), or formula units (for an ionic compound). We saw previously in Section 1.4 that the atomic mass given beneath the symbol for each element in the periodic table is the mass in grams for 1 mol of atoms (or the mass in amu for 1 atom) for the element, with the molar mass defined as this mass (in grams) per mol of the element. For example, the molar mass of carbon is 12.01 g/mol (expressed using four digits of precision). Similarly, the concept of molar mass applies for compounds: the molar mass of a compound is the sum of the molar mass for all of the atoms in the molecular formula (for a molecule) or formula unit (for an ionic compound).

Example 3.1: Detection of TNT

TNT (trinitrotoluene) has a molecular formula $C_7H_5N_3O_6$.

a) What is the molar mass (in g/mol) for TNT?

b) TNT can be detected at levels as low as 2.02 ng in a 2-L volume (Pushkarsky *et al.*, 2006). What is the minimum number of molecules of TNT in a 2-L volume that can be detected?

Solutions

a) Determine the molar of mass of TNT based on molar masses of 12.01 g/mol for C, 1.01 g/mol for H, 14.01 g/mol for N, and 16.00 g/mol for O as:

$$M_{TNT} = M_{C_7H_5N_3O_6} = 7M_C + 5M_H + 3M_N + 6M_O$$
$$= 7(12.01\text{ g/mol}) + 5(1.01\text{ g/mol}) + 3(14.01\text{ g/mol}) + 6(16.00\text{ g/mol})$$
$$M_{TNT} = 227.15\text{ g/mol}$$

b) Apply the factor-label method to successively convert ng TNT to g TNT to mol TNT to molecules TNT using the molar mass for TNT and Avogadro's number:

$$2.02\,\text{ng TNT} \times \frac{1\text{ g TNT}}{10^9\text{ ng TNT}} \times \frac{1\text{ mol TNT}}{227.15\text{ g TNT}}$$
$$\times \frac{6.022\times 10^{23}\text{ molecules TNT}}{1\text{ mol TNT}} = 5.36\times 10^{12}\text{ molecules TNT}$$

3.4 Characterizing Compounds Based on Their Percent Composition

Chemically 1 mole ofTNT is equivalent to six moles of carbon atoms, five moles of hydrogen atoms, three moles of nitrogen atoms, and six moles of oxygen atoms, such that we can make the following statements:

$$227.15\text{g } C_7H_5N_3O_6 = 7(12.01\text{ g C}) + 5(1.01\text{ g H}) + 3(14.01\text{ g N}) + 6(16.00\text{ g O})$$
$$227.15\text{g } C_7H_5N_3O_6 = \{84.07\text{ g C, } 5.05\text{ g H, } 42.03\text{ g N, } 96.00\text{ g O}\}$$

We generalize this statement by defining the mass percent of an element in a compound as:

$$\% - mass \equiv \frac{\{\text{Mass of element in compound}\}}{\{\text{Mass of compound}\}} \times 100\,\% \qquad [3.1]$$

Analytical techniques, such as mass spectrometry, can provide experimental data for the mass percent composition of a compound, useful for a variety of problems including identification of contaminants in water sources, impurities in semiconductors, and material leakages from medical devices.

Example 3.2: Percent Composition for TNT

What are the mass percentages of carbon, hydrogen, nitrogen, and oxygen for TNT?

Solutions

Our general strategy is to define the mass percent for each element based on the number of moles of the element per mole of the compound:

$$\% - C = \frac{7\text{ mol C}}{1\text{ mol } C_7H_5N_3O_6} \times 100\,\% = \frac{7(12.01\text{ g C})}{(227.15\text{g } C_7H_5N_3O_6)} \times 100\,\% = 37.01\,\%\text{ C}$$
$$\% - H = \frac{5\text{ mol C}}{1\text{ mol } C_7H_5N_3O_6} \times 100\,\% = \frac{5(1.01\text{ g H})}{(227.15\text{g } C_7H_5N_3O_6)} \times 100\,\% = 2.22\,\%\text{ H}$$
$$\% - N = \frac{3\text{ mol N}}{1\text{ mol } C_7H_5N_3O_6} \times 100\,\% = \frac{7(14.01\text{ g N})}{(227.15\text{g } C_7H_5N_3O_6)} \times 100\,\% = 18.50\,\%\text{ N}$$
$$\% - O = 100\% - (\% - C + \% - H + \% - N)$$
$$= 100\,\% - [(37.01\,\%\text{ C}) + (2.22\,\%\text{ H}) + (18.50\,\%\text{ N})]$$
$$\% - O = 42.27\%\text{ O}$$

Example 3.3: Composition of a Superconducting Ceramic

A 5.3764 g ceramic disc of $YBa_2Cu_3O_7$ becomes superconducting when it is cooled below -183 °C with liquid N_2. Superconducting materials repel magnetic fields, cause magnetic materials to levitate, and have applications in high-speed transportation systems.

a) What is the molar mass of $YBa_2Cu_3O_7$?

b) How many moles of $YBa_2Cu_3O_7$ are in the disc?

c) How many moles of yttrium (Y) are in the disc?

d) How many grams of yttrium are in the disc?

e) What is the mass percent of yttrium in $YBa_2Cu_3O_7$?

Figure 3.2. Levitating magnet above a $YBa_2Cu_3O_7$ disc cooled by liquid nitrogen.

Solutions

a) Determine the molar mass as based on the molar masses of yttrium, Ba (barium), Cu (copper), and oxygen and their corresponding mole numbers per formula unit:

$$M_{YBa_2Cu_3O_7} = M_Y + 2M_{Ba} + 3M_{Cu} + 7M_O$$
$$= (88.91\text{ g/mol}) + 2(137.33\text{ g/mol}) + 3(63.55\text{ g/mol}) + 7(16.00\text{ g/mol})$$
$$M_{YBa_2Cu_3O_7} = 666.22\text{ g/mol}$$

b) Determine the number of moles of $YBa_2Cu_3O_7$ using the mass and molar mass:

$$5.3764\text{ g YBa}_2\text{Cu}_3\text{O}_7 \times \frac{1\text{ mol YBa}_2\text{Cu}_3\text{O}_7}{666.22\text{ g YBa}_2\text{Cu}_3\text{O}_7} = 8.0700 \times 10^{-3}\text{ mol YBa}_2\text{Cu}_3\text{O}_7$$

c) Determine the number of moles of $YBa_2Cu_3O_7$ based on 1 mole of yttrium per 1 mole of compound:

$$8.0700 \times 10^{-3}\ \text{mol}\ YBa_2Cu_3O_7 \times \frac{1\ \text{mol Y}}{1\ \text{mol}\ YBa_2Cu_3O_7} = 8.0700 \times 10^{-3}\ \text{mol Y}$$

d) Determine the mass of yttrium as based on the number of moles of yttrium and its molar mass:

$$8.0700 \times 10^{-3}\ \text{mol Y} \times \frac{88.91\ \text{g Y}}{1\ \text{mol Y}} = 7.1750 \times 10^{-1}\ \text{g Y}$$

e) Determine the %-Y based on the mass of yttrium and the mass of the compound using Equation [3.1]:

$$\%-\text{Y} = \frac{m_Y}{m_{YBa_2Cu_3O_7}} \times 100\,\% = \frac{(7.1750 \times 10^{-1}\ \text{g Y})}{(5.3764\ \text{g}\ YBa_2Cu_3O_7)} \times 100\,\% = 13.345\,\%\ \text{Y}$$

3.5 Determining a Chemical Formula for a Substance Based on its Percent Composition

Consider the reverse of determining the mass percent composition given a formula: how can we determine the formula for a compound given its mass percent composition?

Example 3.4: Determining the Molecular Formula for an Unknown Explosive from its Percent Composition and Molar Mass

Residue from an explosion was analyzed to identify the explosive used, and a composition of 15.85% C, 2.22% H, 18.51% N, and 63.41% O was determined. What is the molecular formula for the explosive if it has a molar mass of 227.11 g/mol?

Solutions

Based on its percent composition and Example 3.2, we know the explosive is *not* TNT. To identify the empirical formula we apply the following strategy:

Step 1: Choose 100 g of the compound as a "basis."

100.00 g compound = 15.86 g C = 2.22 g H = 18.51 g N = 63.41 g O

Step 2: Determine the number of moles for each element in this 100-g basis.

$$15.86\ \text{g C} \times \frac{1\ \text{mol C}}{12.01\ \text{g C}} = 1.321\ \text{mol C}$$

$$2.22\ \text{g H} \times \frac{1\ \text{mol H}}{1.01\ \text{g H}} = 2.20\ \text{mol H}$$

$$18.51\ \text{g N} \times \frac{1\ \text{mol N}}{14.01\ \text{g N}} = 1.321\ \text{mol N}$$

$$63.41\ \text{g O} \times \frac{1\ \text{mol O}}{16.00\ \text{g O}} = 3.963\ \text{mol O}$$

Step 3: Write a ***tentative empirical*** formula based on the moles for each element: $C_{1.321}H_{2.20}N_{1.321}O_{3.93}$. We immediately recognize that this formula is incorrect because the ratio between moles for each element must be whole numbers (a compound can't have fractional amounts of an atom!).

Step 4: Convert the formula to a form with only integer coefficients for each element. We first divide each coefficient by the smallest coefficient, here 1.321, to produce a revised tentative empirical formula:

$$C_{\frac{1.321}{1.321}}H_{\frac{2.20}{1.321}}N_{\frac{1.321}{1.321}}O_{\frac{3.963}{1.321}} = C_{1.000}H_{1.66}N_{1.000}O_{3.000}$$

We use the guideline that if our coefficients were within ±0.05 of being integers, we could round off to integers. Sometimes, however, this first adjustment is insufficient, as we see for this example. We then consider multiplying our revised formula by small integers 2×, 3×, 4× ... stopping when we obtain coefficients that are sufficiently close to integers:

2×: $C_{2.000}H_{3.32}N_{2.000}O_{6.000}$ (*insufficiently* close to integer-valued coefficients)

3×: $C_{3.000}H_{4.99}N_{3.000}O_{9.000}$ (*sufficiently* close to integer-valued coefficients)

We conclude our empirical formula is $C_3H_5N_3O_9$. Note that very rarely do we need to multiply by 5× or larger integers!

Step 5: Determine the molecular formula by comparing the molar mass based on the empirical formula with the molar mass for the compound. The molar mass for the empirical formula is determined as:

$$\begin{aligned} M_{\text{C H N O}} &= 3M_{\text{C}} + 5M_{\text{H}} + 3M_{\text{N}} + 9M_{\text{O}} \\ &= 3(12.01\ \text{g/mol}) + 5(1.01\ \text{g/mol}) + 3(14.01\ \text{g/mol}) + 9(16.00\ \text{g/mol}) \\ M_{\text{C H N O}} &= 227.11\text{g/mol} \end{aligned}$$

Because the given molar mass matches the molar mass for the compound, the empirical and molecular formulas are identical. Note that the actual molar mass for the compound always will be a positive-integer multiple of the molar mass corresponding to the empirical formula.

4: Chemical Bonding and Structures of Molecules and Polyatomic Ions

Learning Objectives

By the end of this chapter you should be able to:

1) Evaluate the ionic character of a bond and identify the polarity of a covalent bond based on differences in electronegativity between its constituent atoms.
2) Identify the Lewis *e*- dot symbol for an atom or monatomic ion.
3) Draw the Lewis structure for a molecule or polyatomic ion given its molecular formula.
4) Identify relationships between bond order, bond length, and bond energy.
5) Name inorganic compounds based on their formulas and write formulas for inorganic compounds given their name.
6) Predict the geometry of a molecule or polyatomic ion using the VSEPR model.
7) Determine whether a molecule or polyatomic ion is polar or non-polar.
8) Apply the valence-bond method to identify hybridized orbitals and geometry and polarity of molecules with multiple central atoms.
9) Name simple organic compounds from their chemical formulas and write chemical formulas for simple organic compounds given their names.
10) Describe the structural characteristics of complex organic molecules, including organic polymers.

4.1 Characterizing Bonds Based on Differences in Electronegativity

The physical and chemical properties associated with a compound depend on the compound's structure. Because strategies for identifying structural formulas of molecular and ionic compounds differ profoundly, we need to consider how to distinguish whether a bond is covalent or ionic. This problem is complicated because assuming a bond is either ionic or covalent is an oversimplification: most bonds between atoms of non-identical elements are *neither* purely covalent nor purely ionic (in fact, a purely ionic bond does not exist!) but have *mixed* characteristics (*i.e.*, involve a mixture of transfer and sharing of electrons).

We can address this issue by considering how atoms of different elements differ in their ability to compete for electrons, much like a tug-of-war. We assess the affinity of an atom for electrons based on its electronegativity (EN) and apply a scale for electronegativity devised by Linus Pauling based on quantum mechanics and bond energies, signing a value $0.7 \leq EN \leq 4.0$ for each elment (Table 4.1). Note that electronegativities are not assigned to the noble gases (of which only xenon, Xe, even rarely forms bonds). Also, the electronegativities of metals are less than for non-metals, with electronegativity increasing with decreasing atomic number in a group and with increasing atomic number in a period. This behavior can be understood by considering that core (non-valence) electrons shield valence electrons from the nucleus, and orbitals associated with higher energy levels hold electrons further, on average, from the nucleus (*i.e.*, more core electrons, more weakly held valence electrons), but orbitals with the same principle and orbital angular momentum quantum numbers n and l hold electrons with approximately the same energy (*i.e.*, an increase in nuclear charge along a period results in additional valence electrons held more tightly). Consequently, we observe that fluorine and francium (Fr) have the highest and lowest electronegativities of $EN_F = 4.0$ and $EN_{Fr} = 0.7$, respectively.

Table 4.1. Electronegativities for Elements

H 2.1																
Li 1.0	Be 1.5											B 2.0	C 2.5	N 3.0	O 3.5	F 4.0
Na 0.9	Mg 1.2											Al 1.5	Si 1.8	P 2.1	S 2.5	Cl 3.0
K 0.8	Ca 1.0	Sc 1.3	Ti 1.5	V 1.6	Cr 1.6	Mn 1.5	Fe 1.8	Co 1.8	Ni 1.8	Cu 1.9	Zn 1.6	Ga 1.6	Ge 1.8	As 2.0	Se 2.4	Br 2.8
Rb 0.8	Sr 1.0	Y 1.2	Zr 1.4	Nb 1.6	Mo 1.8	Tc 1.9	Ru 2.2	Rh 2.2	Pd 2.2	Ag 1.9	Cd 1.7	In 1.7	Sn 1.8	Sb 1.9	Te 2.1	I 2.5
Cs 0.8	Ba 0.9	La* 1.1	Hf 1.3	Ta 1.5	W 2.4	Re 1.9	Os 2.2	Ir 2.2	Pt 2.2	Au 2.4	Hg 1.9	Tl 1.8	Pb 1.8	Bi 1.9	Po 2.0	At 2.2
Fr 0.7	Ra 0.9	Ac‡ 1.1														

*Lanthanides: 1.1-1.3
‡Actinides: 1.3-1.5

We evaluate characteristics of bonds based on differences in electronegativity between the pair of atoms forming the bond:

$$\Delta EN = |EN_{atom\,1} - EN_{atom\,2}| \qquad [4.1]$$

As ΔEN increases, the %-ionic character of the bond increases (*i.e.*, the bonds becoms more ionic, less covalent). We identify that if $\Delta EN = 0$, the bond electrons are shared equally, and we have a non-polar covalent bond. The bonds in homonuclear diatomics, such as H_2 and Cl_2, are non-polar covalent, as are bonds between unlike atoms with equal electronegativities in molecules such as NCl_3. If $0 < \Delta EN \leq 1.7$, however, the bond electrons are shared unequally, such that a greater percentage of the bond electrons are associated with the more electronegative atom, and we have a polar covalent bond. The presence of unequal sharing results in the formation of partial positive and negative charges, denoted with the symbols δ^+ and δ^-, respectively, on the less and more electronegative atoms, respectively (Figure 4.1). We observe that less equal sharing occurs as ΔEN increases. If $\Delta EN > 1.7$, the sharing of bond electrons is so unequal that electron transfer dominates, and we have a bond that is primarily ionic; however, only in the limit of infinite separation between a pair of atoms is a bond 100% ionic.

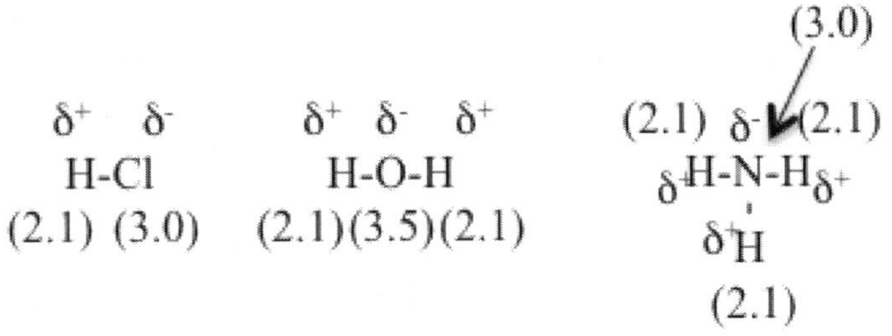

δ^+ δ^-
H-Cl
(2.1) (3.0)

δ^+ δ^- δ^+
H-O-H
(2.1)(3.5)(2.1)

(3.0)
(2.1) δ^- (2.1)
δ^+H-N-Hδ^+
δ^+H
(2.1)

Figure 4.1. Examples of polar covalent bonding. Electronegativities for atoms given in parentheses.

Example 4.1: Evaluating Bond Characteristics

Characterize the bonds in the following diatomic species as either nonpolar covalent, polar covalent, or ionic based on differences in electronegativity between the participating atoms.

a) LiH

b) NaCl

c) O_2

d) NO

e) MgO

Solutions

As each of the compounds contains a single bond, we evaluate the difference in electronegativity for the elements in the bond using Equation 4.1 to determine whether the bond is non-polar covalent ($\Delta EN = 0$), polar covalent ($0 < \Delta EN \leq 1.7$), or ionic ($\Delta EN > 1.7$).

a) LiH has a polar covalent bond because $\Delta EN = |\, 2.1 - 1.0 \,| = 1.1$. Note that this bond has significant ionic characteristics because of its intermediate value for ΔEN.

b) NaCl has a primarily ionic bond because $\Delta EN = |\, 3.0 - 0.9 \,| = 2.1$ is greater than the covalent/ionic bond threshold of 1.7.

c) O_2, like all diatomic compounds of a single element, has a non-polar covalent bond because $\Delta EN = 0$.

d) NO has a polar covalent bond because $\Delta EN = |\, 3.5 - 3.0 \,| = 0.5$. Note that this bond has minimal ionic characteristics because of its low value for ΔEN.

e) MgO has an ionic bond because $\Delta EN = |\ 3.5 - 1.2\ | = 2.3$ is greater than the covalent/ionic bond threshold of 1.7. Note that this bond is predicted to be more ionic than the Na-Cl bond.

4.2 Lewis Theory

Further examination of chemical bonding requires introducing a qualitative model for how atoms share electrons. The Lewis theory (developed by G. N. Lewis from 1916–1919) proposes that the valence electrons for each atom play a fundamental role in how the atom forms bonds, and that atoms transfer electrons to form ionic bonds or share electrons to form covalent bonds such that each atom (with few exceptions) has eight valence electrons—termed an octet—and assume the electron configuration for a noble gas. We apply Lewis theory using Lewis e^- dot symbols as a fourth alternative for representing electron configurations besides orbital diagrams, condensed spdf notation, and valence shell notation. The Lewis e^- dot symbol for an element (Table 4.2) is generated based on the following guidelines:

- The number of valence electrons is the group/family for the element in the periodic table. Note that because this strategy does not work for transition metals, for which the valence shell includes electrons in the d subshell, we will not consider writing Lewis e^- dot symbols for transition metals.
- Depict the number of valence electrons using dots arranged around the atomic symbol for the element. Note that there is no sidedness in the depiction of these electrons.

Tab 4.2. Lewis e^- Dot Symbols

Group/Family	Generic Lewis Symbol		Examples	
	Atom	Ion	Atom	Ion
1A (alkali metals)	$X\cdot$	X^+	$Na\cdot$	Na^+
2A (alkaline metals)	$\cdot X\cdot$	X^{2+}	$\cdot Ca\cdot$	Ca^{2+}
3A	$\cdot\dot{X}\cdot$	X^{3+}	$\cdot\dot{Al}\cdot$	Al^{3+}
4A	$\cdot\underset{\cdot}{\dot{X}}\cdot$	Not observed	$\cdot\underset{\cdot}{\dot{C}}\cdot$	Not observed
5A	$\cdot\underset{\cdot}{\dot{X}}:$	$[\cdot\underset{\cdot}{\dot{X}}:]^{3-}$	$\cdot\underset{\cdot}{\dot{N}}:$	$[\cdot\underset{\cdot}{\dot{N}}:]^{3-}$
6A	$:\underset{\cdot}{\dot{X}}:$	$[:\underset{\cdot}{\dot{X}}:]^{2-}$	$:\underset{\cdot}{\dot{O}}:$	$[:\underset{\cdot}{\dot{O}}:]^{2-}$
7A (halogens)	$:\underset{\cdot}{\ddot{X}}:$	$[:\underset{\cdot}{\ddot{X}}:]$	$:\underset{\cdot}{\ddot{F}}:$	$[:\underset{\cdot}{\ddot{F}}:]$
8A (noble gases)	$:\underset{\cdot\cdot}{\ddot{X}}:$	Not observed	$:\underset{\cdot\cdot}{\ddot{Ne}}:$	Not observed

4.3 Drawing Lewis Structures for Molecules and Polyatomic Ions

We generate structural formulas for molecules and polyatomic ions based on Lewis structures as combinations of Lewis e^- dot symbols linked by shared pairs of electrons as representations of covalent bonds. Any valid Lewis structure must satisfy four requirements:

1. All valence electrons *must* appear in the structure.

2. All valence electrons are *usually* paired. Exceptions to this rule include some nitrogen-containing compounds and other free radicals. We will focus solely on polyatomic species in which all of the electrons are paired.

3. Each atom *usually* has a valance octet. Exceptions that we will consider include: hydrogen and lithium (Li), which have two valence electrons (a duet); beryllium (Be), which has four valence electrons (a quartet); and boron (B) and aluminum (Al), which have six valence electrons (a sextet). Note that non-metals in the 3rd third and higher periods (*e.g.*, sulfur, S, and phosphorus, P) that are bonded to highly electronegative atoms (*e.g.*, oxygen) sometimes have ten or twelve valence electrons, forming expanded valence shells (expanded octets). We will not consider further compounds containing atoms with expanded valence shells.

4. Some bonds involve sharing more than one pair of electrons. In particular, double and triple bonds involve two and three pairs of shared electrons, respectively.

We apply the following seven-step strategy for drawing valid Lewis structures:

Step 1: Count the total number of valence electrons. For polyatomic ions add/subtract the charge. Determine the number of valence electron pairs as half the number of valence electrons.

Step 2: Draw a skeletal structure, arranging the atoms in the alignment in which they are connected by bonds. Central atoms are atoms bonded to two or more other atoms; terminal atoms are atoms bonded to only one other atom. Carbon (except in carbon monoxide, CO), silicon, and germanium (Ge) are always central atoms; hydrogen, lithium, or beryllium are always terminal atoms. Atoms (other than H, Li, and Be) with lowest electronegativity are central atoms; atoms with highest electronegativity are terminal atoms. Oxygen,however, occurs as a central atom in hydroxy (-O-H) and peroxy (-O-O-) linkages. In general, compact, symmetrical structures are favored (except for polymers, large chain-like macromolecules).

Step 3: Place two valence electrons on each bond, subtracting the number of electrons or pairs placed from the totals determined in Step 1.

Step 4: Complete octets for terminal atoms (other than the exceptions such as H noted above, for which their special rules apply), subtracting the number of electrons or pairs placed from the running totals determined in Step 3.

Step 5: Place the remaining valence electrons on the central atom(s), attempting to complete octets as possible (noting the exceptions for B and Al) and subtracting the number of electrons or pairs placed from the running totals determined in Step 3.

Step 6: If central atoms without filled valence shells remain, introduce additional sharing in the form of double and/or triple bonds as necessary.

Step 7: Place brackets and the charge around the structure for a polyatomic ion.

Note that a valid Lewis structure, by itself, does *not* depict the shape of the molecule or polyatomic ion!

As we examine a series of examples, we will defer learning how to name compounds that don't have familiar names until Sections 4.8 and 4.9 for molecular compounds containing carbon, and Section 4.5 for all other compounds.

Guided Example 4.1: H_2O (Water)

The O atom contributes six valence electrons and each H atom two valence electrons, for a total of eight valence electrons (four valence electron e^- pairs). Draw a skeletal structure with O as the central atom, H – O – H, such that two e^- pairs remain. Assign these remaining pairs to complete the octet for the central O atom as $\mathrm{H-\ddot{\underset{..}{O}}-H}$.

Guided Example 4.2: CF_2Cl_2 (Difluorodichloromethane)

This molecule, a very stable chlorinated refrigerant also known as Freon-12, Refrigerant 12, or R-12 but now no longer used due to its link as a contributor to the depletion of the ozone layer, has a total of 32 valence electrons (four from the C atom and seven from each fluorine and chlorine atom), for a total of 16 valence e^- pairs.

C is the central atom, with twelve e^- pairs remaining.

```
  F
  |
F-C-Cl
  |
  Cl
```

Assign three e^- pairs to each terminal F and Cl atom.

```
    ..
   :F:
 ..  |  ..
:F-C-Cl:
 ..  |  ..
   :Cl:
    ..
```

Guided Example 4.3: C_2F_4 (Tetrafluoroethene)

This molecule, a precursor to the polymer tetrafluoroethylene (trademarked as Teflon® by DuPont), has thirty-six valence electrons (four from each C atom and seven from each fluorine), for a total of eighteen valence e^- pairs.

Each C is a central atom, with three F atoms bonded to each C atom and thirteen e^- pairs remaining.

```
  F
  |
F-C-C-F
    |
    F
```

Assign three e^- pairs to each terminal F atom and one e^- pair to a central C atom.

```
   ..
  :F:
..  | .. ..
:F-C-C-F:
..     | ..
      :F:
       ..
```

Because we have run out of valence e^-, create a double bond between the C atoms.

```
   ..
  :F:
..  |    ..
:F-C=C-F:
..     | ..
      :F:
       ..
```

Guided Example 4.4: OH- (the Hydroxide Ion)

This polyatomic ion has eight valence electrons (six from the O atom, one from the H atom, and one for the -1 charge), for a total of four valence e^- pairs. Draw a skeletal structure as O – H, such that three e^- pairs remain. Assign these remaining pairs to complete the octet for the O atom, $:\ddot{\underset{..}{O}}-\mathrm{H}$, and place brackets around the structure and label its charge as $\left[:\ddot{\underset{..}{O}}-\mathrm{H}\right]$.

Guided Example 4.5: SO_2 (Sulfur dioxide)

This molecule, a pollutant generated by burning sulfur-bearing coal, has eighteen valence electrons (six each from the S atom and pair of O atoms), for a total of nine valence e^- pairs.

S (with a lower electronegativity) is the central atom, with seven e^- pairs remaining.

O – S – O

Assign three e^- pairs to each terminal O atom and one e^- pair to the central S atom.

:Ö – S̈ – Ö:

Because we have run out of valence e^-, create a double bond between the S atom and an O atom.

:Ö = S̈ – Ö: -or- :Ö – S̈ = Ö:

Both structures appear correct but neither by itself is complete because the actual bonds are resonance hybrids of single and double bonds (*i.e.*, a "1½" bond). This phenomenon is termed resonance: identical arrangements of atoms differing only in arrangement of electrons. Consider that we can distinguish between individual atoms for an element using different isotopes for this element. We depict resonance using a double arrow.

:Ö = S̈ – Ö: ↔ :Ö – S̈ = Ö:

Guided Example 4.6: CH_3COOH (Ethanoic acid)

This molecule, found in vinegar and commonly known as acetic acid, contains a –COOH group, termed a carboxyl group, one of several functional groups for molecules containing carbon that we will consider in Section 4.9 Because this formula contains elements of the structure, it is known as a condensed structural formula. CH_3COOH has twenty-four valence electrons (four from each C atom, one from each H atom, and six from each O atom), for a total of twelve valence e^- pairs.

Each C is a central atom, with a carboxyl group -COOH) not being a peroxy (-O-O-) linkage, such that five e^- pairs remain:

```
 H
H-C-C-O-H
 H O
```

Assign three e^- pairs to the terminal O atom and two e^- pairs to the central O atom:

```
 H    ..
H-C-C-O-H
 H :O:
    ..
```

Because we have run out of valence e^-, we need to create a double bond between a C atom and one of the O atoms. Two possibilities exist (Figure 4.2), but are both correct? Only one?

```
 H     ..           H     ..
H-C-C-O-H          H-C-C=O-H
 H ||              H :O:
   O:                 ..
   ..
(a)                 (b)
```

Figure 4.2. Correct (a) and incorrect (b) Lewis structures for ethanoic acid, CH_3COOH.

We distinguish between these possibilities based on the concept of formal charges, *apparent* (not real!) charges arising because atoms do *not* contribute equal numbers of electrons to bonds. We calculate the formal charge, denoted FC, for any individual atom based on the following formula:

$$FC = \{\#\text{ of valence } e^- \text{ on free atom}\} - \{\#\text{ of lone pair } e^-\} - \tfrac{1}{2}\{\#\text{ of bond pair } e^-\} \quad [4.2]$$

Each atom in the correct structure has a formal charge $FC = 0$. For the incorrect structure, however, one O atom has $FC = +1$ and one O atom has $FC = -1$:

O atom on upper right: $FC = (6) - (2) - \tfrac{1}{2}(6) = +1$
O atom on lower left: $FC = (6) - (6) - \tfrac{1}{2}(2) = -1$

We present these results using the following scheme: depict any non-zero formal charges for an atom above that atom using "+" for +1, "+2" for +2, … and "-" for -1, "-2" for -2, … We then apply the following guidelines for distributing formal charges:

- The sum of the individual formal charges is the overall charge.
- We seek to minimize the number of non-zero formal charges, such that structures with fewer non-zero formal charges are preferred.
- If the presence of non-zero formal charges is unavoidable, preferred structures have negative formal charges on atoms with higher electronegativity and positive formal charges on atoms with lower electronegativity.
- Preferred structures do *not* have formal charges with the same sign on adjacent atoms.

Applying these guidelines allows us to identify the correct structure.

```
    H                    H
    |    ..              |    .+
H-C-C-O-H           H-C-C=O-H
  | ||  ..             |  |
  H O:                 H:O:
    ..                   ..
   (a)                  (b)
```

Figure 4.3. All formal charges for the correct Lewis structure for CH_3COOH (a) are zero; incorrect Lewis structure (b) has multiple atoms with non-zero formal charge.

Note that just as it is possible for multiple compounds to have the same empirical formula but different molecular formulas, it is possible for multiple compounds to have the same molecular formula but different structural formulas. Such molecules are called isomers and are discussed in Section 4.9.

4.4 Other Properties of Bonds

Consider three properties associated with individual bonds. Bond order (BO) has a value of 1 for a single bond, 2 for a double bond, and 3 for a triple bond. We observe that as bond order increases (*i.e.*, more electrons are shared), the nuclei connected by the bond are held together more tightly. Bond length (l_B) is the *equilibrium* distance between the pair of nuclei joined by a covalent bond. We observe that as bond order increases, the bond length decreases. Bond dissociation energy (D) is the energy that must be added to break one mole of bonds in the vapor phase. We observe that as bond order increases, bond dissociation energy increases.

Example 4.2: Comparing O-O Bonds Among Different Molecules

Consider the following set of molecules: H_2O_2, O_2 (molecular oxygen), and O_3 (ozone).

a) For each molecule draw a valid Lewis structure, including any resonance hybrids.

b) Rank the molecules in order of increasing length of their O-O bonds.

c) Rank the molecules in order of increasing dissociation energy of their O-O bonds.

Solutions

a) Valid Lewis structures are given below. Note that only O_3 (ozone) shows resonance.

O=O   O=O—O ↔ O—O=O   H—O—O—H

b) Bond order for the O-O bond varies from 1 for H_2O_2 to 2 for O_2 and 1.5 for O_3. Accordingly, we predict O_2 has the shortest O-O bond and HOOH the longest O-O bond. Experimental measurements of O-O bond length concur: l_B = 1.21 Å for O_2, 1.272 Å for O_3, and 1.452 Å for HOOH. We can predict these values (obtained experimentally using advanced forms of spectroscopy) using computational chemistry with *Gaussian*, a software package that solves the Schrödinger equation numerically for multi-electron atoms and molecules.

c) Based on bond order for the O-O bond, we predict O_2 has the strongest O-O bond and HOOH the weakest O-O bond.

Because covalent bonds behave like (and can be described as) springs connecting a pair of nuclei, the length of any covalent bond is not constant but rather varies as the bond vibrates; the equilibrium bond length is merely the distance between the nuclei with the lowest energy. Species with only one covalent bond, such as molecular oxygen, can solely experience stretches, but species with multiple covalent bonds can experience complex vibrational motion, including potentially both symmetric and asymmetric stretching, rocking, scissoring, and wagging, as animated at http://en.wikipedia.org/wiki/Molecular_vibration.

Recall from Section 4.1 that polar covalent bonds involve partial separations of charge, in contrast with non-polar covalent bonds that involve no partial separations of charge. When sets of complete or partial charges move asymmetrically, their motion distorts the electric fields associated with the bond. Light can interact with these electric fields if the frequency of the light matches the frequency of vibration for the bond. Because bonds vibrate with frequencies corresponding to the infrared (IR) region of the electromagnetic spectrum, molecules absorb IR light if they have polar covalent bonds and can vibrate asymmetrically. CO_2 is an example of such a molecule. CO_2 exhibits four modes of vibrational motion: symmetric and asymmetric stretches and a pair of bending modes oriented perpendicular to each other in 3-D. Only the asymmetric stretch and the bends are IR active. In contrast, molecules with non-polar covalent bonds, such as O_2 and N_2, are IR inactive (*i.e.*, do not absorb IR radiation). Greenhouse gases (which include water and CH_4, known as methane) are molecules that can absorb IR light radiated from the Earth's surface, preventing the escape of this radiation into outer space.

4.5 Inorganic Nomenclature

Now that we have learned how to identify structures for molecules and polyatomic ions, we can consider how to name these structures. Nomenclature is the use of a formal system to match names and formulas for different compounds. The commonly accepted system for chemical nomenclature established by the IUPAC (the International Union of Pure and Applied Chemistry) is based on dividing compounds into two categories as alternatives to classifying a compound as either molecular or ionic: organic compounds, molecular compounds formed by carbon with other elements and the compounds of life and fossil fuels, and inorganic compounds, all other molecular and all ionic compounds. We will examine organic nomenclature in Sections 4.8 and 4.9.

Nomenclature for inorganic compounds is based on identifying the number of elements in an inorganic compound. Binary compounds are inorganic compounds consisting of only two elements. Binary ionic compounds, formed by a metal cation and a non-metal anion, have a two-part name in which the first name is that of the metal and the second name is the base name of the non-metal with the suffix "-ide" (Table 4.3). For metals not from Group 1/1A or 2/2A we also enclose within parentheses the metal's positive charge in Roman numerals after the metal's name. For example, CaF_2 is "calcium fluoride," and Fe_2O_3 is "iron (III) oxide."

Table 4.3. Names of Common Monatomic Non-metal Anions

Anion	Name
F^-	fluoride
Cl^-	chloride
Br^-	bromide
I^-	iodide

Anion	Name
O^{2-}	oxide
S^{2-}	sulfide
N^{3-}	nitride
H^-	hydride

Binary inorganic compounds formed between a pair of nonmetals also have a two-part name: the first name is the name of the first element, and the second name is the base name of the second element with the suffix "-ide"; however, we append a Greek prefix (Table 4.4) for the number of atoms in the element to its name, with the exception that no prefix is used if there is only one atom for the first element. For example, N_2O_5 is "dinitrogen pentoxide" and SiO_2 is "silicon dioxide."

Table 4.4 Greek Prefixes for Inorganic Nomenclature

Number	Prefix
1	mono-
2	di-
3	tri-
4	tetra-
5	penta-
6	hexa-

Binary acids, formed by hydrogen paired with a halogen (a Group 17/7A element), have a two-part name in which the first name is the base name of the halogen with the prefix "hydro-" and suffix "ic" and the second name "acid." For example, HF is "hydrofluoric acid."

Inorganic compounds with three or more elements are frequently compounds with polyatomic ions. Table 4.5 lists commonly encountered polyatomic ions, including the hydroxide ion introduced in Section 4.3. Note that the ammonium ion is the only common polyatomic cation and that only the hydroxide and cyanide ions have names that end with the suffix "-ide. Oxygen is a common element in polyatomic ions, forming oxoanions. All common oxoanions with chlorine or nitrogen have a charge of -1. Some non-metals, including chlorine, nitrogen, and sulfur, form a series of oxoanions with different numbers of oxygen atoms but constant charge (Figure 4.4). The prefix "thio-'" is used when sulfur is substituted for one or more oxygen atoms (*e.g.*, compare sulfate and thiosulfate). Many oxoanions also have oxoacid counterparts. For example, H_2SO_4 can be called "dihydrogen sulfate" or its more common name of "sulfuric acid" and H_3PO_4 can be called "trihydrogen phosphate" or its more common name of "phosphoric acid."

Table 5. Common Polyatomic Ions

Formula	Name
NH_4^+	ammonium
OH^-	hydroxide
CN^-	cyanide
ClO^-	hypochlorite
ClO_2^-	chlorite
ClO_3^-	chlorate
ClO_4^-	perchlorate

Formula	Name
NO_2^-	nitrite
NO_3^-	nitrate
SO_3^{2-}	sulfite
SO_4^{2-}	sulfate
$S_2O_3^{2-}$	thiosulfate
CO_3^{2-}	carbonate
PO_4^{3-}	phosphate

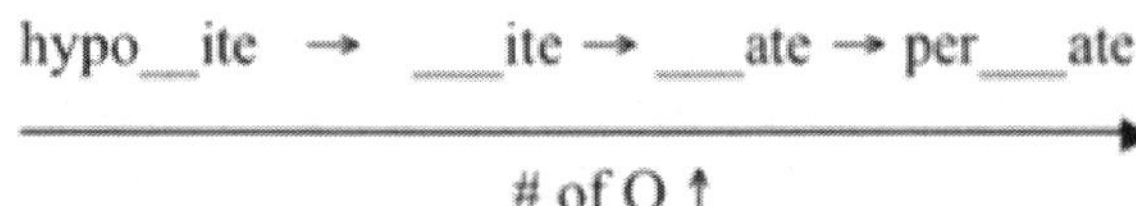

Figure 4.4. Trends in nomenclature for oxoanions of chlorine, nitrogen, and sulfur.

One additional type of inorganic compound with more than two elements is the hydrates, compounds with one or more chemically associated water molecules. These compounds are named by appending to the name of the core compound the third name "hydrate" with the Greek prefix for the number of associated water molecules per formula unit. For example, $CaSO_4 \bullet 2\ H_2O$ is "calcium sulfate dihydrate," commonly known as the building material gypsum.

Example 4.3: Naming Inorganic Compounds

Write the names for $CaCO_3$, FeO, NO_2, HBr, and GaAs.

Solutions

$CaCO_3$ is "calcium carbonate," the major component of limestone. FeO is "iron (II) oxide," a pigment used in some tattoos. NO_2 is "nitrogen dioxide," a major component in air pollution. HBr is "hydrobromic acid" (or, alternatively, "hydrogen bromide"). GaAs is "gallium (III) arsenide," a semiconductor used in some modern electronics.

Example 4.4: Identifying Formulas for Inorganic Compounds from Their Names

Identify the formulas for hydrogen cyanide, sodium hydroxide, ammonium nitrate, potassium iodide, and manganese (IV) oxide.

Solutions

Hydrogen cyanide has the formula HCN and is used in the electroplating of gold and silver. Sodium hydroxide has the formula NaOH and is a strong base used as a catalyst in the production of biodiesel. Ammonium nitrate has the formula NH_4NO_3 and is a major component of synthetic fertilizer. Potassium iodide has the formula KI and is a nutritional supplement and component of some dye-sensitized solar cells. Manganese (IV) oxide has the formula MnO_2 and is a component of dry cell batteries.

4.6 Molecular Geometry Based on the VSEPR Model

We can identify the shape of a molecule or polyatomic ion based on further consideration of its Lewis structure. We define molecular shape (geometry) as the equilibrium three-dimensional spatial arrangement of atoms connected by covalent bonds. Specifying this arrangement requires identifying both equilibrium bond lengths, l_B, and equilibrium bond angles (*i.e.*, the angle between a pair of bonds connected to a central atom), θ_B. When we consider possible molecular geometries, we encounter three types of possibilities:

1. The molecule has only one bond and is linear, with no associated bond angles. Such a molecule has no central atom. Examples include diatomic molecules, such as molecular hydrogen (H-H), molecular nitrogen ($\ddot{N} \equiv \ddot{N}$), and hydrochloric acid ($H - \ddot{\underset{..}{Cl}}:$).

2. The molecule has two bonds, with two possibilities for the equilibrium bond angle and associated geometry: $\theta_B = 180°$, such that the molecule is linear, or $\theta_B < 180°$, such that the molecule is nonlinear. We shall consider some examples for each possibility shortly.

3. The molecule has three or more bonds, with equilibrium bond angles other than the 180° generally observed.

Note that the second and third possibilities both involve molecules with at least one central atom.

One strategy to predict molecular shape is based on a set of concepts known as valence-shell electron-pair repulsion (VSEPR) theory. The basic tenet of VSEPR theory is that valence electron pairs orient themselves at equilibrium to minimize repulsions with each other, regardless of whether these pairs occur in bonds or as lone pairs. Application of VSEPR theory is based on first counting the number of electron groups associated with a central atom using the following formula:

$$\begin{Bmatrix} \text{\# of} \\ e^- \text{ groups} \end{Bmatrix} = \begin{Bmatrix} \text{\# of} \\ \text{bond groups} \end{Bmatrix} + \begin{Bmatrix} \text{\# of} \\ \text{lone pairs} \end{Bmatrix} \quad [4.3]$$

In this formula single, double, and triple bonds each count as a single bond group! For example, for the molecule CO_2, with Lewis structure $:\ddot{O} = C = \ddot{O}:$, the central C atom has two bond groups and no lone pairs, such that it has two electron groups.

We apply VSEPR theory using a four-step strategy:

Step 1: Draw a plausible Lewis structure and identify the central atom(s).

Step 2: For each central atom determine whether it has two, three, or four electron groups and whether it has zero, one, or two lone pairs.

Step 3: Use the number of electron groups to establish an electron group geometry for each central atom. In particular, the central atom will be associated with a linear geometry if it has two electron groups, a trigonal planar geometry if it has three electron groups, and a tetrahedral geometry if it has four electron groups.

Step 4: Establish the molecular geometry based on the electron group geometry and the number of lone pairs using Table 4.6.

Table 4.6. VSEPR Table for Molecular Geometries

# of e^- Groups	e^- Group Geometry	# of Lone Pairs	VSEPR Notation (Molecular Geometry)	Ideal Bond Angles
2	Linear	0	AX_2 (Linear)	180°
3	Trigonal-planar	0	AX_3 (Trigonal-planar)	120°
		1	AX_2E (Bent)	
4	Tetrahedral	0	AX_4 (Tetrahedral)	109.5°
		1	AX_3E (Trigonal pyramidal)	
		2	AX_2E_2 (Bent)	

We note the following trends in repulsion between electron pairs:

- Repulsion between electron pairs increases as the bond angle decreases. In particular, repulsion between electron pairs separated by ~109° is greater than repulsion between electron pairs separated by 120°, which in turn is greater than repulsion between electron pairs separated by 180°.
- Lone pair electrons repel more than bond pair electrons. In particular, repulsion between two lone pairs is greater than repulsion between a lone pair and a bond pair, which in turn is greater than repulsion between two bond pairs. This behavior results in the observation that the bond angle of ~104.5° for H_2O is less than the bond angle of ~107.5° for NH_3 (ammonia), which in turn is less than the ideal bond angle of 109.5° for CH_4.

Note that VSEPR theory and the above rules explain why cyclic structures consisting of three- or four-member rings are uncommon: the angles between the bond pairs would be 60° for a three-member ring and 90° for a four-member ring, which introduces enormous repulsion between electron pairs around central atoms on the ring. We will not consider such highly strained structures further other than recognizing that they do exist (*e.g.*, epoxides are three-member rings of one O atom and two C atoms).

Example 4.5: Determining Molecular Shape Using the VSEPR Table

Determine the VSEPR notation and geometry associated with each central atom for:

a) SiF_4 (a by-product of etching silicon wafers or glass with HF)

b) CO_3^{2-} (formed when CO_2 dissolves in water)

c) N_2O (a relatively nonreactive species known as the "Teflon" of the atmosphere)

d) CH_3COOH

Solutions

a) SiF_4, silicon tetrafluoride, has the Lewis structure depicted below, an AX_4 VSEPR notation for its central silicon atom, and a tetrahedral molecular geometry.

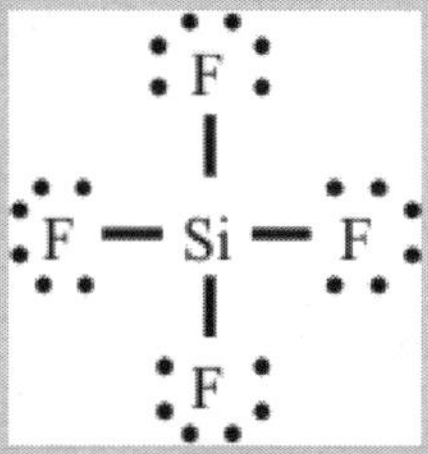

b) CO_3^{2-}, the carbonate ion, has the Lewis structure depicted below as three resonance hybrids, an AX_3 VSEPR notation for its central carbon atom, and a trigonal planar molecular geometry.

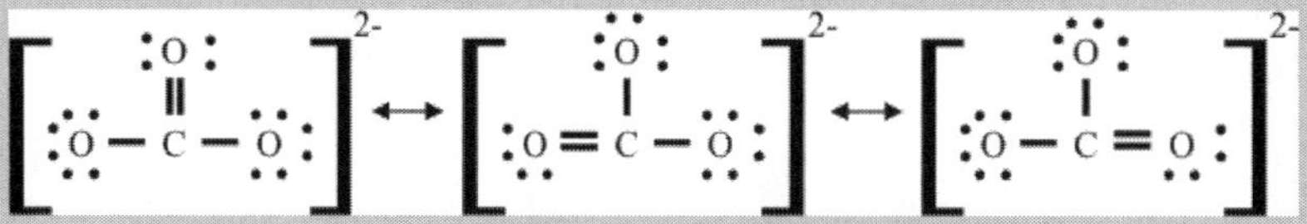

c) N_2O, dinitrogen monoxide, has the Lewis structure depicted below as two resonance hybrids, an AX_2 VSEPR notation for its central nitrogen atom, and a linear molecular geometry.

d) The Lewis structure for CH_3COOH, ethanoic acid, which was determined previously in Guided Example 4.6 in Section 4.3 as depicted below, has an AX_4 VSEPR notation and tetrahedral molecular geometry for its left-most carbon atom, an AX_3 VSEPR notation and trigonal planar molecular geometry for its right-most carbon atom, and an AX_2E_2 VSEPR notation and bent molecular geometry for its right-most oxygen atom. The overall shape of this molecule is not easily described, other than that it is clearly neither linear nor planar.

$$\mathrm{H_3C{-}C({=}\ddot{O}{:}){-}\ddot{O}{-}H}$$

4.7 Molecular Polarity

Now that we have established procedures for identifying molecular geometry, we can consider how to identify the polarity of a molecule or polyatomic ion. The overall polarity of a polyatomic species with multiple covalent bonds depends on the polarity of the individual bonds comprising the polyatomic species as well as the spatial arrangement of these bonds. For each bond we can define a dipole moment, $\vec{\mu}$, as the vector pointing from the atom on the bond with a partial positive charge (δ^+) to the atom on the other end of the bond with a partial negative charge (δ^-). The magnitude of the dipole moment, $\|\vec{\mu}\|$, depends on the difference in electronegativity, ΔEN, for the bond: for non-polar covalent bonds $\Delta EN =$ 0 and $\|\vec{\mu}\| = 0$, and for polar covalent bonds $\Delta EN > 0$ and $\|\vec{\mu}\| > 0$, with $\|\vec{\mu}\|$ increasing as ΔEN increases. We can also assign a dipole moment to lone pairs of electrons where the dipole moment $\vec{\mu}_{lone\,pair}$ points towards the lone pair and $\|\vec{\mu}_{lone\,pair}\| > \|\vec{\mu}_{bond\,pair}\|$. The magnitude of the dipole moment is expressed most conveniently in units of debyes (D), where 1 D = 3.336×10^{-30} C-m.

We apply the concept of dipole moment to molecules and polyatomic ions by considering the overall dipole moment for a molecule or polyatomic ion as the vector sum of the dipole moments for individual bonds and lone pairs:

$$\vec{\mu}_{molecule} = \sum \vec{\mu}_i \qquad [4.4]$$

where the set $\{\vec{\mu}_i\}$ consists of all the dipole moments for individual bonds and lone pairs. We then identify that if $\|\vec{\mu}_{molecule}\| = 0$, the overall molecule or polyatomic ion is non-polar, but if $\|\vec{\mu}_{molecule}\| > 0$, the overall molecule or polyatomic ion is polar. The vector addition is simplified by considering the following pair of guidelines concerning symmetry. Symmetric molecules always are non-polar, regardless of the type and/or number of bonds. For example, CO_2, AlH_3 (aluminum hydride), and CH_4 are symmetric and non-polar molecules. In contrast, asymmetric molecules with polar bonds and/or lone pairs are polar. For example, O_3, NH_3, and H_2O are asymmetric with polar bonds.

Computational chemistry using *Gaussian* allows us to evaluate the magnitude of the dipole moment—where this magnitude has units of debyes (D)—and to visualize bond and molecular polarities using electrostatic potential (ESP) maps in which regions of relatively high electron density are depicted in red (medium gray regions in black and white), and relatively low electron density in blue (darker regions in black and white), as depicted in Figure 4.5.

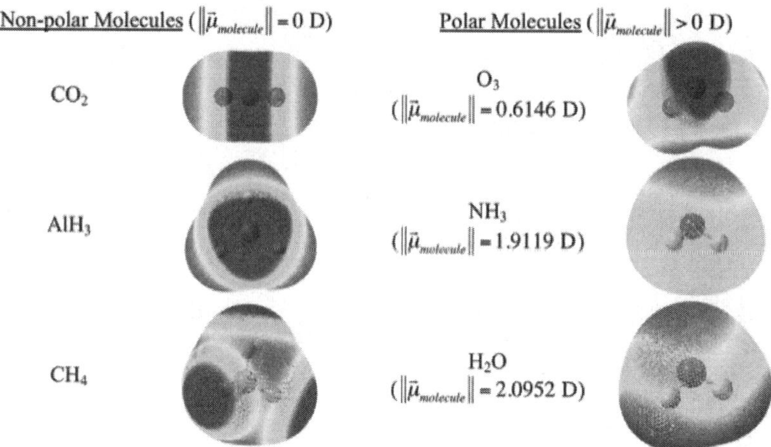

Figure 4.5. Comparison of non-polar and polar molecules, with associated dipole moments and ESP maps generated using *Gaussian* 03.

Example 4.6: Determining Molecular Polarity

Identify whether the species examined in Example 4.5 are polar or non-polar.

Solutions

a) SiF_4 is non-polar because it has a perfectly symmetric arrangement of polar bonds.

b) CO_3^{2-} is non-polar because it has a perfectly symmetric arrangement of polar bonds.

c) N_2O is polar because it has an asymmetric arrangement of polar bonds.

d) CH_3COOH is polar because it has an asymmetric arrangement of polar bonds.

4.8 Describing the Geometry of Hydrocarbons Based on Hybridized Orbitals

To make unambiguous predictions for the geometry and polarity of molecules with multiple central atoms requires a more sophisticated model for chemical bonding than the combination of Lewis theory and VSEPR theory. Valence-bond (VB) theory, based on the concept that covalent bonds occur as the result of overlap between atomic orbitals, offers one such description. VB theory is illustrated most clearly by its application to the structure of hydrocarbons, organic compounds (i.e., molecules containing carbon other than carbon monoxide and carbon dioxide) consisting only of carbon and hydrogen atoms. Hydrocarbons are the basis for fossil fuels, such as natural gas, gasoline, and petroleum-based diesel.

The name for a hydrocarbon consists of two parts. The first part of the name is a Greek-based prefix for the number of carbon atoms. These prefixes, listed in Table 4.7, differ somewhat from the prefixes introduced for inorganic compounds in Section 4.5. The second part of the name is a suffix for the type of C-C bonds (Table 4.8). Carbon-carbon double (C=C) bonds and carbon-carbon triple (C≡C) bonds are types of functional groups, individual or groups of atoms that give an organic molecule its characteristic properties. For example, we encountered our first functional group, the carboxyl functional group, -COOH (ethanoic acid), in Section 4.3. C-C single bonds tend to be more stable and less reactive than C=C double bonds and C≡C triple bonds.

Table 4.7. Greek Prefixes for Organic Nomenclature

Number of C Atoms	Greek Name	Greek Prefix
1	methyl	meth-
2	ethyl	eth-
3	propyl	prop-
4	butyl	but-
5	pentyl	pent-
6	hexyl	hex-

Table 4.8 Suffixes for Nomenclature of Hydrocarbons

Type of C-C Bond	Suffix	Generic Molecular Formula
C-C (single bond)	-ane	C_nH_{2n+2}
C=C (double bond)	-ene	C_nH_{2n} (assuming only 1 C=C bond)
C≡C (triple bond)	-yne	C_nH_{2n-2} (assuming only 1 C≡C bond)

Let's now explore the application ofVB theory to describe the shape of three hydrocarbons with different bonding: methane (CH_4), ethene (C_2H_4, commonly known as ethylene and with a Lewis structure similar to that of tetrafluoroethene, examined in Guided Example 4.3 in Section 4.3), and ethyne (C_2H_2, commonly known as acetylene and with a Lewis structure of $H - C \equiv C - H$).

Guided Example 4.7: Methane

Methane is formed between a C atom with a $[He]2s^22p^2$ ground-state electron configuration and four H atoms with $1s^1$ ground-state electron configurations. Consider the corresponding orbital diagrams for valence electrons:

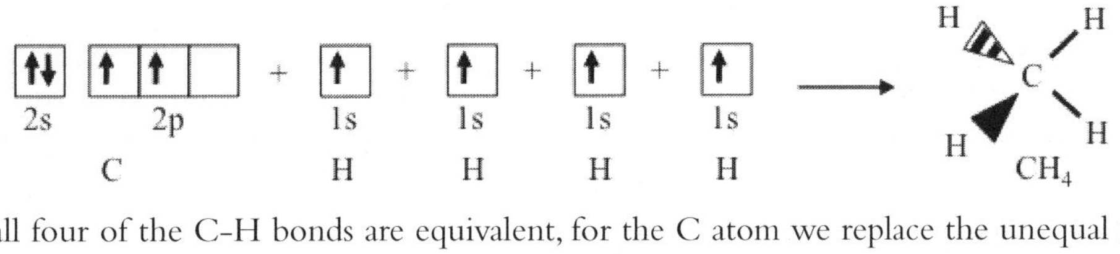

2s 2p (C) + 1s (H) + 1s (H) + 1s (H) + 1s (H) ⟶ CH_4

Because all four of the C-H bonds are equivalent, for the C atom we replace the unequal 2s and 2p atomic orbitals—which have different energies and two valance electrons each—with a single sp^3 hybridized orbital, such that the interaction between the four valence electrons in a sp^3 hybridized orbital for the C atom with the single valence electrons for each H atom produces four equivalent sigma (σ) bonds oriented 109.5° apart (as predicted by VSEPR theory):

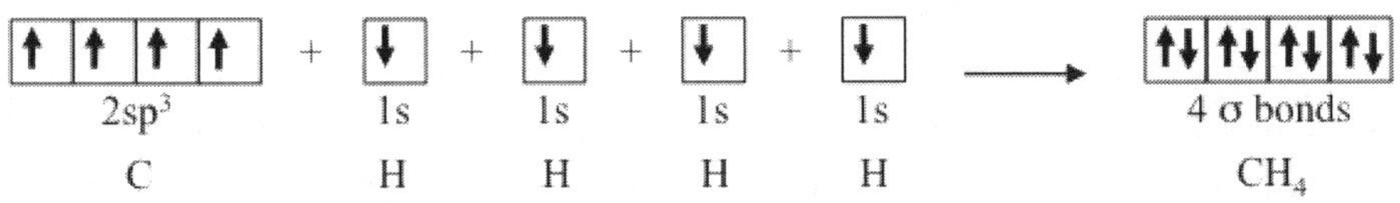

$2sp^3$ (C) + 1s (H) + 1s (H) + 1s (H) + 1s (H) ⟶ 4 σ bonds (CH_4)

Note that these σ bonds, produced by end-to-end overlap of atomic orbitals, are free to rotate.

Guided Example 4.8: Ethene

Ethene is formed between a pair of C atoms bonded to each other through a double bond (*i.e.*, two pairs of shared electrons), with each C atom also bonded to a pair of H atoms. Consider the corresponding orbital diagrams for valence electrons:

[↑↓] [↑|↑|] + [↑↓] [↑|↑|] + [↑] + [↑] + [↑] + [↑] → $H_2C=CH_2$

2s 2p (C) + 2s 2p (C) + 1s (H) + 1s (H) + 1s (H) + 1s (H) → C_2H_4

In order to produce the observed orientation of 120° between the pair of C-H and the C-C bond around each C atom (based on a VSEPR notation of AX_3), we exchange three of the four valence electrons for each C atom from unequal 2s and 2p atomic orbitals into a single sp^2 hybridized orbital, but leave one valence electron for each C atom in a higher-energy 2p atomic orbital:

[↑|↑|↑] [↑] + [↓|↓|↓] [↓] + [↓] + [↓] + [↑] + [↑] → [↑↓|↑↓|↑↓|↑↓|↑↓] [↑↓]

$2sp^2$ 2p (C) + $2sp^2$ 2p (C) + 1s (H) + 1s (H) + 1s (H) + 1s (H) → 5 σ bonds, 1 π bond (C_2H_4)

We find that the interaction among the three valence electrons for each C atom in a sp^2 hybridized orbital and the set of single valence electrons for the four H atoms produces five equivalent σ bonds oriented 120° apart (as predicted by VSEPR theory). We also observe that the out-of-plane, side-to-side overlap between the 2p orbitals for the pair of C atoms results in a pi (π) bond. Note that π bonds (and, therefore, double bonds) are *not* free to rotate because rotation would break these bonds!

VSEPR theory alone tells us that each C atom for ethene has a trigonal planar geometry but not whether the two pairs of H atoms lie in the same or different planes. In contrast VB theory tells us that all of the C and H atoms lie in a single plane, as depicted in Figure 4.6.

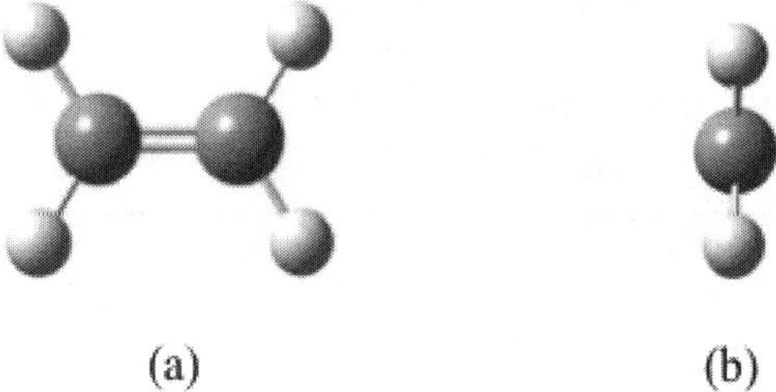

(a) (b)

Figure 4.6. Face-on (a) and side-on (b) views of the planar shape of ethene. Images generated using *Gaussian* 03.

Guided Example 4.9: Ethyne

Ethyne is formed between a pair of C atoms bonded to each other through a triple bond (*i.e.*, three pairs of shared electrons), with each C atom also bonded to an H atom. Consider the corresponding orbital diagrams for valence electrons:

[↑↓] [↑|↑|] + [↑↓] [↑|↑|] + [↑] + [↑] → H—C≡C—H

2s 2p (C) + 2s 2p (C) + 1s (H) + 1s (H) → C_2H_2

In order to produce the observed orientation of 180° between the C-H and the C-C bond around each C atom (based on a VSEPR notation of AX_2), we exchange two of the four valence electrons for each C atom from unequal 2s and 2p atomic orbitals for a single sp hybridized orbital, but leave two valence electrons for each C atom in a pair of higher-energy 2p orbitals:

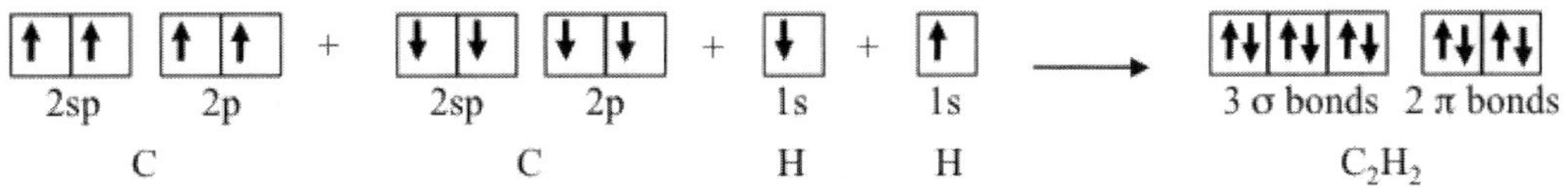

↑ ↑	↑ ↑	+	↓ ↓	↓ ↓	+	↓	+	↑	⟶	↑↓ ↑↓ ↑↓	↑↓ ↑↓
2sp	2p		2sp	2p		1s		1s		3 σ bonds	2 π bonds
C			C			H		H		C_2H_2	

We find that the interaction between the two valence electrons for each C atom in a sp hybridized orbital and the set of single valence electrons for the two H atoms produces three equivalent σ bonds oriented 180° apart (as predicted by VSEPR theory). We also observe that the out-of-plane, side-to-side overlap between a pair of 2p orbitals for the pair of C atoms results in two bonds. Note again that the π bonds are not free to rotate!

The conclusions we have reached for our set of hydrocarbons can be generalized. Table 4.9 combines the summary of the predictions for VSEPR theory from Table 4.5 in Section 4.6 with the corresponding orbital hybridizations predicted by VB theory.

Table 4.9. VSEPR Table for Molecular Geometries with Orbital Hybridization from VB Theory

# of e^- Groups	e^- Group Geometry	# of Lone Pairs	VSEPR Notation (Molecular Geometry)	Ideal Bond Angles	Orbital Hybridization
2	Linear	0	AX_2 (Linear)	180°	sp
3	Trigonal-planar	0	AX_3 (Trigonal-planar)	120°	sp^2
		1	AX_2E (Bent)		
4	Tetrahedral	0	AX_4 (Tetrahedral)	109.5°	sp^3
		1	AX_3E (Trigonal pyramidal)		
		2	AX_2E_2 (Bent)		

Example 4.7: Determining Orbital Hybridization from VB Theory

Assign each central atom in the species examined in Examples 4.5 and 4.6 a hybridization of sp, sp^2, or sp^3.

Solutions

a) The central Si atom in SiF_4 has an AX_4 VSEPR notation and a sp^3 hybridization.

b) The central C atom in CO_3^{2-} has an AX_3 VSEPR notation and a sp^2 hybridization.

c) The central N atom in N_2O has an AX_2 VSEPR notation and a sp hybridization.

d) CH_3COOH has multiple central atoms. As depicted below, the left-most C atom has an AX_4 VSEPR notation and a sp^3 hybridization, the right-most C atom has an AX_3 VSEPR notation and a sp^2 hybridization, and the right-most O atom has an AX_2E_2 VSEPR notation and a sp^3 hybridization.

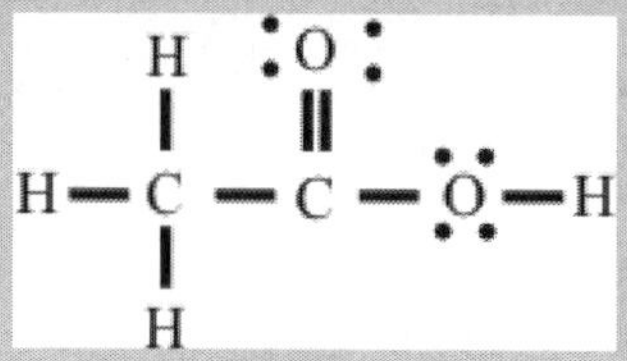

Note that VB theory has limitations, including an inability to explain why molecular oxygen actually has two unpaired electrons and is paramagnetic. A more powerful and sophisticated description, molecular orbital (MO) theory—based on the concept that when atoms combine to form a molecule, atomic orbitals combine to form molecular orbitals—is needed to explain this behavior. We will apply qualitative predictions from MO theory in Section 5.2 to explain some of the properties of metals and semiconductors.

4.9 Nomenclature and Structure of Organic Compounds with Functional Groups

Many organic compounds occur as isomers: multiple structural formulas exist for the same molecular formula. For example, the three Lewis structures depicted in Figure 4.7 correspond to alkenes that share the molecular formula C_4H_8 but differ in the placement of the C=C double bond and/or H atoms. These three isomers are not in resonance, as one would need to change the connectivity between atoms to convert among isomers. To distinguish between these isomers we number the carbon atoms, starting from the end of the molecule closest to the functional group, and append the smallest number for the C atom closest to the functional group as a prefix followed by a hyphen to the molecule's name.

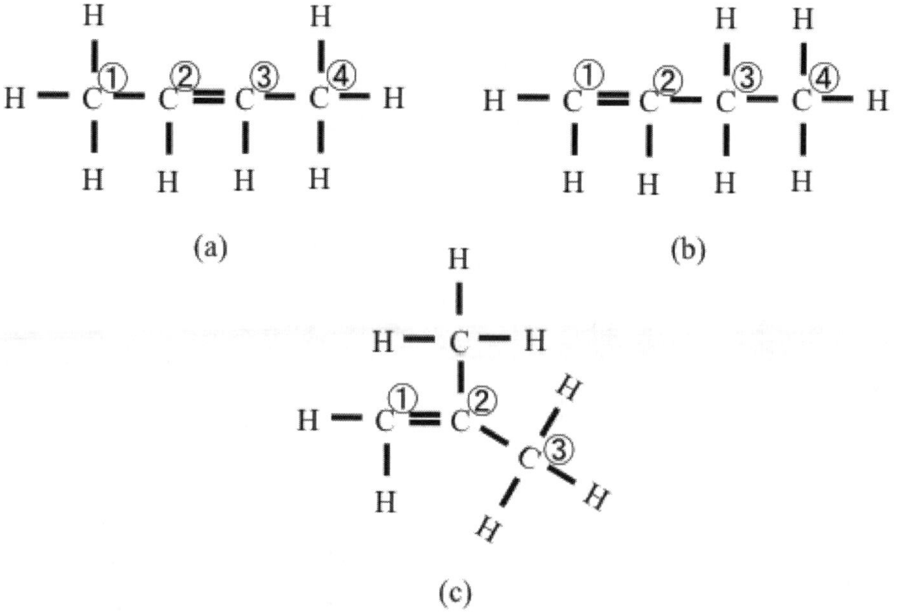

Figure 4.7. Isomers of C_4H_8. (a) but-1-ene; (b) but-2-ene; (c) 2-methyl-prop-1-ene. Numbering enables distinguishing different isomers.

Consider the structure and nomenclature for organic compounds more broadly, including organic compounds that contain other elements (*e.g.*, many organic compounds also contain oxygen, nitrogen, sulfur, and phosphorus). These elements appear as parts of functional groups that give an organic molecule its characteristic properties. We have previously encountered three types of functional groups: C=C double and C≡C triple bonds in Section 4.8, and carboxyl groups (-COOH) in Section 4.3. Organic molecules with functional groups other than C=C or C≡C bonds can be described as hydrocarbons for which one or more hydrogens (-H), methylene groups (-CH_2-), or methyl groups (-CH_3) have been replaced—we use the phrase "substituted"—with one or more atoms of other elements (other than only C or H). In considering examples of functional groups we will frequently employ condensed structural formulas—first introduced in Section 4.3 to represent CH_3COOH as a specific isomer of $C_2H_4O_2$—as a means to depict structural elements more succinctly. We also note that many important organic compounds contain multiple functional groups.

Carboxylic (organic) acids are formed by substituting a carboxyl group for a methyl group (Figure 4.8a). To name a carboxylic acid we append the suffix "-oic" and the separate word "acid" to the base name of the hydrocarbon on which the substitution occurs, using the Greek prefix for the number of carbon atoms, including the carbon atom in the carboxyl group. Note that the C=O double bond also sometimes itself is identified as a carbonyl group (-CO-). For example, ethanoic acid is based on replacing a methyl group on ethane, CH_3CH_3, with a carboxyl group. Similarly, butanoic acid, the origin of the foul smell associated with rotten eggs, is based on replacing a methyl group on butane, $CH_3(CH_2)_2CH_3$, with a carboxyl group.

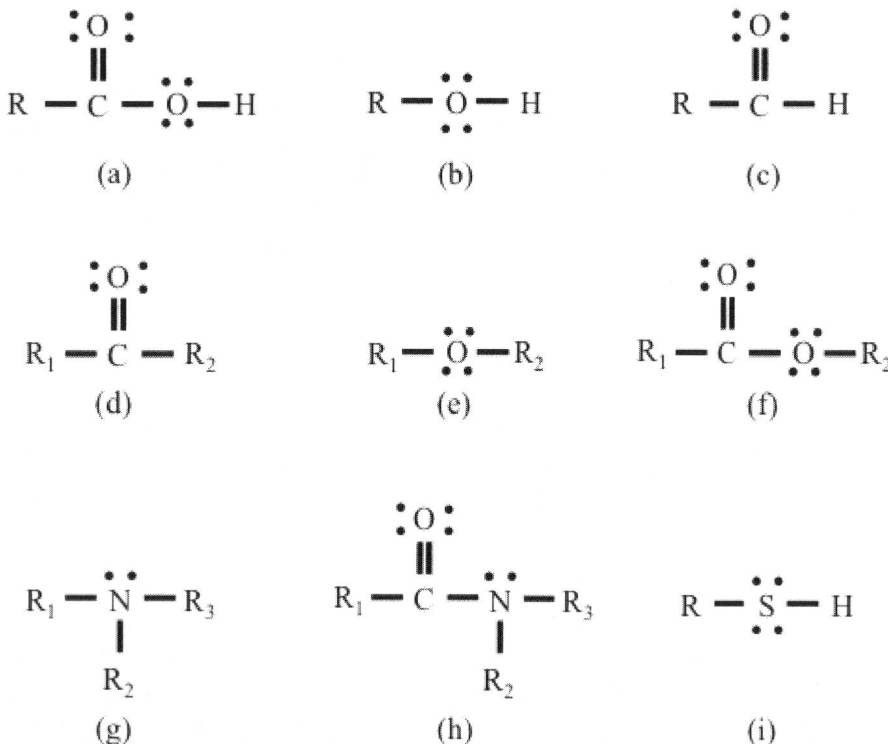

Figure 4.8. Generic structures of organic molecules with functional groups. The R groups are hydrocarbon chains that may contain other functional groups. (a) Carboxylic acids; (b) alcohols; (c) aldehydes; (d) ketones; (e) ethers; (f) esters; (g) amines; (h) amides; (i) thiols.

Alcohols are formed by substituting a hydroxyl (-OH) group for a hydrogen atom (Figure 4.8b). To name an alcohol we append the suffix "-ol" to the base name of the hydrocarbon on which the substitution occurs. For example, ethanol, CH_3CH_2OH, is based on replacing an H atom on ethane with

CH_3CH_3 with a hydroxyl group; and 2-propanol, $CH_3CHOHCH_3$, is based on replacing an H atom on the middle C atom (carbon #2 from either end) with a hydroxyl group.

Halogenated hydrocarbons (also known as alkyl halides) are formed by halo substitution—substituting a halogen (-X, where X = F, Cl, Br, or I) for an H atom. We previously encountered halogenated hydrocarbons in Guided Examples 4.2 and 4.3 as difluorodichloromethane and tetrafluoroethene, CF_2Cl_2 and C_2F_4, respectively. To name a halogenated hydrocarbon we append the halogen root as a prefix. For example, CH_3CH_2F is fluoroethane, based on replacing an H atom on ethane with an F atom. The now-discontinued refrigerant Freon 113, Cl_2FCCCl_2F, has the formal name 1,1,2,2,-tetrachloro-1,2-difluorethane, based on replacing all six H atoms on ethane with two Cl atoms and one F atom on each C atom.

Aldehydes are formed by substituting a -CHO group (a carbonyl group with an H atom) for a methyl group (Figure 4.8c). To name an aldehyde we append the suffix "-al" to the base name of the hydrocarbon on which the substitution occurs, using the Greek prefix for the total number of C atoms in the molecule. For example, CH_3CHO is "ethanal."

Ketones are formed by substituting a carbonyl group for a methylene group (Figure 4.8d). To name a ketone we append the suffix "-one" to the base name of the hydrocarbon on which the substitution occurs, using the Greek prefix for the total number of C atoms in the molecule. If there are more than three carbon atoms we also append the number for the C atom in the carbonyl group, based on counting from the nearest end, as a prefix followed by a hyphen to the name. For example, CH_3COCH_3 is "propanone" (commonly known as acetone), and $CH_3CH_2CO(CH_2)_2CH_3$ is "3-hexanone."

Ethers are formed by substituting an oxygen atom for a methylene group (Figure 4.8e). To name an ether we append the Greek prefix for the number of carbon atoms in the smaller of the groups R_1 and R_2, followed by "oxy" to the name for the larger of the groups $R_1$and R_2. For example, $CH_3OCH_2CH_3$ is "methoxyethane" (commonly known as methyl ethyl ether).

Esters are formed by substituting a –COO- group for a methylene group (Figure 4.8f). Esters are produced by reacting an alcohol with a carboxylic acid. Esters have two-part names: the Greek name for the R_2 group as a first name, and a second name as the suffix "-oate" appended to the Greek prefix for the R group. For example, $CH_3(CH_2)_2COOCH_2CH_3$ is "ethyl butanoate," which is a fruit fragrance. Biodiesel—a fuel for diesel engines produced from vegetable oil—is another example of an ester.

Amines (also called amino groups) are formed when a nitrogen atom joins one or more alkyl groups (*i.e.*, hydrocarbon chains) (Figure 4.8g). Primary amines occur when R_2 and R_3 are hydrogen atoms, secondary amines when only R_3 is a hydrogen atom, and tertiary amines when neither R group is a hydrogen atom. Nomenclature for amines can be complicated!

Amides are formed by substituting a $–CONR_2R_3$ group for a methyl group (Figure 4.8h). Amides are produced by reacting an amine with a carboxylic acid. Nomenclature for amides also is complicated!

Thiols are formed by substituting a thiol group, -SH, for a hydrogen atom (Figure 4.8i). To name a thiol we append the suffix "-thiol" to the name of the hydrocarbon on which the substitution occurs. For example, CH_3CH_2SH is "ethanethiol," a vile-smelling molecule detectable by the human nose at relatively low levels and doped into otherwise odorless natural gas.

Aromatic groups are formed by substituting a phenyl group ($-C_6H_5$) for an H atom. The phenyl group is a ring of six carbon atoms exhibiting resonance: each C-C bond is intermediate between a single and a double bond in character (Figure 4.9a). From the perspective ofVB theory this ring is a planar structure in which electrons in the out-of-plane π bonds are delocalized and able to move freely within a π-bonded "conjugated" network. Benzene, C_6H_6 (depicted below), is the smallest molecule with an aromatic group and is formed by adding a hydrogen atom to a phenyl ring. This structure, which is the basis for a wide variety of important molecules, is often depicted using shorthand representations (Figure 4.9b).

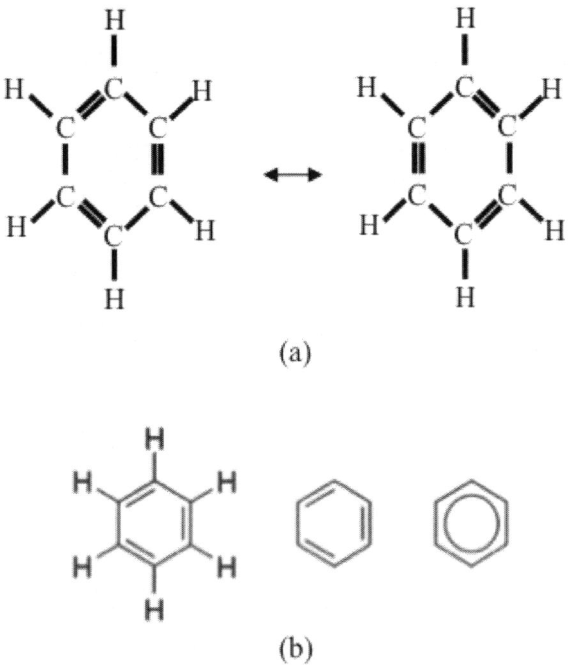

(a)

(b)

Figure 4.9. Aromatic functional group. (a) Phenyl ring exhibiting resonance; (b) shorthand representations for benzene, C_6H_6.

Example 4.7: Naming Organic Compounds

Write the names for C_5H_{12}, C_5H_{10}, C_5H_8, HCOOH, and CF_2CF_2.

Solutions

C_5H_{12} is pentane; C_5H_{10} is pentene; C_5H_8 is pentyne (or penadiene); HCOOH is methanoic acid; CF_2CF_2 is 1,1,2,2-tetrafluoroethene.

Example 4.8: Identifying Formulas for Organic Compounds from Their Names

Identify the condensed structural formulas for hexane, propene, methanol,1,2-difluoroethene, and 1,2-ethandiol.

Solutions

Hexane is C_6H_{14}; Propene is C_3H_6; Methanol is CH_3OH; 1,2-difluoroethene is CHFCHF; 1,2-ethandiol is CH_2OHCH_2OH.

Example 4.9: Identifying Organic Functional Groups

Identify the functional group(s) for and name the condensed structural formulas CH_3CH_2CHO, $CH_3(CH_2)_2CONH_2$, $CH_3(CH_2)_3COO(CH_2)_4CH_3$, and $CH_3OC_6H_5$.

Solutions

CH_3CH_2CHO is an aldehyde named propanal; $CH_3(CH_2)_2CONH_2$ is an amide named butanamide; $CH_3(CH_2)_3COO(CH_2)_4CH_3$ is an ester named pentyl pentanoate; $CH_3OC_6H_5$ is an ether with an aromatic ring named methyl phenyl ether.

Example 4.10: Drawing Lewis Structures for Organic Molecules

Identify the functional group(s) and draw the Lewis structure for:

a) 1-hexanethiol

b) Butanone (also known as methyl ethyl ketone)

c) Aminoethanoic acid (the simplest amino acid, glycine)

d) 2-hydroxybenzoic acid (also known as salicylic acid, the active ingredient in aspirin)

Solutions

a) 1-hexanethiol has a thiol group.

```
   H   H   H   H   H   H
   |   |   |   |   |   |     ..
H— C — C — C — C — C — C  —  S — H
   |   |   |   |   |   |     ..
   H   H   H   H   H   H
```

b) Butanone has a carbonyl group.

```
     H  :O:  H   H
     |   ||  |   |
H —  C — C — C — C — H
     |       |   |
     H       H   H
```

c) Aminoethanoic acid has an amino group and a carboxyl group.

d) 2-hydroxybenzoic acid has a hydroxyl, a carboxyl, and an aromatic group. The proper representation for the Lewis structure for the molecule indicates that there are two resonance hybrids.

4.10 Polymers

Polymers are giant molecules—also called macromolecules—occurring as long chains of many identical (or nearly identical) repeat units with very high molar masses (the term molecular weight is used for polymers), and with physical and chemical properties distinct and unusual compared with the small molecules. Polymers have an almost limitless range of applications, including plastics, textiles, electronics, and medical devices. Table 4.10 presents the structure of monomers (precursors for polymers) and repeat units for some common organic polymers as well as the type of functional group that links repeat units. The size of a polymer is frequently represented as its degree of polymerization, *n*, defined as:

$$n = \frac{M_{polymer}}{M_{monomer}} \sim 10^2 - 10^5 \quad [4.5]$$

Table 4.1 Structure of common synthetic organic polymers

Polymer (Acronym/Trade Name)	Monomer(s)	Repeat Unit	Bond Connecting Repeat Units
Polyethylene* (HDPE & LDPE)	$H_2C{=}CH_2$	$-CH_2-$	C-C
Polypropylene (Type 5 recyclable)	$H_2C{=}CH(CH_3)$	$-CH_2-CH(CH_3)-$	C-C
Polyvinyl chloride (PVC) (Type 3 recyclable)	$H_2C{=}CHCl$	$-CH_2-CHCl-$	C-C
Polystyrene (Type 6 recyclable)	$H_2C{=}CH(C_6H_5)$	$-CH_2-CH(C_6H_5)-$	C-C
Polytetrafluoroethylene (PTFE/Teflon®)	$F_2C{=}CF_2$	$-CF_2-$	C-C
Polyethylene terephthalate (PET) (Type 1 recyclable)	$HOCH_2CH_2OH$ + $HOOC-C_6H_4-COOH$	$-O-CH_2CH_2-O-C(=O)-C_6H_4-C(=O)-$	Ester
Poly(hexamethyl adipamide) (Nylon 6,6)	$H_2N(CH_2)_6NH_2$ + $ClOC(CH_2)_4COCl$	$-NH-(CH_2)_6-NH-C(=O)-(CH_2)_4-C(=O)-$	Amide
Polycarbonate (PC/Lexan®)	$HO-C_6H_4-C(CH_3)_2-C_6H_4-OH$ + $Cl-C(=O)-Cl$	$-O-C_6H_4-C(CH_3)_2-C_6H_4-O-C(=O)-$	Carbonate

5: Intermolecular Forces and the Macroscopic Properties of Pure Substances

Learning Objectives

By the end of this chapter you should be able to:

1) Identify the types of intermolecular forces and their effects on the structure of a compound.
2) Explain the behavior of metals and semiconductors using band theory.
3) Interpret the pressure-temperature phase diagram for a typical substance and identify key features such as phases, phase transitions, and critical and triple points.
4) Calculate the relative rates of effusion through a small aperture in a vessel wall for a gaseous mixture of light and heavy molecules.
5) Use the van der Waals equation of state to relate volumetric properties for a fluid experiencing non-negligible intermolecular forces.
6) Relate the effects of different types of intermolecular forces to properties of liquids, such as boiling point, viscosity, and surface tension.
7) Compare and contrast properties of different types of crystalline solids, including metals, ceramics, and network covalent solids, and of non-crystalline solids, including liquid crystals and amorphous polymers.
8) Relate composition, atomic radius, dimensions of a unit cell and density for crystals with cubic units cells.
9) Determine the empirical formula for a crystalline solid from the structure of its unit cell.

5.1 The Intermolecular Forces Experienced by a Pure Substance Depends on its Structure

We are familiar with the concept that there are two forms for energy: kinetic energy associated with motion, and potential energy associated with stored energy. This concept applies at both the macroscopic level, where we commonly associate potential energy (PE) with the energy stored in a piece of matter due to gravity, and the molecular level (*i.e.*, for individual atoms, ions,and molecules). Molecular kinetic energy is the sum of the contributions to kinetic energy from translation, vibration, and rotation of a particle. Molecular potential energy is the sum of the energy stored in covalent bonds, in excited electrons (*i.e.*, when the electronic configuration is not the ground state), and due to intermolecular (IM) forces between atoms not connected by covalent bonds. IM forces arise due to interactions between the electronic structures of neighboring atoms, ions, and molecules and can be either attractive or repulsive. From physics we can relate an IM force to its corresponding intermolecular (IM) potential energy, Γ_{IM}, as $F_{IM} = -\nabla\Gamma_{IM}$, where ∇ is the gradient operator from multi-dimensional calculus. We observe that attractive IM forces correspond to $\Gamma_{IM} < 0$ and repulsive IM frces to $\Gamma_{IM} > 0$. Attractive forces decrease PE and stabilize matter, and repulsive forces increase PE and *destabilize* matter.

We can specify the types of attractive IM forces experienced between pairs of isolated atoms, ions, or molecules or within an individual molecule (*i.e.*, intramolecularly) and their corresponding IM PE by considering its Lewis structure and polarity and identifying if and how charges are separated. For example, electrostatic interactions occur between pairs of point charges (*i.e.*, monatomic and polyatomic ions), pairs of particles experiencing *permanent* and *complete* separation of charge. Electrostatic interactions between ions of opposing charge are attractive, with a strength described by Coulomb's law as $F_{coulombic} = \frac{Q_1 Q_2}{4\pi\varepsilon_o r^2}$, where Q_1 and Q_2 are the ionic charges (in C), *r* is the separation distance between the ions, and $\varepsilon_o = 8.8542 \times 10^{-12}\ C^2/J\text{-}m$ is the dielectric permittivity of a vacuum. The corresponding IM PE per pair of ions (in J/pair) then is represented as:

$$\Gamma_{coulombic} = \frac{Q_1 Q_2}{4\pi\varepsilon_o r} \qquad [5.1]$$

The magnitude of the IM PE associated with electrostatic interactions increases as the charge number on either ion increases. Consequently, electrostatic interactions are weaker between monovalent ions (*i.e.*, ions with charges of +1 or -1) than between divalent ions (*i.e.*, ions with charges of +2 or -2) and between trivalent ions (*i.e.*, ions with charges of +3 or -3).

Dipole-dipole interactions occur between pairs of polar molecules, pairs of particles experiencing permanent but incomplete separation of charge (*i.e.*, molecules with non-zero net dipole moments). The relative orientation of a pair of dipoles affects whether dipole-dipole interaction for the pair is attractive or repulsive, as depicted in Figure 5.1. If the molecules are free to rotate, the attractive orientation occurs with an enormously greater frequency, and individual dipole-dipole interactions are typically attractive. We find that the average IM PE per pair of dipoles, $\overline{\Gamma}_{dipole}$ (in J/pair), is negative (*i.e.*, the average of these interactions is attractive) and depends on the magnitude of the molecular dipole moment, $\|\vec{\mu}_{molecule}\|$, separation, *r*, and temperature as:

$$\overline{\Gamma}_{dipole} = -\frac{\|\vec{\mu}_{molecule}\|^4}{24\pi^2\varepsilon_o^2 k_B T r^6} \qquad [5.2]$$

where $k_B = \frac{R}{N_A} = 1.38 \times 10^{-23}$ J/K is the Boltzmann constant. The strength of dipole-dipole interactions increases as the magnitude of the dipole moment for the molecule increases. Tabulated data for molecular dipole moments are available in the *CRC Handbook of Chemistry and Physics* (available on-line at http://www.hbcpnetbase.com/).

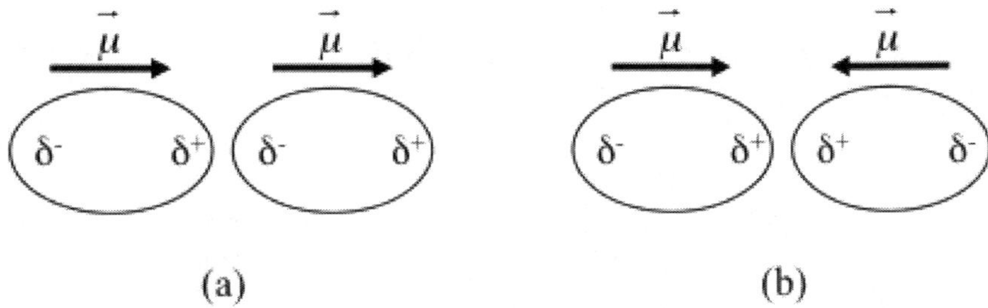

Figure 5.1. Possible dipole-dipole interactions between a pair of polar molecules. (a) Attractive; (b) repulsive.

Dispersion forces (also known as London forces) arise as temporary and partial separations of charge that occur as a consequence of electrons not being fixed in space (consider the Heisenberg uncertainty principle introduced in Section 2.5). We visualize distributions of electrons as "electron clouds" because quantum mechanics tells us we can only state the probability of finding electrons in given sets of positions. When a pair of particles approaches each other, they disturb their neighbor's electron clouds, resulting in an instantaneous but short-lived dipole. This temporary dipole, in turn, induces instantaneous, short-lived dipoles with attractive orientations in other neighboring particles that propagate as additional instantaneous, short-lived dipoles. This behavior is experienced by all matter containing electrons.

The IM PE per pair of particles associated with dispersion forces separated by a distance r can be expressed using quantum mechanics as:

$$\Gamma_{dispersion} = -\frac{3\alpha^2 I}{4r^6} \qquad [5.3]$$

where α is the polarizability—characterizing the ease by which the electron cloud for a species is displaced by an electric field—and I is the first ionization potential—characterizing the amount of energy to remove the weakest-held electron for the species. We observe that polarizability increases with atomic/molecular size, as there are more electrons to be disturbed and a larger range of positions over which they may be found. Thus, the strength of dispersion forces increases as molar mass increases and/or species becomes more elongated. Tabulated data for polarizabilities and ionization potentials are available in the *CRC Handbook of Chemistry and Physics* (available on-line at http://www.hbcpnetbase.com/). Data for ionization potentials also are available on-line through the NIST Chemistry Webbook at http://webbook.nist.gov.

Hydrogen bonding is a special, stronger form of dipole-dipole interactions between an H atom covalently bonded to a very electronegative atom (typically fluorine, oxygen, or nitrogen) and a lone electron pair on a different neighboring very electronegative atom (again, typically F, O, or N). Note that to a lesser extent chlorine and sulfur are other relatively high electronegativity atoms that can participate in hydrogen bonding. A hydrogen bond is not a covalent bond but rather a directional interaction. Figure 5.2 depicts the results of a simulation of hydrogen bonding generated using the molecular modeling software *Odyssey*. Note that water is the best hydrogen bonding molecule because its number of H atoms matches its number of lone pairs. Intramolecular hydrogen bonding is critical to stabilizing the structure of biological molecules(*e.g.*, proteins and nucleic acids). Intermolecular hydrogen bonding can also result in the phenomenon of self-assembly, which is an essential feature for nanotechnology. Hydrogen bonding tends to be important only for liquids and solids.

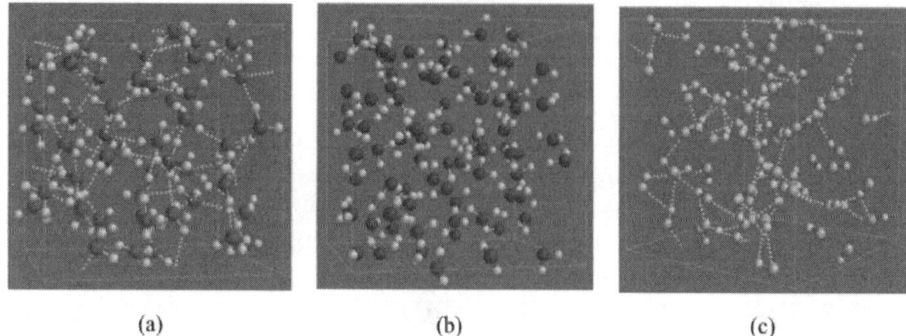

(a) (b) (c)

Figure 5.2. Hydrogen bonding in (a) NH_3 (3 H atoms for every 1 lone pair), (b) H_2O (2 H atoms for every 2 lone pairs), and (c) HF (1 H atom for every 3 lone pairs). Dashed lines depict hydrogen bonds. Images created using the molecular modeling software *Odyssey*.

We rank the relative strengths of attractive IM forces as:

$$\begin{Bmatrix}\text{Electrostatic}\\\text{interactions}\end{Bmatrix} > \begin{Bmatrix}\text{Hydrogen}\\\text{bonding}\end{Bmatrix} > \begin{Bmatrix}\text{Dipole - dipole}\\\text{interactions}\end{Bmatrix} > \begin{Bmatrix}\text{Dispersion}\\\text{forces}\end{Bmatrix}$$

Dipole-dipole interactions and dispersion forces are together known as van der Waals forces, which have an attractive PE proportional to the reciprocal of the separation distance between molecules to the sixth power.

Repulsive IM forces between non-covalently-bonded atoms arise because very close overlap of electron clouds is energetically unfavorable. Note that repulsive forces are very short in range compared to attractive forces. All matter containing electrons experiences repulsive IM forces.

Example 5.1: Identifying Attractive IM Forces Experienced by Different Substances

The compounds $H_3C(CH_2)_4NH_2$ (pentylamine), $H_3C(CH_2)_3SH$ (1-butanethiol), $H_3CCHCH_3CHCH_3CH_3$ (2,3-dimethylbutane), and $(NH_4)_2CO_3$ (ammonium carbonate) have similar molar masses. Rank these compounds in order of increasing strength of attractive intermolecular forces experienced.

Solutions

The compounds in this set differ substantially in types of attractive intermolecular forces experienced.

Compound	M(g/mol)	Types of IM Forces Experienced			
		Dispersion Forces?	Dipole-Dipole Interactions?	Hydrogen Bonding?	Electrostatic Interactions?
$H_3C(CH_2)_4NH_2$	87	Yes	Yes	Yes	No
$H_3C(CH_2)_3SH$	90	Yes	Yes	Weak	No
$H_3CCHCH_3CHCH_3CH_3$	86	Yes	No	No	No
$(NH_4)_2CO_3$	96	Yes	Yes	Yes	Yes

Electrostatic interactions are stronger than hydrogen bonding, which in turn is stronger than other forms of dipole-dipole interactions, which in turn are stronger than dispersion forces.

Thus, we rank these compounds in order of increasing strength of attractive intermolecular forces as 2,3-dimethylbutane, 1-butanethiol, pentylamine and ammonium carbonate.

Example 5.2: Identifying Relative Strengths of Dispersion Forces

Rank the organic molecules dichloromethane (CH_2Cl_2, also known as methylene chloride), trichloromethane ($CHCl_3$, also known as choloroform), methane, tetrachloromethane (CCl_4, also known as carbon tetrachloride), and chloromethane (CH_3Cl, also known as methyl chloride) in order of increasing dispersion forces experienced.

Solutions

Dispersion forces increase with increasing molar mass and/or molecular elongation. The five molecules have similar shapes but differ in molar mass, such that ranking them in order of increasing molar mass provides us their ranking in terms of increasing dispersion forces: CH_4, CH_3Cl, CH_2Cl_2, $CHCl_3$, and CCl_4.

Example 5.3: Quantitatively Comparing the IM PE Associated with Different IM Forces

The following data are available for water: $\|\vec{\mu}_{molecule}\| = 1.85$ D, $\alpha = 14.8\times10^{-31}$ m^3, and I = 12.62 eV. If a pair of water molecules at 400. K are separated by 2.76 Å (an angstrom, Å, is a tenth of a nm), compare the attractive IM PE (in kJ/mol) associated with dipole-dipole interactions, dispersion forces, and gravity, where $\Gamma_{gravity} = -\frac{Gm^2}{r}$ and $G = 6.67\times10^{-11}$ m^3/kg-s^2 is the gravitational constant.

Solution

Determine the average attractive PE associated with dipole-dipole interaction forces as:

$$\Gamma_{dispole} = -\frac{\mu_{H_2O}^{\;4}}{24\pi^2 \varepsilon_o^2 k_B T r^6} = -\frac{\left(1.85\ \text{D} \times \frac{3.336 \times 10^{-30}\ \text{C}-\text{m}}{1\ \text{D}}\right)^4}{\left[24\pi^2\left(8.8542 \times 10^{-12}\ \frac{\text{C}^2}{\text{J}-\text{m}}\right)^2\left(1.38 \times 10^{-23}\ \frac{\text{J}}{\text{K}}\right)(400.\ \text{K}) \times \left(2.76\ \text{A} \times \frac{10^{-10}\ \text{m}}{1\ \text{A}}\right)^6\right]}$$

$$\Gamma_{dispole} = -3.20 \times 10^{-20}\ \text{J/pair} \times \frac{10^{-3}\ \text{kJ}}{1\ \text{J}} \times \frac{6.022 \times 10^{23}\ \text{pairs}}{1\ \text{mol}} = -19.3\ \text{kJ/mol}$$

Determine the PE associated with dispersion forces as:

$$\Gamma_{dispersion} = -\frac{3\alpha_{H_2O}^{\;2} I_{H_2O}}{4r^6} = -\frac{3(14.8 \times 10^{-31}\ \text{m}^3)^2\left(12.62\ \text{eV} \times \frac{1.602 \times 10^{-19}\ \text{J}}{1\ \text{eV}}\right)}{4\left(2.76\ \text{A} \times \frac{10^{-10}\ \text{m}}{1\ \text{A}}\right)^6}$$

$$\Gamma_{dispersion} = -7.51 \times 10^{-21}\ \text{J/pair} \times \frac{10^{-3}\ \text{kJ}}{1\ \text{J}} \times \frac{6.022 \times 10^{23}\ \text{pairs}}{1\ \text{mol}} = -4.52\ \text{kJ/mol}$$

Determine the PE associated with gravitational forces as:

$$\Gamma_{gravity} = -\frac{\left(6.67 \times 10^{-11}\ \frac{\text{m}^3}{\text{kg}-\text{s}^2}\right)\left(18.0\ \text{g/mol} \times \frac{10^{-3}\ \text{kg}}{1\ \text{g}} \times \frac{1\ \text{mol}}{6.022 \times 10^{23}\ \text{molecule}}\right)^2}{\left(2.76\ \text{A} \times \frac{10^{-10}\ \text{m}}{1\ \text{A}}\right)}$$

$$\Gamma_{gravity} = -2.16 \times 10^{-52}\ \text{J/pair} \times \frac{10^{-3}\ \text{kJ}}{1\ \text{J}} \times \frac{6.022 \times 10^{23}\ \text{pairs}}{1\ \text{mol}} = -1.30 \times 10^{-31}\ \text{kJ/mol}$$

Clearly, $|\Gamma_{dipole}| > |\Gamma_{dispersion}| >> |\Gamma_{gravity}|$.

5.2 Models for Metals and Semiconductors

We seek to relate the microscopic structure of matter, based on our descriptions for atoms, ions, and molecules, to observable macroscopic properties We start by considering metals. Atoms in a metallic solid are held together by metallic bonds, a fifth type of attractive IM interaction. Metallic bonding is isotropic (*i.e.*, one-directional) among multiple neighbors and occurs with relatively few electrons—metals have valence-shell orbitals that are relatively unoccupied (*e.g.*, each Na atom has one valence electron but four valence-shell orbitals, 3s, $3p_x$, $3p_y$, and $3p_z$). VB theory is unable to explain this phenomenon. Metallic bonding, however, can be described using an electron sea model (Figure 5.3) in which a network of positively-charged nuclei are immersed in a "sea" of delocalized (*i.e.*, free), valence electrons (recall that metals have relatively low electronegativities and bind their valence electrons relatively weakly).

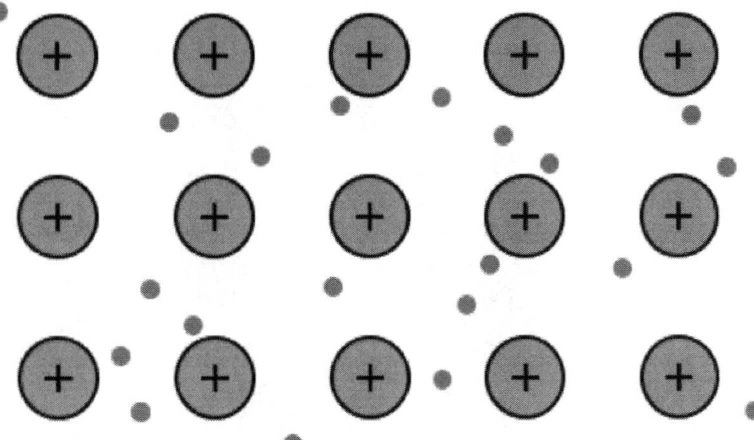

Figure 5.3. Electron sea model for metallic bonding showing delocalized valence electrons (small circles) loosely associated with metal nuclei.

The electron sea model qualitatively explains many properties associated with metals. For example, metals are good conductors because the electron sea is mobile and easily disturbed by external electric fields (*i.e.*, valence electrons can flow when a voltage is imposed). Metals also are opaque and lustrous because the energy of free electrons is not quantized (*i.e.*, valence electrons in metals can absorb and re-radiate a continuous range of wavelengths). Further, metals are malleable and ductile because the bonding is diffuse and the electron sea readily adjusts to mechanical stresses.

Band theory offers a superior description that quantitatively explains the behavior of metals as well as semiconductors and insulators. In particular, band theory provides a basis for quantitative descriptions for materials used in modern electronics such as transistors, LEDs, and photovoltaic (solar) cells. Band theory is an outcome of molecular orbital (MO) theory in which bonding between atoms is described in terms of bonding and anti-bonding molecular orbitals (MOs). When an enormous number of atoms combine to form a solid, an enormous number of MOs overlap in continuous energy bands. MOs occupied by valance electrons form a valence band; unoccupied MOs form a conduction band. These two bands can be separated by a gap termed the band gap.

Figure 5.4 depicts the interpretation of the behavior of materials based on band theory. In electrical conductors, such as metals, there is no band gap—the valence and conduction bands overlap, and electrons move easily between bands. In contrast, in electrical insulators such as non-metallic solids (*e.g.*, diamond), the valence band is fully occupied, but a substantial band gap exists between valence and conduction bands, and substantial increases in temperature are necessary to give electrons enough energy to transition to the conduction band. Semi-conductors fall in between conductors and insulators—the valence band is fully occupied, but the band gap is small enough that only a relatively small amount of energy is necessary for an electron to transition to the conduction band. Increasing the temperature can provide this additional energy and semi-conductors become conductors as the temperature is increased.

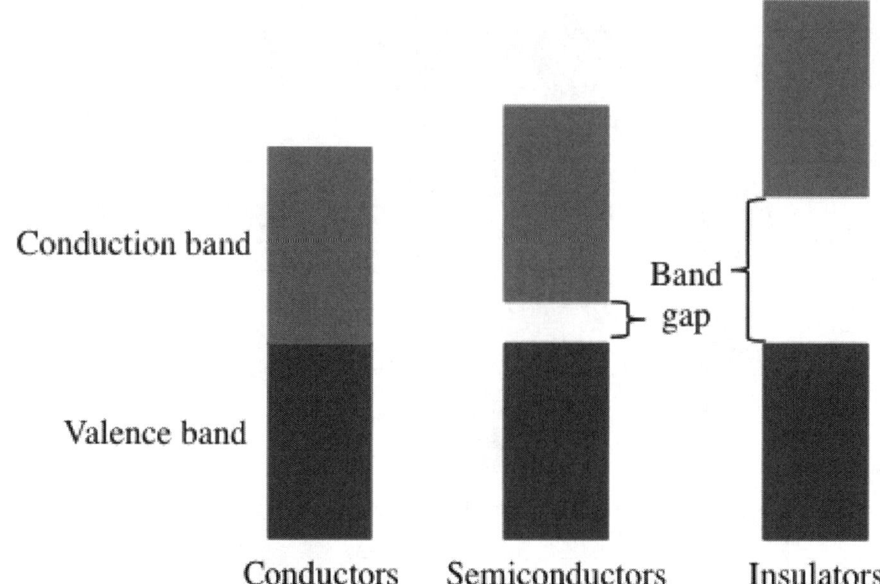

Figure 5.4. Description based on band theory for metals, semi-conductors, and insulators.

The electronic properties of semi-conductors depend on the band gap; intrinsic semi-conductors, such as Si, can be tailored by deliberately adding impurities, termed dopants, which modify the band gap. Doping donor atoms with additional valence electrons (*e.g.*, Group 5A elements for Si) brings the upper level of the valence band closer to the conduction band and forms an *n*-type semiconductor. Doping acceptor atoms with fewer valence electrons (*e.g.*, Group 3A elements for Si) introduces positively-charged "holes" and brings the lower level of the conduction band closer to the valence band and forms a *p*-type semiconductor. The color of LEDs can be controlled and the design of photovoltaic cells can be based on this process.

5.3 Pressure-Temperature Phase Diagrams

Phase diagrams provide graphical representations of the phases present at equilibrium (*i.e.*, when the properties of matter do not change with time and there are no net driving forces for change), which are useful for addressing problems in the formulation of pharmaceuticals, the operation of liquid-crystal displays, the behavior of polymers and ceramics, and the processing of semiconductors. There are multiple types of phase diagrams, including pressure-temperature (*P*–*T*) and pressure-volume (*P*–*V*) phase diagrams. For example, for a typical *P*–*T* phase diagram (Figure 5.5) we observe the following behavior for pure substances:

- Vapors (gases, *g*) are typically present at relatively high temperatures and low pressures.
- Solids (*s*) are typically present at relatively low temperatures and high pressures.
- Liquids (*l*) are typically present at intermediate temperatures and pressures.

Liquids and solids are condensed phases in which matter packs relatively close together. Liquids and vapors are fluids, phases that flow when a force is exerted, and take on the shape of confining vessels.

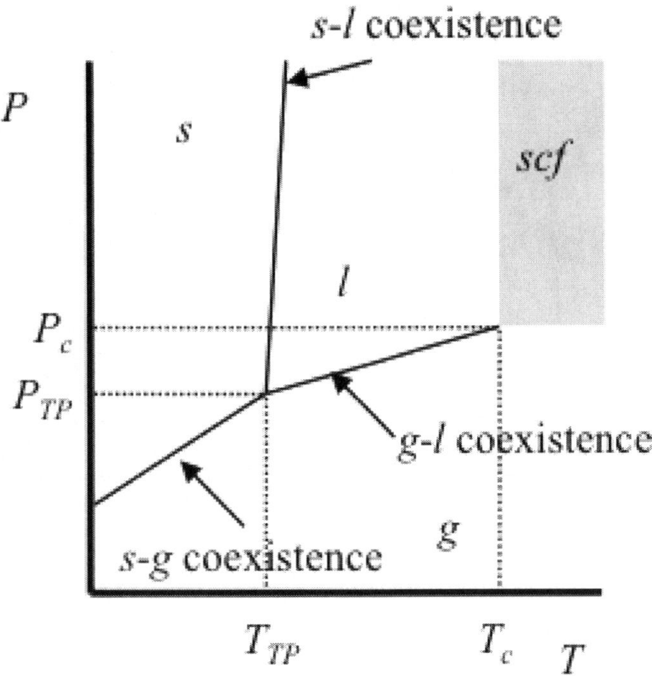

Figure 5.5. Typical pressure-temperature phase diagram for a pure substance

We identify boundaries between pairs of phases as coexistence curves for which two phases are present at equilibrium. We term these phases as "saturated." The coexistence curves intersect at a single point, defined as a triple point, (T_{TP}, P_{TP}), for which the saturated solid, saturated liquid, and saturated vapor coexist. Note that some substances exhibit more than one triple point because they exist in more than one solid phase, a phenomenon known as polymorphism. For example, water has multiple solid phases that differ in the packing of water molecules and the arrangement of hydrogen bonds.

At relatively high temperatures and pressures, liquid and vapor phases are indistinguishable. The critical point, (T_c, P_c), specifies the lowest temperature and pressure, respectively, for which the liquid and vapor are indistinguishable; for ($T \geq T_c$, $P \geq P_c$) the observed phase is a supercritical fluid (*scf*). Supercritical fluids have properties intermediate between those of the liquid and vapor. For example, solids have solubilities in the *scf* phase more like those in liquids than in vapors, but the *scf* phase flows and mixes more easily than do liquids. These properties have led to the use of supercritical CO_2 in the decaffeination of coffee.

Consider the types of phase transitions commonly encountered. Fusion (*i.e.*, melting) is the transformation of a saturated solid to a saturated liquid, while freezing is the reverse process of transforming a saturated liquid into a saturated solid. Vaporization is the transformation of a saturated liquid to a saturated vapor, while condensation is the reverse process of transforming a saturated vapor to a saturated liquid. Sublimation is the direct transformation of a saturated solid to a saturated vapor, with deposition the reverse process of transforming a saturated vapor directly to a saturated solid.

Typically the *s-l* coexistence curve is nearly vertical (*i.e.*, has a very steep slope on a *P-T* phase diagram) with a slight positive tilt (*i.e.*, $\left(\frac{dP}{dT}\right)_{s\text{-}l\,coexistence} > 0$). This behavior is observed because the density of most solids is greater than the density of their corresponding liquids, such that increasing the pressure of a liquid at constant temperature near the *s-l* coexistence curve results in the transformation of the liquid to the solid (Figure 5.6a). Water, however, has a *s-l* coexistence curve with a slight negative tilt (*i.e.*, $\left(\frac{dP}{dT}\right)_{s\text{-}l\,coexistence} < 0$) because the density of ice is *less* than the density of liquid water. Accordingly, increasing the pressure on ice near the *s-l* coexistence curve can result in its transformation to liquid water (Figure 5.6b).

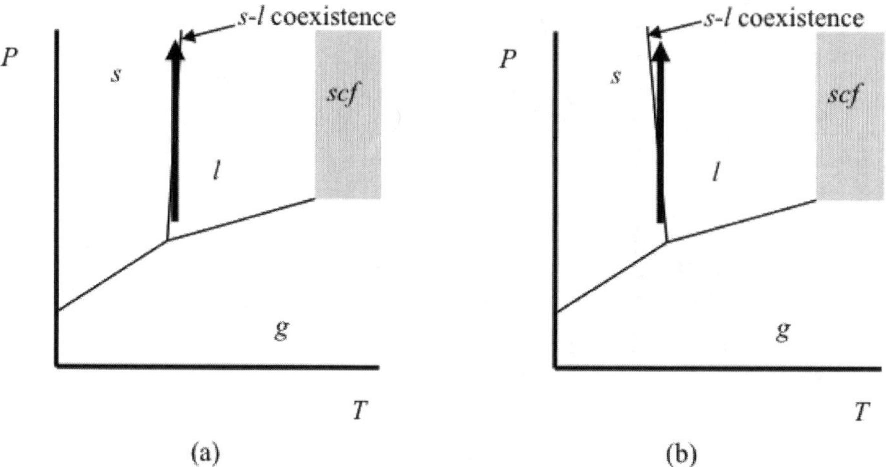

Figure 5.6. The effect of increasing pressure near the triple point depends on the slope of the solid-liquid coexistence curve. (a) Typical substance with liquid less dense than solid; (b) A substance, such as water, with liquid more dense than solid.

Under some conditions liquids actually can exist at temperatures and pressures that don't correspond to equilibrium states. For example, a superheated liquid—a liquid that exists at a temperature greater than the boiling point at a given pressure—and a supercooled liquid—a liquid that exists at a temperature less than the fusion point at a given pressure—are metastable phases. In contrast, plasmas, fluid phases occurring at very high temperatures for which electrons have been stripped from nuclei, are equilibrium states.

5.4 Kinetic Molecular Theory and Effusion of Ideal Gases

In Section 1.6 we introduced the ideal gas law (Equation [1.6]) as an equation of state providing a relationship between pressure, volume, temperature, and the number of moles for a gas based on experimental observations. This relationship also can be derived by considering a statistical, molecular-scale model known as kinetic molecular theory (KMT), which has five key assumptions:

1. A gas is composed of an enormous number of particles in constant, random motion.
2. These particles don't all move at the same speed but rather have a *distribution* of speeds.
3. These particles have negligible volumes.
4. These particles collide with each other and with walls relatively infrequently.
5. These particles exert negligible IM forces on each other.

KMT predicts that the average molecular translational kinetic energy of an ideal gas, $\overline{KE}_{molecular} = \frac{3}{2}k_B T$, is proportional to absolute temperature and independent of what the gas is. This behavior leads to the use of the phrase "thermal energy for the quantity term *RT*. Further, KMT predicts that the root-mean-squared (rms) speed, $u_{rms} = \sqrt{\frac{3RT}{M}}$, depends only on temperature and molar mass.

The dependence of rms speed on molar mass can explain the relative rates of passage of gas particles through pinholes, a process known as effusion. Effusion occurs when a balloon deflates slowly over time, and was used in the Manhattan Project during World War II to separate isotopes of uranium. Note that effusion is distinct from but related to diffusion, the random motion of particles resulting in their spread over time. Consider that the probability of a particle passing through a pinhole is proportional to its rms speed. Thus, larger particles with a smaller rms speed have a lower probability of escaping through a pinhole (Figure 5.7).

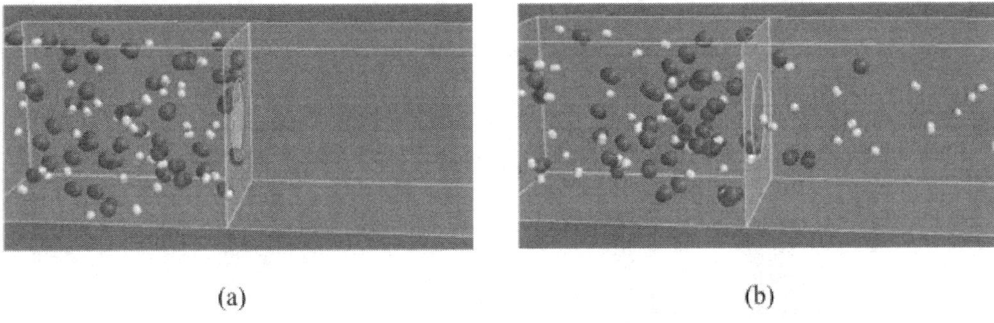

Figure 5.7. Simulation of effusion of H_2 and CO through a pinhole using the molecular modeling software *Odyssey*. (a) Gases contained on left-side of chamber before pinhole opened; (b) Smaller H_2 effuses more rapidly than larger CO through pinhole into right-side of chamber.

Our qualitative observations can be expressed quantitatively in the form of Graham's law, where $\dot{n}_i$ denotes the rate of effusion (in moles per unit time) of species i:

$$\frac{\{\text{Rate of effusion of A}\}}{\{\text{Rate of effusion of B}\}} = \frac{\dot{n}_A}{\dot{n}_B} = \frac{u_{rms,A}}{u_{rms,B}} = \frac{\sqrt{\frac{3RT}{M_A}}}{\sqrt{\frac{3RT}{M_B}}} = \sqrt{\frac{M_B}{M_A}} \qquad [5.4]$$

Example 5.4: Effusion of Gases from Pipelines

Natural gas is primarily methane and is marketed as a relatively clean-burning fossil fuel. But methane is a greenhouse gas, and losses from natural gas systems account for a significant percentage of worldwide methane emissions. Microscopic leakage from pipelines, measured in scfh (ft^3/hr of methane at STP), is an important contributor to these losses.

a) In a study by the U.S. Department of Energy (DOE) a leak rate of 500 scfh—equivalent to a leak rate of 0.175 mol/s—was detected for CH_4 in a pipeline. If hydrogen gas (H_2) were transported in this pipeline, what would be the leak rate of H_2 (in mol/s)?

b) Because methane is odorless, natural gas supplies are commonly odorized with ethanethiol, which has a strong, foul-smelling odor detectable at relatively low levels by the human nose. What is the leakage rate of ethanethiol relative to the leakage rate of methane?

Solutions

a) Apply Graham's law of effusion, Equation [5.4], based on molar masses of 2.02 g/mol for molecular hydrogen and 16.05 g/mol for methane:

$$\frac{\dot{n}_{H_2}}{\dot{n}_{CH_4}} = \sqrt{\frac{M_{CH_4}}{M_{H_2}}}$$

$$\dot{n}_{H_2} = \dot{n}_{CH_4}\sqrt{\frac{M_{CH}}{M_{H_2}}} = (0.175 \text{ mol/s})\sqrt{\frac{(16.05 \text{ g/mol})}{(2.02 \text{ g/mol})}} = 0.495 \text{ mol } H_2/\text{s}$$

Note that the rate of leakage for hydrogen gas is nearly three times as great as for natural gas, suggesting that the conversion of natural gas pipelines for transport of hydrogen gas for a hydrogen-based economy may require more stringent measures for reducing leaks.

b) Apply Graham's law of effusion again, now based on a molar mass of 62.15 g/mol for ethanethiol:

$$\frac{\dot{n}_{CH_3CH_2SH}}{\dot{n}_{CH_4}} = \sqrt{\frac{M_{CH_4}}{M_{CH_3CH_2SH}}} = \sqrt{\frac{(16.05 \text{ g/mol})}{(62.15 \text{ g/mol})}} = 0.5082$$

We see that the odorant leaks at about half the rate of the natural gas.

5.5 Volumetric Descriptions for Non-Ideal Gases

The ideal gas law and KMT are inappropriate for describing the behavior of matter at relatively low temperatures—for which the assumption that particles exert negligible forces on each other is invalid—and/or relatively high pressures—for which the assumption that particles have negligible volumes is invalid. How do we describe gases at relatively low temperatures and/or relatively high pressures? Consider the results depicted in Figure 5.8 from a simulation of how the pressure for ammonia at 25.0 °C affects its compressibility factor, defined as:

$$z \equiv \frac{PV}{nRT} = \frac{Pv}{RT} \qquad [5.5]$$

where the molar volume is defined as $v \equiv \frac{V}{n}$ We observe that NH_3 behaves ideally (*i.e.*, $z = 1$) at very low pressures; the compressibility factor for NH_3, however, decreases to a minimum at ~50 atm as the pressure initially increases, but that further increases in pressure increase the compressibility factor, with the compressibility factor equal to ~1 at ~500 atm.

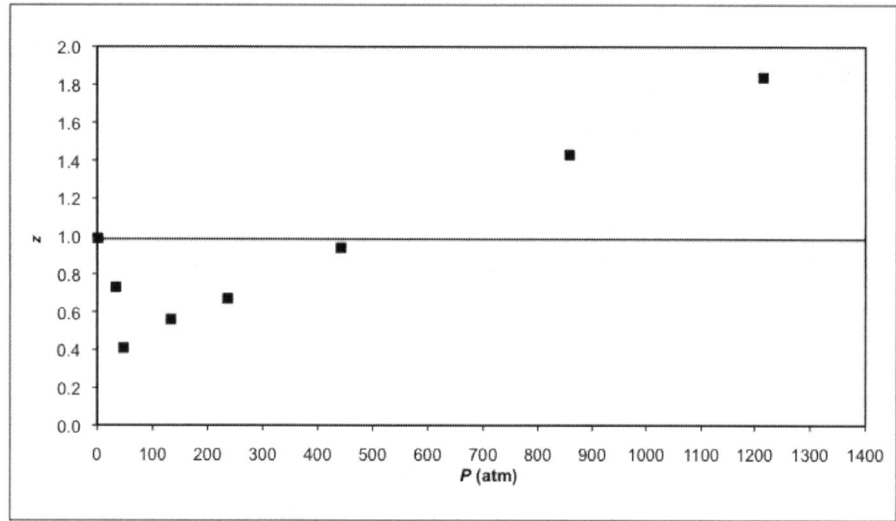

Figure 5.8. Results from a simulation of dependence of compressibility factor on pressure for ammonia at room temperature, using the molecular modeling software *Odyssey*.

We can explain this behavior by considering the roles of attractive and repulsive IM forces as introduced in Section 5.1. Attractive IM forces pull particles together, such that the particles occupy less volume than predicted by the ideal gas law. Moderate increases in pressure from very low pressure allow particles to interact more strongly and longer-range attractive IM forces to dominate. Repulsive IM forces, however, push particles apart, such that the particles occupy more volume than predicted by the ideal gas law. Large increases in pressure bring particles into very close proximity, such that shorter-range repulsive IM forces dominate. We summarize this interpretation as:

$$z = \left(\frac{Pv}{RT}\right)\left(\frac{v_{IG}}{v_{IG}}\right) = \left(\frac{Pv_{IG}}{RT}\right)\left(\frac{v}{v_{IG}}\right) = \frac{v}{v_{IG}} \quad [5.6]$$

where we applied the ideal gas law rearranged as $\frac{Pv_{IG}}{RT} = 1$ and identify that $z < 1$ when attractive IM forces dominate, $z > 1$ when repulsive IM forces dominate, and $z = 1$ when attractive and repulsive IM forces balance.

Example 5.5: Comparing the Predictions of Different Equations of State for Water with Experimental Observations

Experimental data for the molar volume of water for two different pairs of pressures and temperatures are provided below.

a) Compare the molar volume based on the ideal gas law for each temperature-pressure pair with the experimental observations.

b) Interpret your results from part a based on evaluating the compressibility factor.

Property	Condition A	Condition B
Pressure (bar)	1.00	100.
Temperature (°C)	1000.	500.
v_{exp} (m^3/mol)	1.059×10^{-1}	5.909×10^{-4}

Solutions

a) Apply the ideal gas law (Equation [1.6]) rearranged as $v^{IG} = \frac{RT}{P}$ to each condition, expressing the temperature in Kelvin and pressures in Pa:

$$v_{IG}^{A} = \frac{(8.314 \text{ J/mol - K})(1073.\text{ K})}{\left(1.00 \text{ bar} \times \frac{10^5 \text{ Pa}}{1 \text{ bar}}\right)} = 1.06 \times 10^{-1} \text{ m}^3/\text{mol}$$

$$v_{IG}^{B} = \frac{(8.314 \text{ J/mol - K})(773.\text{ K})}{\left(100.\text{ bar} \times \frac{10^5 \text{ Pa}}{1 \text{ bar}}\right)} = 6.43 \times 10^{-4} \text{ m}^3/\text{mol}$$

The ideal gas law compares most favorably with the experimental observations for Condition A (a relatively low pressure and relatively high temperature) for which we would expect IM interactions to be less important.

b) Evaluate the compressibility factor for each condition using Equation [5.6] with $v = v_{exp}$:

$$z^A = \frac{(1.059 \times 10^{-1} \text{ m}^3/\text{mol})}{(1.06 \times 10^{-1} \text{ m}^3/\text{mol})} = 1.00$$

$$z^B = \frac{(5.909 \times 10^{-4} \text{ m}^3/\text{mol})}{(6.43 \times 10^{-4} \text{ m}^3/\text{mol})} = 0.919$$

We find that the compressibility factor is unity for Condition A, consistent with negligible IM forces, and that attractive IM forces dominate over repulsive IM forces for Condition B. Recall that in Example 5.3 we characterized the contributions of dipole-dipole interactions and dispersion forces to the attractive IM potential between pairs of water molecules.

We can modify the ideal gas law, expressed as $P = \frac{nRT}{V}$, to introduce the presence non-negligible IM forces through separate corrections for the effect of repulsive and attractive interactions, respectively. In a first modification we introduced repulsive interactions as a reduction in volume, $V_{actual} = V_{macroscopic} - nb$, where V_{actual} is the net or "free" volume available for motion for gas particles, $V_{macroscopic}$ is the measured macroscopic volume, and b is the actual volume occupied per mole of gas particles (*i.e.*, excluded from the available space for motion), termed the excluded volume parameter and reflecting the strength of repulsive IM forces. The resulting modification to the ideal gas law is $P = \frac{nRT}{V - nb}$.

In a second modification we introduce attractive interactions as a reduction in pressure, $P_{macroscopic} = \frac{RT}{V - nb} - \left\{\begin{matrix}\text{Correction for} \\ \text{attractive interactions}\end{matrix}\right\}$. Because the attractive IM PE associated with van der Waals' interactions is proportional to the separation between particles to r^{-6}, equivalent to V^{-2}, we expect this correction for attractive interactions also to scale as V^{-2}. Introducing a parameter a reflecting the strength of attractive IM forces leads to an overall modification to the ideal gas law in the form of van der Waals equation of state:

$$P = \frac{RT}{v-b} - \frac{a}{v^2} \quad [5.7]$$

where a and b are termed the van der Waals parameters. The values for the van der Waals parameters can be determined based on the critical temperature and critical pressure using multi-variable calculus by evaluating the inflection point that occurs in the isotherm (*i.e.*, curve of constant of temperature) for $T = T_c$ on a P-v phase diagram, $\left(\frac{\delta P}{\delta v}\right)_{T_c} = \left(\frac{\delta^2 P}{\delta v^2}\right)_{T_c} = 0$, as:

$$a = \frac{27(RT_c)^2}{64P_c} \quad [5.8a]$$

$$b = \frac{RT_c}{8P_c} \quad [5.8b]$$

Tabulated data for the critical temperature and pressure are available in the *CRC Handbook of Chemistry and Physics* (available at http://www.hbcpnetbase.com/).

Example 5.6: Evaluating the van der Waals Parameters for Water

Determine the van der Waals parameters a (in J-m^3/mol^2) and b (in m^3/mol) for water, for which T_c = 647.3 K and P_c = 220.48 bar.

Solutions

Apply Equations [5.8a] and [5.8b] using the gas constant expressed in units of J/mol-K and converting the critical pressure to Pa, recognizing 1 J = 1 Pa-m^3:

$$a = \frac{27[(8.314 \text{ J/mol} - \text{K})(647.3 \text{ K})]^2}{64\left(220.48 \text{ bar} \times \frac{10^5 \text{ Pa}}{1 \text{ bar}}\right)} = 0.5542 \text{ J} - \text{m}^3/\text{mol}^2$$

$$b = \frac{(8.314 \text{ J/mol} - \text{K})(647.3 \text{ K})}{8\left(220.48 \text{ bar} \times \frac{10^5 \text{ Pa}}{1 \text{ bar}}\right)} = 3.051 \times 10^{-5} \text{ m}^3/\text{mol}$$

Figure 5.9 compares a collection of isotherms for water based on the ideal gas law with the van der Waals equation of state in plots of pressure versus molar volume. For relatively high temperatures—for which molecules possess sufficient thermal energy to overcome attractive forces—and relatively high molar volumes—corresponding to relatively low pressures and for which the molecules will be too far apart to experience significant repulsive forces—the two volumetric descriptions are similar. But for relatively low temperatures or relatively high pressures the models differ. In particular, for $T < T_c$ at relatively low and relatively high pressures only a single phase—the gas phase at low pressures and the liquid phase at high pressures—is observed and the pressure and molar volume map 1:1, but two phases—the saturated liquid and the saturated vapor—can exist in equilibrium at intermediate pressures.

Thus, the van der Waals equation of state provides a qualitative description for all three types of simple fluids (*i.e.*, liquids and supercritical fluids as well as gases), unlike the ideal gas law that is appropriate only for gases.

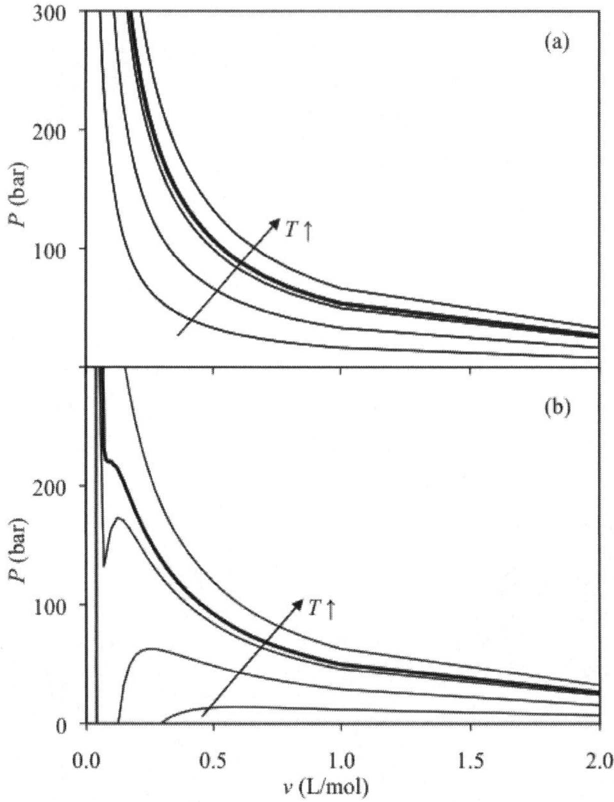

Figure 5.9. Comparison of *P-v* isotherms for water at 200 K, 400 K, 600 K, 647.3 K (Tc), and 800 K for (a) ideal gas law and (b) van der Waals equation of state. Isotherms at critical temperature are in bold. Note the inflection point in the isotherm for the van der Waals model at the critical temperature.

Example 5.7: Applying the van der Waals Equation of State to Water

Compare the pressure (in bar) corresponding to the temperature and experimentally observed molar volume for Condition B in Example 5.5 based on the van der Waals equation of state versus the ideal gas law.

Solutions

Apply Equation [5.7] with T = 773. K, v = 5.909×10^{-4}, and the values for the van der Waals parameters from Example 5.6:

$$P_{vdW} = \frac{(8.314\ \text{J/mol}-\text{K})(773.\ \text{K})}{(5.909\times10^{-4}\ \text{m}^3/\text{mol})-(3.051\times10^{-5}\ \text{m}^3/\text{mol})} - \frac{(0.5542\ \text{J}-\text{m}^3/\text{mol}^2)}{(5.909\times10^{-4}\ \text{m}^3/\text{mol})^2}$$

$$P_{vdW} = 9.88\times10^6\ \text{Pa}\times\frac{1\ \text{bar}}{10^5\ \text{Pa}} = 98.8\ \text{bar}$$

In comparison, application of the ideal gas law predicts:

$$P_{IG} = \frac{(8.314\ \text{J/mol}-\text{K})(773.\ \text{K})}{(5.909\times10^{-4}\ \text{m}^3/\text{mol})} = 1.09\times10^{7}\ \text{Pa}\times\frac{1\ \text{bar}}{10^{5}\ \text{Pa}} = 109.\ \text{bar}$$

The van der Waals equation of state does a much better job describing the experimental observations.

Unfortunately, although the van der Waals model can be physically satisfying and applied to gases, liquids, and supercritical fluids, in practice it encounters difficulties with the type of quantitative calculations needed for process design because it assumes that all particles in a fluid are spheres, and that the strength of attractive IM forces is independent of temperature, either or both being inappropriate for substances such as water and most hydrocarbons.

5.6 The Strength of Intermolecular Forces Affects the Properties of Liquids

Consider now the effects of attractive IM forces on characteristic properties of liquids. The normal boiling point ($T_{b,n}$), defined as the boiling point of a liquid at a pressure of 1 atm, increases as the strength of attractive IM forces increases because IM forces oppose the transformation of matter from the liquid to the vapor phase. This same behavior explains why the vapor pressure (P_{vap}), defined as the pressure of the vapor in equilibrium with the liquid, decreases with increasing strength of attractive IM forces—there is a lower probability that a molecule can escape from the liquid into the vapor with larger attractive IM forces. Note that the boiling point of a liquid depends on the pressure and the vapor pressure on the temperature.

Two other properties of liquids, its viscosity and surface tension, are also affected by IM forces. The viscosity (η) of a liquid, defined qualitatively for both liquids and gases as its resistance to flow, increases as the strength of attractive IM forces increases because IM forces oppose the relative motion of particles past each other. Surface tension (γ), defined as the energy per unit area required to increases the surface area of a liquid, similarly increases as the strength of attractive IM forces increases. We can understand the underlying physical origin of surface tension and the role of IM forces by considering that a particle at a vapor-liquid interface experiences different strengths of IM forces than a particle in the bulk of the liquid (Figure 5.10). We commonly observe that the adsorption of anything at a vapor-liquid interface decreases the surface tension of the liquid. For example, we can determine the purity of water most sensitively by measuring its surface tension (γ_{H_2O} = 72 erg/cm^2 = 7.2 x 10^{-2} J/m^2 for pure water at room temperature). Gradients in surface tension result in the phenomenon of Marangoni effects.

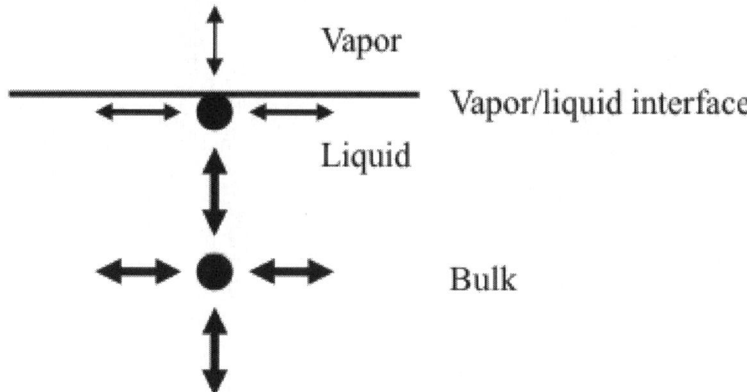

Figure 5.10. Comparison of IM forces for a molecule at the vapor-liquid interface vs. within the bulk of the liquid.

Example 5.8: Ranking the Properties of Organic Liquids Based on their Relative Strengths of Dispersion Forces

Rank the organic molecules examined in Example 5.2—dichloromethane, trichloromethane, methane, tetrachloromethane, and chloromethane—in order of increasing normal boiling point, vapor pressure at room temperature, viscosity, and surface tension.

Solutions

We need to consider effects of changes in both dispersion forces and dipole-dipole interactions because CH_4 and CCl_4 are non-polar molecules experiencing only dispersion forces, but CH_3Cl, CH_2Cl_2, and $CHCl_3$ are polar molecules experiencing both dipole-dipole interactions and dispersion forces. Because the normal boiling point increases with increasing strength of IM forces, trends within these two groups are straightforward to identify based on differences in molar mass: $T_{b,n}(CH_4) < T_{b,n}(CCl_4)$ and $T_{b,n}(CH_3Cl) < T_{b,n}(CH_2Cl_2) < T_{b,n}(CHCl_3)$. Rankings in order of increasing viscosity and surface tension are expected to be identical to the rankings in order of increasing normal boiling point. As vapor pressure decreases with increasing strength of IM forces, we predict that the rankings in order of increasing vapor pressure should be the reverse of the trends for normal boiling point, surface tension,and viscosity. Note that trends across these two groups are more difficult to predict because both the types of IM forces experienced as well as molar mass are varying. Experimental data for the normal boiling point are consistent with our qualitative predictions: $T_{b,n}(CH_4)$ = -162 °C, $T_{b,n}(CCl_4)$ = 77 °C, $T_{b,n}(CH_3Cl)$ = -24 °C, $T_{b,n}(CH_2Cl_2)$ = 40 °C, and $T_{b,n}(CHCl_3)$ = 62 °C.

5.7 Types of Solids and their Properties

Solids are condensed phases of matter in which the particles (*i.e.*, atoms in an atomic solid, ions in an ionic solid, molecules in a molecular solid) are packed closely together, such that IM forces are extremely important. Solids differ from liquids in that liquids are completely disordered phases in which the particles typically are spaced further apart than in solids. Solids can be classified into three broad classes, based on the degree of ordering they exhibit. Usually when we think of a "solid," we are visualizing crystalline solids, solids for which all of the particles pack in regularly-ordered, repeating (termed periodic) structures (termed lattices). Crystalline solids exhibit a melting point (T_m) that increases as the strength of attractive IM forces increases because IM forces oppose the transformation from the solid to the liquid phase.

Example 5.9: Ranking the Melting Points of Different Solids Based on their Relative Strengths of Attractive IM Forces

Rank the compounds examined in Example 5.1—pentylamine, 1-butanethiol, 2,3-dimethylbutane, and ammonium carbonate—in order of increasing melting point.

Solutions

In our solutions for Example 5.1 we ranked these compounds in order of increasing strength of attractive intermolecular forces as 2,3-dimethylbutane, 1-butanethiol, pentylamine, and ammonium carbonate. We expect this same ranking to apply in terms of increasing melting point. Experimental data support our prediction: $T_m(H_3CCHCH_3CHCH_3CH_3)$ = -128 °C, $T_m(H_3C(CH_2)_3SH)$ = -116 °C, $T_m(H_3C(CH_2)_4NH_2)$ = -50 °C, and $T_m((NH_4)_2CO_3)$ > 58 °C (ammonium carbonate decomposes at this temperature before it melts).

We classify types of crystalline solids based on the types of IM forces that dominate (Table 5.1). Although you probably can identify examples of many of these types of solids from your everyday experiences, you may not recognize network covalent solids. These types of materials can be visualized as incredibly large molecules. Two important examples of network covalent solids are solid carbon in the form of graphite and diamond. Graphite (Figure 5.11a) and diamond (Figure 5.11b) are examples of allotropes of carbon, defined as multiple forms of an element that differ in bonding and molecular structure. Note that we previously have encountered a pair of allotropes: compare the bonding and structure of molecular oxygen with its allotrope ozone in Example 4.2. Graphite is elemental carbon with C atoms covalently bonded as sheets of atoms with AX_3 VSEPR notation, trigonal-planar geometry, and sp^2 hybridization. In contrast, diamond is elemental carbon with C atoms covalently bonded in a three-dimensional structure with AX_4 VSEPR notation, tetrahedral geometry, and sp^3 hybridization. Note that only relatively weak dispersion forces hold sheets of graphite together, such that these sheets shear easily (*e.g.*, as in pencil lead). However, the tetrahedral arrangement of carbon atoms in diamond means there are no planes of relatively weak shear strength. In combination with the relatively strong bonding between each carbon atom and four others, pure diamond is the hardest substance known.

Table 5.1. Properties of Different Types of Crystalline Solids

Type of Solid	Components	Primary/Strongest IM Forces	Representative Properties
Metallic	Metal atoms (atoms and delocalized e^-)	Metallic bonds	Lustrous, ductile, malleable, good conductors, high T_m
Ionic ("Ceramics")	Cations and anions	Electrostatic interactions	Hard, nonconducting solids/ conducting melts, high T_m
Network Covalent	Atoms	Covalent bonds	High T_m
Molecular	Hydrogen-bonding molecules	Hydrogen bonds	Moderate T_m
	Polar molecules	Dipole-dipole interactions	Low T_m
	Nonpolar molecules	Dispersion forces	Very low T_m

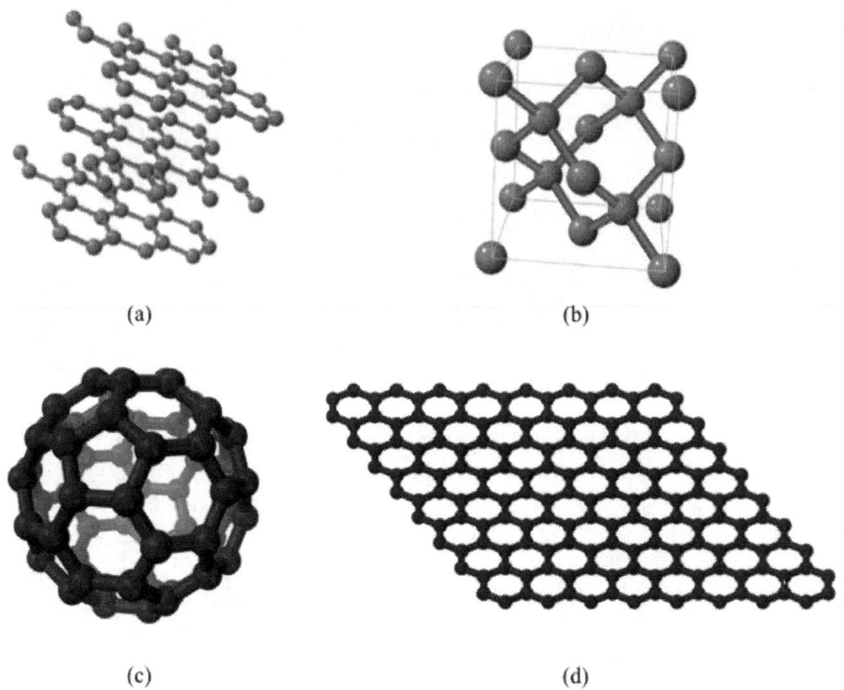

Figure 5.11. Structures of allotropes of carbon. (a) Graphite; (b) diamond; (c) buckminsterfullerene; (d) graphene.

Until the past few decades graphite and diamond were the only known allotropes of carbon. The discovery of buckminsterfullerene, with a molecular formula C_{60} and known colloquially as "buckyball" (Figure 5.11c), offered an additional allotrope of solid carbon and stimulated the search for others. Fullerenes in the form of spheres have been examined as carriers for drug delivery, and fullerenes formed into cylindrical carbon nanotubes and graphene sheets (essentially, single sheets of graphite, Figure 5.11d) are being studied for potential applications in advanced high-speed electronics.

Solids also can exist in non-crystalline forms. Liquid crystalline solids (also known as liquid crystals) are composed of highly asymmetric (*i.e.*, rod- and disc-like) molecules with large dipole moments that can exist in phases of matter with spatial ordering intermediate between crystalline solids—which have regularly-ordered, periodic molecular packing—and liquids (which have completely disordered packing. Multiple types of liquid crystalline phases can occur as a function of temperature and/or applied electric field, with different phases varying in the degree of microscopic ordering exhibited (Figure 5.12).

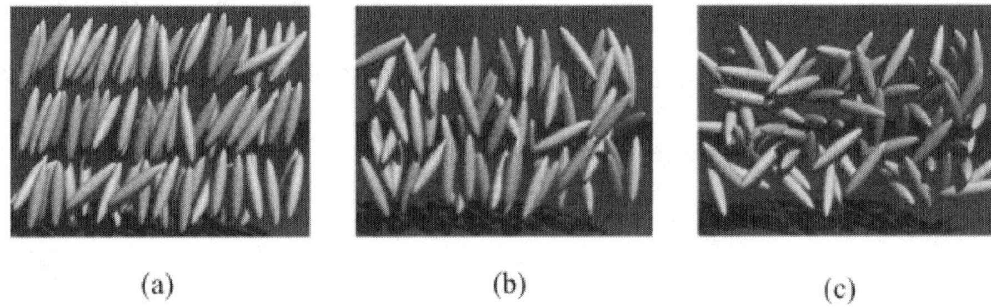

Figure 5.12. Liquid crystalline and ordinary liquid phases. (a) In smectic phase molecules are aligned roughly in a common direction (termed the director) within a layer. (b) In nematic phase molecules are aligned in the direction of the director but not in layers. (c) In the isotropic liquid phase no spatial alignment exists in any direction.

In amorphous solids not all of the particles pack in regularly-ordered, periodic structures, such that either no or only negligible spatial ordering is observed (Figure 5.3). Many polymers occur as completely amorphous or partially amorphous (*i.e.*, some regions, termed domains, are amorphous; other domains are crystalline) solids. Amorphous solids do not melt but rather experience a phase transition known as a glass transition between a hard, brittle "glassy" solid phase and a soft, rubbery solid phase. This transition occurs at the glass transition temperature (T_g): for $T < T_g$, molecules have relatively restricted motion, but for $T > T_g$ molecules have much less restricted motion. As a consequence the physical behavior of an amorphous solid depends on its temperature relative to its glass transition temperature. The glass transition temperature is a function of molecular structure.

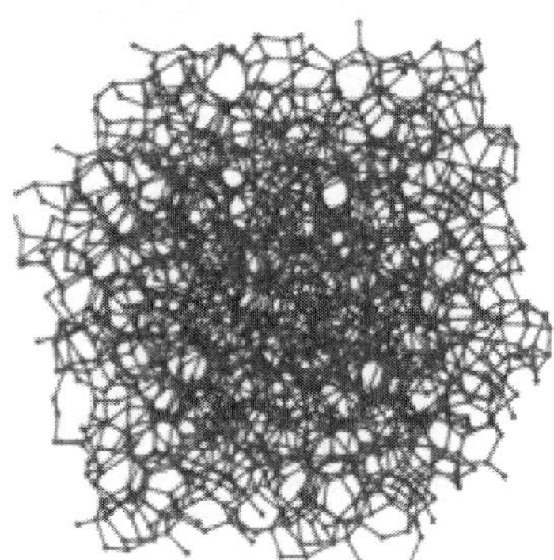

Figure 5.13. Example of structure for an amorphous form of carbon; compare with Figure 5.11. (Credit: Mstroek)

5.8 Models for the Structure of Simple Crystalline Solids

Let's now examine some possible structures of crystal lattices describing the ordered, periodic three-dimensional packing of atoms, molecules, or ions in a crystalline solid. Note that the particles may be identical (*e.g.*, for a pure metal) or of multiple types (*e.g.*, for a metallic alloy or ionic solid consisting of cations and anions). In fact, we can consider crystal lattices for ionic compounds as their equivalent of a "structural formula."

We represent a crystal lattice most concisely based on its unit cell—the simplest repeating arrangement of particles. In practice, a wide variety of unit cells are encountered, and the U.S. Naval Research Labs provides a tremendous website (http://cst-www.nrl.navy.mil/lattice/index.html) at which you can interactively examine this variety. As an introduction to this subject we focus on three types of cubic unit cells (*i.e.*, unit cells based on a cubic geometry), as depicted in Figure 5.14. Simple cubic (denoted *sc*) unit cells consist of eight particles, one at each corner of cube. Body-centered cubic (denoted *bcc*) unit cells consist of nine particles, one at each corner of a cube as well as one in the center of the cube. Lastly, face-centered cubic (denoted *fcc*) unit cells consist of fourteen particles, one at each corner of a cube as well as one on each face of the cube. Many metals crystallize as an *fcc* unit cell, and this type of packing is also known as a cubic close-packed unit cell. An interactive visualization of an *fcc* unit cell is available at http://cst-www.nrl.navy.mil/lattice/struk/a1.html.

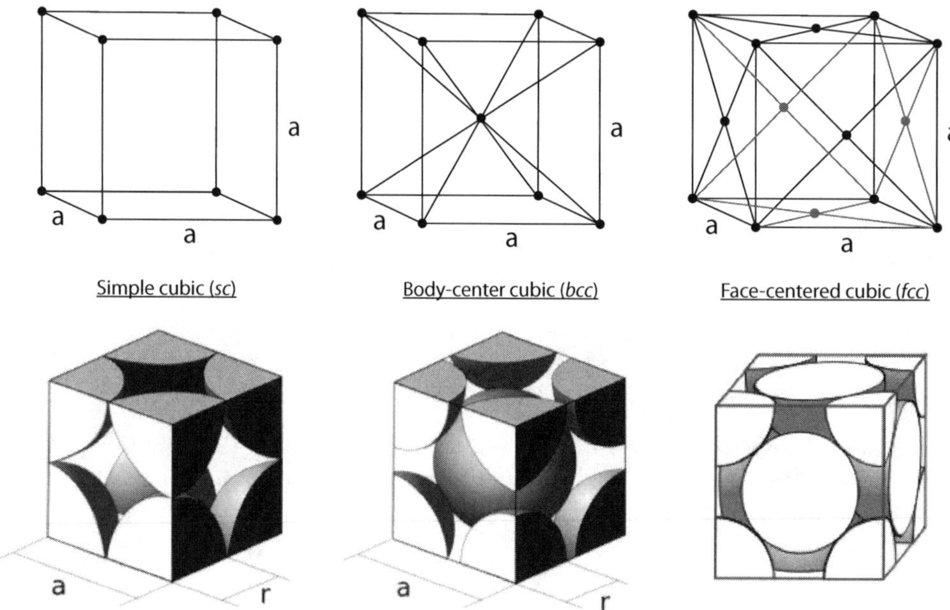

Figure 5.14. Lattice (top row) and cut-away space-filling depictions of cubic unit cells.

Table 5.2 summarizes some of the characteristic properties of these different cubic unit cells. We assume that each particle contributing to the unit cell has an identical radius, *r*. The coordination number is the number of particles with which any given particle is in contact. The net number of particles per unit cell, however, is not identical to the coordination number, but rather depends on where the particles are located within/at the boundary of a unit cell and, if a particle is at a boundary, how many other unit cells share this article. Consider how sharing occurs: particles at corners are shared among eight different unit cells, particles in the center are contained entirely within a given unit cell, and particles at faces are shared between two different unit cells. The *sc* unit cell has the lowest coordination number and number of particles per unit cell; the *fcc* unit cell the highest coordination number and number of particles per unit cell. Yet packing more particles into the *fcc* unit cell has consequences regarding the size of this unit cell relative to the *sc* and *bcc* unit cells for a given particle radius. The size of a unit cell can be represented either by the length of its sides (a) or by its volume ($V_{unitcell} = a^3$). Relationships between volume and particle radius can be determined from geometric considerations.

Table 5.2. Properties of Cubic Unit Cells

Property	Type of Crystal Lattice		
	Simple Cubic (*sc*)	Body-Centered Cubic (*bcc*)	Face-Centered Cubic (*fcc*)
Coordination number	6 (1 in plane above, 4 in-plane, 1 in plane below)	8 (4 in plane above, 0 in-plane, 4 in plane below)	12 (4 in plane above, 4 in-plane, 4 in plane below)
Particles/unit cell	1 (1/8 of each of 8 corner particles)	2 (1 center particle + 1/8 of each of 8 corner particles)	4 (½ of each of 6 face particles + 1/8 of each of 8 corner particles)
Length of side (a)	$2r\ 2r$	$\frac{4\sqrt{3}}{3}r$	$2\sqrt{2}(r)$
Volume (a^3)	$8r^3\ 8r^3$	$\frac{64\sqrt{3}}{9}r^3$	$16\sqrt{2}(r^3)$

Note that the actual volume occupied by particles, in contrast with the volume of the unit cell, is based on the number of particles per unit cell and the volume per particle:

$$V_{particles} = \{\text{particles/unitcell}\}\{\text{volume/particle}\} = \frac{4\pi\{\text{particles/unitcell}\}r^3}{3} \qquad [5.8]$$

We then can determined the amount of empty space with a unit cell, its void volume, as:

$$V_{Void} = V_{unitcell} - V_{particles} > 0 \qquad [5.9]$$

We predict the density of a crystalline solid based on the properties of its unit cell as:

$$\rho = \frac{\{\text{particles/unitcell}}{V_{unitcell}} \qquad [5.10]$$

where $m_{particle}$ is the mass per particle.

5.9 Structures and Formula Units for Ionic Solids

We can apply our models for cubic unit cells to determine the formula unit—the equivalent of an empirical formula—for ionic solids and metal alloys Examples 5.12 and 5.13 illustrate the general technique.

Example 5.10: Predicting the Density of a Metal Based on its Unit Cell and Metallic Radius

Iridium (Ir) has a metallic radius of 135.5 pm and crystallizes as an *fcc* unit cell. What is its density (in g/cm^3)?

Solutions

First, determine the mass per Ir atom in units of g/atom based on its molar mass of 192.2 g/mol:

$$m_{Ir atom} = 192.2 \text{ g/mol Ir} \times \frac{1 \text{ mol Ir}}{6.022 \times 10^{23} \text{ atoms}} = 3.192 \times 10^{-22} \text{ g/atom}$$

Next, determine the volume per unit cell based on the metallic radius and the relationship between metallic radius and volume for an *fcc* unit cell from Table 5.2:

$$V_{fcc unit cell} = 16\sqrt{2}(r_{Ir}^3) = 16\sqrt{2}\left(135.5 \text{ pm} \times \frac{1 \text{ m}}{10^{12} \text{ pm}} \times \frac{10^2 \text{ cm}}{1 \text{ m}}\right)^3 = 5.629 \times 10^{-23} \text{ cm}^3\text{/unit cell}$$

Recognize that there are four Ir atoms per *fcc* unit cell from Table 5.2, such that applying Equation [5.10] we find:

$$\rho = \frac{m_{unit cell}}{V_{unit cell}} = \frac{\{\text{\# of Ir atoms per } bcc \text{ unit cell}\}\{\text{mass per Ir atom}\}}{\{\text{volume per } fcc \text{ unit cell}\}}$$

$$= \frac{(4 \text{ atoms/unit cell})(3.192 \times 10^{-22} \text{ g/atom})}{(5.629 \times 10^{-23} \text{ cm}^3\text{/unit cell})}$$

$$\rho = 22.69 \text{ g/cm}^3$$

This value compares favorably with experimental measurements.

Example 5.11: Determining a Metallic Radius from the Density of a Metal and its Unit Cell

Iron crystallizes in a body-centered cubic (*bcc*) unit cell. If the density of iron is 7.06 g/cm^3, what is the metallic radius (in pm) of an atom of iron?

Solutions

We can relate the density of iron to its unit cell structure using Equation [5.10]:

$$\rho = \frac{m_{unitcell}}{V_{unitcell}} = \frac{\{\#\text{ of Fe atoms per } bcc \text{ unit cell}\}\{\text{mass per Fe atom}\}}{\{\text{volume per } bcc \text{ unit cell}\}}$$

where from Table 5.2 there are two atoms per *bcc* unit cell. The mass of an Fe atom can be determined from its molar mass of 55.85 g/mol as:

$$m_{Featom} = 55.85 \text{ g/mol} \times \frac{1 \text{ mol Fe}}{6.022 \times 10^{23} \text{ Fe atoms}} = 9.274 \times 10^{-23} \text{ g/atom}$$

The volume of a *bcc* unit cell depends on the metallic radius of an Fe atom based on Table 5.2:

$$V_{bccunitcell} = \frac{64\sqrt{3}}{9}(r_{Fe})^3$$

We then solve for the metallic radius of an Fe atom by rearrangement:

$$V_{unitcell} = \frac{\{\#\text{ of Fe atoms per } bcc \text{ unit cell}\}\{\text{mass per Fe atom}\}}{\rho} = \frac{64\sqrt{3}}{9}(r_{Fe})^3$$

$$(r_{Fe})^3 = \frac{9\{\#\text{ of Fe atoms per } bcc \text{ unit cell}\}\{\text{mass per Fe atom}\}}{(64\sqrt{3})\rho}$$

$$r_{Fe} = \left[\frac{9\{\#\text{ of Fe atoms per } bcc \text{ unit cell}\}\{\text{mass per Fe atom}\}}{(64\sqrt{3})\rho}\right]^{1/3}$$

$$= \left[\frac{9(2 \text{ atoms/unit cell})\{9.274 \times 10^{-23} \text{ g/atom}\}}{64\sqrt{3}(7.86 \text{ g/cm}^3)}\right]^{1/3} = 1.242 \times 10^{-8} \text{ cm} \times \frac{1 \text{ m}}{10^2 \text{ cm}} \times \frac{10^{12} \text{ pm}}{1 \text{ m}}$$

$$r_{Fe} = 124.2 \text{ pm}$$

Example 5.12: Determining the Formula Unit for CsCl Based on its Unit Cell

Show that the unit cell for cesium chloride—depicted below in a lattice representation (visualized interactively at http://cst-www.nrl.navy.mil/lattice/struk/b2.html) on the left and a space-filling representation on the right—with cesium ions occupying corner positions and a chloride ion occupying the center position—is consistent with a formula unit of CsCl.

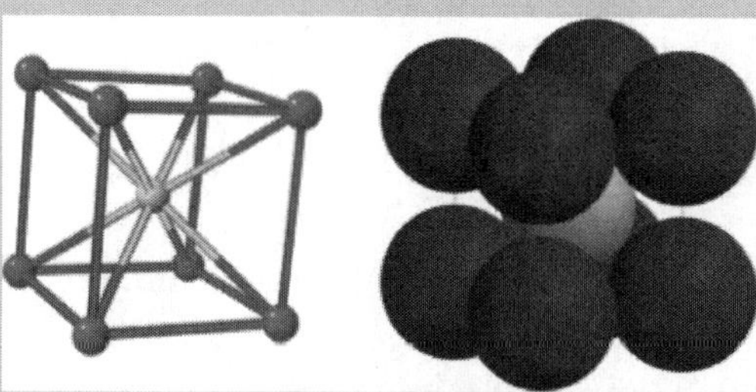

Solution

Count the number of cesium and chloride ions separately:

Cesium: 1 cesium ion per unit cell (8 ions on corners × 1/8 of each corner is within the unit cell)

Chlorine: 1 chloride ion per unit cell

The formula unit is the simplest ratio of 1 Cs: 1 Cl, CsCl. Note that we could shift each ion by a ½ unit cell, such that the cesium ions would occupy center positions and the chloride ions corner positions, without changing any of the properties of the structure.

Example 5.13: Determining the Formula Unit for NaCl Based on its Unit Cell

Show that the unit cell for sodium chloride—depicted below in a lattice representation (visualized interactively at http://cst-www.nrl.navy.mil/lattice/struk/b1.html) on the left and a space-filling representation on the right, with sodium ions occupying corner and face positions and chloride ion occupying the center and edge positions—is consistent with a formula unit of NaCl.

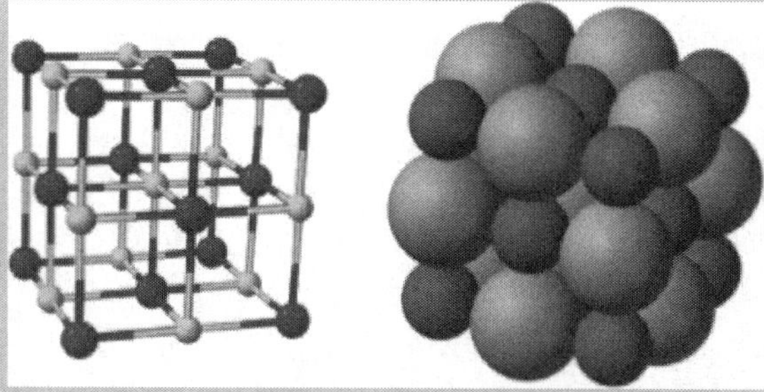

Solution

Count the number of sodium and chloride ions separately:

Sodium: 4 sodium ions per unit cell (8 ions on corners × 1/8 of each corner is within the unit cell + 6 ions on faces × ½ of each face is within the unit cell)

Chlorine: 4 chloride ions per unit cell (12 ions on edges × ¼ of each edge is within the unit cell + 1 central ion)

The formula unit is the simplest ratio of 4 Na: 4 Cl, NaCl.

Example 5.14: Determining the Formula Unit for a Mineral Based on a Verbal Description of its Unit Cell

Perovskite is a mineral containing only calcium, titanium (Ti), and oxygen. The unit cell for perovskite consists of a cube formed with one Ca ion at each vertex (corner), one Ti ion in the center, and one O atom at the center of each face. What is the formula unit for perovskite?

Solutions

Count the numbers of each ion separately; a lattice representation (visualized interactively at http://cst-www.nrl.navy.mil/lattice/struk/e2_1.html) and a space-filling representation of the unit cell are depicted below on the left and right, respectively. There is one calcium ion per unit cell based on 8 vertices × 1/8 of each vertex per unit cell, one titanium ion per unit cell based on the center position lying entirely within the unit cell, and three oxygen ions per unit cell based on 6 faces × ½ of each face per unit cell, such that the formula unit is $CaTiO_3$ (*i.e.*, one Ca^{+2} ion and one Ti^{+4} for every three O^{2-} ions). Note that it would have been impossible to identify the formula unit simply based on balancing ionic charges because titanium can form ions with charge +2, +3, or +4.

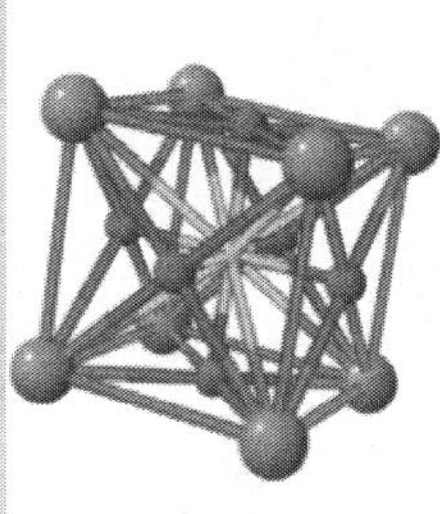

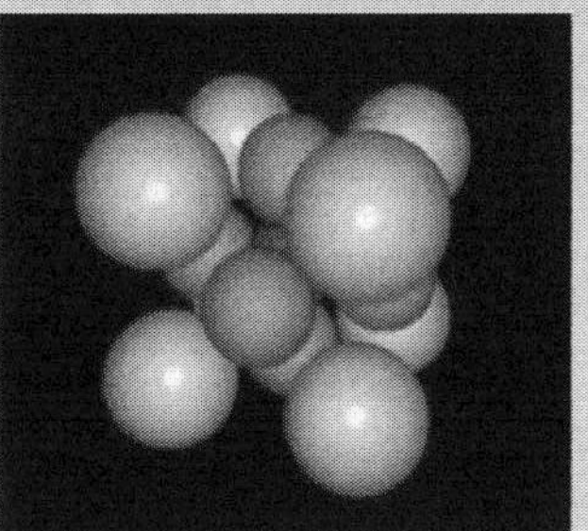

Example 5.15: Determining the Formula Unit for a Mineral Based on a Unit Cell with Filled Holes

Zinc blende, one of the two crystalline structures observed for the compound composed of zinc and sulfur ions, consists of zinc ions on an *fcc* lattice and sulfur ions occupying half the eight tetrahedral holes per unit cell in the lattice, as depicted below in a lattice representation (visualized interactively at http://cst-www.nrl.navy.mil/lattice/struk/b3.html) on the left and a space-filling representation on the right. Determine the formula unit for zinc blende.

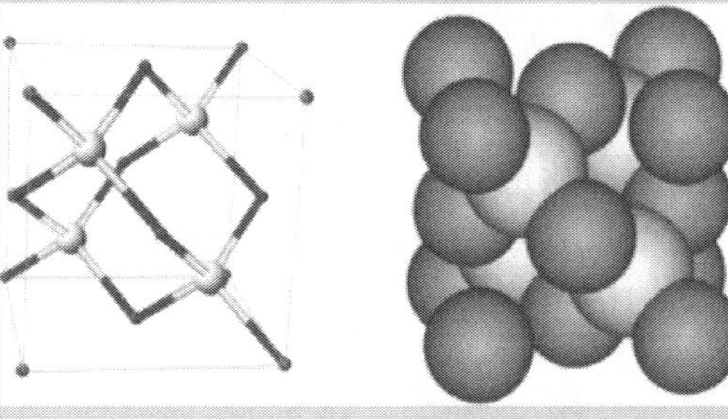

Solutions

Count the numbers of each ion separately. There are four zinc ions per unit cell based on 8 vertices × 1/8 of each vertex per unit cell and 6 faces × ½ of each face per unit cell, and four sulfur ions per unit cell based on 8 tetrahedral holes per unit cell × ½ of the holes occupied by sulfur ions, such that the formula unit is the simplest ratio of 4 Zn: 4 S, ZnS (*i.e.*, one Zn^{2+} ion for every S^{2-} ion). Note that zinc only forms ions with a charge of +2 and sulfur with a charge of -2 and that the positions for the zinc and sulfur ions are equivalent (*i.e.*, the same structure results if the ions are switched).

6: Physical Properties of Mixtures

Learning Objectives

By the end of this chapter you should be able to:

1) Use Dalton's law to calculate partial pressures in gas mixtures.
2) Relate the amount of a compound or ion (in units of moles or mass) and the volume of a solution of that compound or ion with the molarity of the solution.
3) Distinguish between strong electrolytes, weak electrolytes, and nonelectrolytes in solution.
4) Determine the concentrations of gases dissolved in liquids.
5) Relate the amount of a compound or ion (in units of moles or mass) and the mass of a solvent with the molality of a solution of the compound or ion in that solvent.
6) Evaluate the colligative properties of freezing point depression, boiling point elevation, and osmotic pressure for dilute solutions.

6.1 Properties of Mixtures of Gases

Our primary focus in Chapter 5 was explaining the physical properties of pure substances in terms of the microscopic structure. Frequently, however, we are interested in the properties of mixtures of multiple pure substances. We start by considering the mixing of a pair of pure gases, CO and H_2. Figure 6.1a depicts the situation before mixing: the gases are separated by an impermeable barrier. When the barrier is removed, we expect to observe the situation depicted in Figure 6.1b: gases always mix. We seek to understand relationships between the pressure exerted by the mixture, the individual contributions to measurable pressure from CO and H_2, the amounts of CO and H_2 in the mixture, and other overall properties of the mixture (*e.g.*, the volume and temperature).

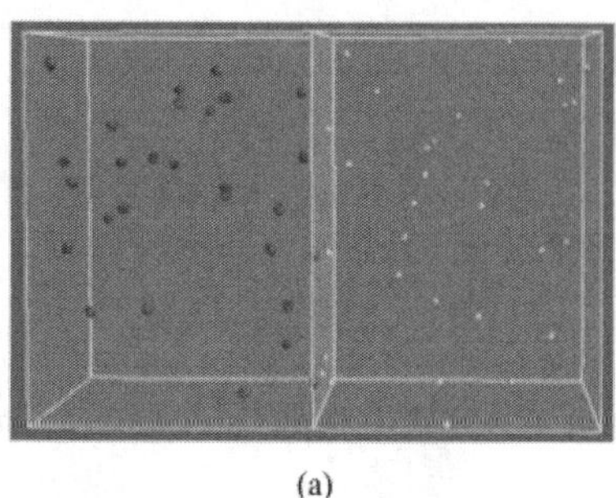

(a)

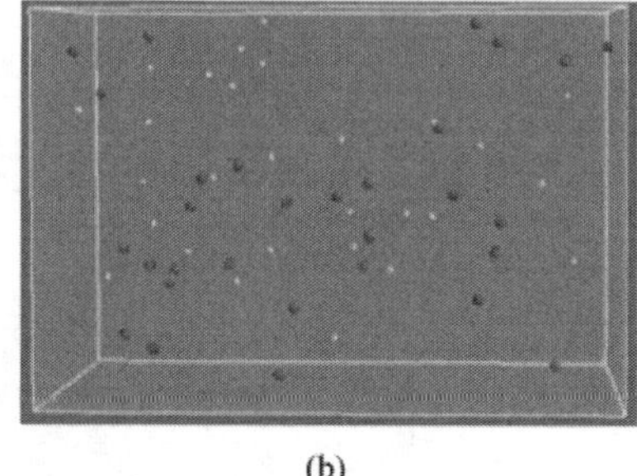

(b)

Figure 6.1. Mixing of the pure gases H_2 (in white) and CO (in gray). (a) Initial state with CO in left chamber and H_2 in right chamber separated by an impermeable barrier; (b) final state with H2 and CO forming a homogeneous mixture after removal of barrier.

Assuming CO and H_2 behave as ideal gases, they do not interact with each other (other than by relatively infrequent collisions) and exert pressures on the walls of the container independent of the presence of each other. We define the pressure exerted by each species as its partial pressure, P_{CO} and P_{H_2}, which represent the respective pressures that would be exerted if *only* pure CO or only pure H_2 were present in the container. We can relate these partial pressures to the actual (denoted as "total") pressure we measure for the mixture, P_{total}, by considering that the total number of moles of gas in our mixture is $n_{total} = n_{CO} + n_{H_2}$ and both gases occupy the same apparent volume (V) and are at the same temperature (T). Applying the rearranged ideal gas law to both the overall mixture as $n_{total} = \frac{P_{Total}V}{RT}$ and to the individual species as $n_i = \frac{P_iV}{RT}$, where P_i is the partial pressure of species i, we find:

$$\frac{P_{total}V}{RT} = n_{total} = n_{CO} + n_{H_2} = \frac{P_{CO}V}{RT} + \frac{P_{H_2}V}{RT} = \left(\frac{V}{RT}\right)(P_{CO} + P_{H_2}) \qquad [6.1]$$

We conclude that the observed pressure is the sum of the partial pressures (*i.e.*, $P_{total} = P_{CO} + P_{H_2}$), a result that we generalize for a mixture of N gases as Dalton's law of partial pressures:

$$P_{total} = \sum_{i=1}^{N} P_i \qquad [6.2]$$

Alternatively, we can express the partial pressure for each species in terms of the total pressure for the mixture as:

$$P_i = \frac{n_iRT}{V} = \left(\frac{n_i}{n_{total}}\right)\left(\frac{n_{total}RT}{V}\right) = \left(\frac{n_i}{n_{total}}\right)P_{total} \qquad [6.3]$$

and express Dalton's law of partial pressure as:

$$P_i = x_iP_{Total} \qquad [6.4]$$

where we define $x_i \equiv \frac{n_i}{n_{Total}}$ as the mole fraction of species i. For example, air is a gas mixture with $x_{N_2} = 0.781$, $x_{O_2} = 0.209$, $x_{Ar} = 0.009$, an $x_{other} = 0.001$ (*e.g.*, CO_2). Note that:

$$\sum_{i=1}^{N} x_i = \sum_{i=1}^{N}\left(\frac{n_i}{n_{Total}}\right) = \frac{\sum_{i=1}^{N} n_i}{n_{Total}} = \frac{n_{Total}}{n_{Total}} = 1 \qquad [6.5]$$

Example 6.1: Humidifying a Tank of Pure Oxygen

A tank of oxygen is humidified by opening a valve connecting a 4.50-L tank of $O_2(g)$ at 1.00 atm and 400 K with a 0.50-L tank of $H_2O(g)$ at 2.00 atm and 400. K.

a) What is the mole fraction of H_2O in the mixture?

b) What is the pressure (in atm) measured after the valve is opened?

c) What is the partial pressure (in atm) of H_2O in the mixture?

Solutions

The process is <u>isothermal</u> (*i.e.*, occurs at a constant temperature of $T = 400.$ K) and produces a mixture with a volume of $V_{total} = V_{O_2,o} + V_{H_2O,o} = (4.50\text{ L}) + (0.50\text{ L}) = 5.00\text{ L}$. We determine the number of moles of O_2 and H_2O in the mixture based on applying the ideal gas law to their individual tanks prior to opening the valve:

$$n_{O_2} = \frac{P_{O_2,o}V_{O_2,o}}{RT} = \frac{(1.00\text{ atm})(4.50\text{ L})}{(8.206\times10^{-2}\text{ atm}-\text{L/mol}-\text{K})(400.\text{ K})} = 0.137\text{ mol}$$

$$n_{H_2O} = \frac{P_{H_2O,o}V_{H_2O,o}}{RT} = \frac{(2.00\text{ atm})(0.50\text{ L})}{(8.206\times10^{-2}\text{ atm}-\text{L/mol}-\text{K})(400.\text{ K})} = 3.05\times10^{-2}\text{ mol}$$

such that:

$$n_{total} = n_{O_2} + n_{H_2O} = (0.137\text{ mol}) + (0.0305\text{ mol}) = 0.168\text{ mol}$$

$$x_{H_2O} = \frac{n_{H_2O}}{n_{total}} = \frac{(3.05\times10^{-2}\text{ mol})}{(0.168\text{ mol})} = 0.182$$

The pressure that we measure for the mixture and the partial pressure of water are then:

$$P_{total} = \frac{n_{total}RT}{V_{total}} = \frac{(0.168\text{ mol})(8.206\times10^{-2}\text{ atm}-\text{L/mol}-\text{K})(400.\text{ K})}{(5.00\text{ L})} = 1.10\text{ atm}$$

$$P_{H_2O} = x_{H_2O}P_{total} = (0.182)(1.10\text{ atm}) = 0.200\text{ atm}$$

Note that relative humidity is defined with respect to the temperature as the ratio of the partial pressure of water to its vapor pressure at that temperature, expressed on a percent basis.

6.2 Composition of Solutions

In Section 3.1 we introduced solutions as homogeneous mixtures in which one or more substances are in large excess. Solutions of metals in the solid state are known as alloys. For example, steel is an alloy consisting primarily of iron, with lesser amounts of chromium, carbon, and other elements. Bronze is an alloy of copper and tin. The physical properties of these alloys depend on their composition.

For solutions we identify two types of components: the solvent(s) is/are the compound(s) in the solution in large excess and that determine(s) the phase of the solution, and the solute(s) is/are the compound(s) in the solution that are *not* in large excess. We are frequently interested in how concentrated a solute is in a solution. One basis is to represent the concentration of the solute, C_{solute} (alternatively denoted in bracket notation as [solute]), in terms of molarity, the number of moles of solute (n_{solute}) per volume of the solution ($V_{solution}$):

$$C_{solute} = [\text{solute}] \equiv \frac{n_{solute}}{V_{solution}} \quad [6.6]$$

where the unit 1 mol/L is defined as 1 molar (M). Here "M" is a unit, not a molar mass.

We encounter two types of problems when considering concentrations of solutions. In Type I problems we seek to determine the amount of solute, the volume of solution, or the concentration of solute given values for two out of three of these properties. These problems require algebraic rearrangement of Equation [6.6] and, if the amount of solute is given or needed in terms of a mass, the use of the molar mass for the solute to relate the number of moles and mass (m_{solute}) of the solute.

Example 6.2: Determining the Mass of Solute in a Solution

How many grams of NaCl are in 100 mL of a room temperature-saturated solution of 5.420 M NaCl(*aq*)?

Solutions

Rearrange Equation [6.6] to solve for the number of moles of the solute NaCl and then convert to the equivalent mass of NaCl using its molar mass of $M_{NaCl} = 58.45$ g/mol:

$$n_{NaCl} = C_{NaCl}V_{solution} = (5.420\ \text{mol/L})\left(100.\ \text{mL} \times \frac{1\ \text{L}}{10^3\ \text{mL}}\right)$$

$$= 0.542\ \text{mol NaCl} \times \frac{58.45\ \text{g NaCl}}{1\ \text{mol NaCl}}$$

$$m_{NaCl} = 31.7\ \text{g NaCl}$$

In Type II problems we seek to determine the result of a dilution, the process of transforming a *more* concentrated solution to a *less* concentrated solution. For dilutions we define the stock as the more concentrated solution with which we start, and the diluent as the pure solvent or less concentrated solution that is added to the stock to create a diluted solution.

Example 6.3: Determining the Volume of Pure Solvent for a Dilution

What volume of pure water is needed to dilute 2.500 L of a 5.420 M NaCl(*aq*) solution to form a 100.0 mM NaCl(*aq*) solution?

Solutions

Because pure solvent is to be used as the diluent, this dilution involves no change in the amount of NaCl, such that using the subscripts *i* and *f* to denote the initial and final states (*i.e.*, before and after the dilution) we find:

$$n_{\mathrm{NaCl},i} = n_{\mathrm{NaCl},f}$$

$$C_{\mathrm{NaCl},i} V_{solution,i} = C_{\mathrm{NaCl},f} V_{solution,f}$$

Rearranging with $C_{\mathrm{NaCl},i} = 5.420$ M, $V_{solution,i} = 2.500$ L, and $C_{\mathrm{NaCl},f} = 100.0$ mM we find the unknown final volume for the solution as:

$$V_{solution,f} = \frac{C_{\mathrm{NaCl},i} V_{solution,i}}{C_{\mathrm{NaCl},f}} = \frac{(5.420\ \mathrm{M})(2.500\ \mathrm{L})}{\left(100.0\ \mathrm{mM} \times \frac{1\ \mathrm{M}}{10^3\ \mathrm{mM}}\right)} = 135.5\ \mathrm{L}$$

We then determine the volume of pure water, $V_{\mathrm{H_2O}}$, added as the difference between the final and initial volumes of the solution:

$$V_{solution,f} = V_{solution,i} + V_{\mathrm{H_2O}}$$

$$V_{\mathrm{H_2O}} = V_{solution,f} - V_{solution,i} = (135.5\ \mathrm{L}) - (2.500\ \mathrm{L}) = 133.0\ \mathrm{L}$$

Example 6.4: Using a Less-Concentrated Solution as a Diluent

It is 5 p.m., and you urgently need to prepare an aqueous solution of 0.25 M hydrochloric acid for an assay testing water quality. A 100.00-L stock solution of 1.00 M HCl is available, but distilled, deionized (DDI) water is unavailable because a new cartridge must be installed. Tap water is insufficiently pure for your analytical needs, but a 0.50-L solution of 0.10 M HCl is available in the lab. Can you prepare 1.00 L of the required solution with the available materials?

Solutions

The total number of moles of hydrochloric acid in your final solution with $V_{final} = 1.00$ L and $C_{\mathrm{HCl},final} = 0.25$ M will be the sum of the number of moles contributed from the stock solution with $C_{\mathrm{HCl},stock} = 1.00$ M and the number of moles contributed from the less-concentrated diluent with $C_{HCl,diluent}$ = 0.10 M. Mathematically, we express this relationship as a material balance on the amount of HCl as:

$$n_{final} = n_{stock} + n_{diluent}$$

$$V_{final} C_{HCl,final} = V_{stock} C_{HCl,stock} + V_{diluent} C_{HCl,diluent} \quad \text{[E6.1]}$$

Assuming there are no changes in volume of the stock and diluent upon mixing, the volume of the final solution is the sum of the volumes of the stock and diluent:

$$V_{final} = V_{stock} + V_{diluent} \quad \text{[E6.2]}$$

We have two independent relationships with two unknowns (the volumes required for the stock and diluent), such that this problem can be solved using linear algebra. Rearrange Equation [E6.2] as $V_{stock} = V_{final} - V_{diluent}$, substitute into Equation [E6.1], and solve for the volume of diluent needed:

$$V_{final} C_{HCl,final} = (V_{final} - V_{diluent}) C_{HCl,stock} + V_{diluent} C_{HCl,diluent}$$

$$V_{final} C_{HCl,final} = V_{final} C_{HCl,stock} + V_{diluent} (C_{HCl,diluent} - C_{HCl,stock})$$

$$V_{diluent} (C_{HCl,diluent} - C_{HCl,stock}) = V_{final} (C_{HCl,final} - C_{HCl,stock})$$

$$V_{diluent} = \left(\frac{C_{HCl,final} - C_{HCl,stock}}{C_{HCl,diluent} - C_{HCl,stock}} \right) V_{need} = \left[\frac{(0.25 \text{ M}) - (1.00 \text{ M})}{(0.10\text{M}) - (1.00 \text{ M})} \right] (1.00 \text{ L})$$

$$V_{diluent} = 0.83 \text{ L}$$

You will not be able to perform your dilution using the available 0.50 L of 0.10 M HCl, as you need 0.83 L of this solution.

6.3 Behavior of Solutes in Aqueous Solutions

Water is a ubiquitous solvent that, as a polar molecule, dissolves ionic compounds and polar molecules better than non-polar species. In contrast, non-polar solvents (*e.g.*, hydrocarbons) dissolve non-polar species better than polar species. Solutions of water are termed aqueous solutions. Compounds that dissolve in water, termed electrolytes, can be categorized as strong, weak, or nonelectrolytes based on whether and how much the compound dissociates into ions. We can distinguish between solutions formed by different types of electrolytes readily by measuring their conductivity. Note that ultrapure water is a poor conductor of electricity, but water as we typically encounter it (*e.g.*, from the tap or in bottled form) has appreciable amounts of dissolved substances.

Strong electrolytes are ionic compounds that dissolve in water by dissociating completely into individual ions For example, solid sodium chloride dissolves in water to form sodium and chloride ions:

$$NaCl(s) \xrightarrow{H_2O} Na^+(aq) + Cl^-(aq)$$

Solutions of strong electrolytes are excellent conductors Note that not all ionic compounds dissolve appreciably in water, as we will examine in Chapter 10 Further, no ionic compound is infinitely soluble (*e.g.*, the maximum concentration of sodium chloride in water at room temperature is 5.420 M, as encountered in Example 6.2).

Weak electrolytes are molecular compounds—and specifically organic acids (introduced in Section 4.9) and organic bases (which we will discuss in Chapter 10)—that dissolve in water by dissociating partially into individual ion. For example, ethanoic (acetic) acid dissolves in water by partially dissociating into ethanoate (the acetate ion), CH_3COO^-, and the hydronium ion, H_3O^+:

$$CH_3COOH(aq) + H_2O(l) \leftrightarrows CH_3COO^-(aq) + H_3O^+(aq) \qquad \text{Rxn [6.1]}$$

The arrows in opposite directions tell us this process is reversible. We will identify how to evaluate the extent of ionization (*i.e.*, how much of the ions are formed given an amount of ethanoic acid) in Chapter 0. Solutions of weak electrolytes are mediocre conductors. Nonelectrolytes are molecular compounds, such as the gases $O_2(g)$ and $N_2(g)$, alcohols, and sugars, as well as noble gases that dissolve in water with negligible dissociation into individual ons. For example, ethanol dissolves in water without dissociating:

$$CH_3CH_2OH(l) \xrightarrow{H_2O} CH_3CH_2OH(aq)$$

Solutions of nonelectrolytes are very poor conductors.

Example 6.5: Mixing Streams of Dissolved Salts

In a waste-water treatment plant 1.0000×10^3 L of 3.0000×10^{-2} M $Na_2CO_3(aq)$ are mixed with 2.0000×10^3 L of 6.0000×10^{-3} M $Ca(NO_3)_2(aq)$ at room temperature. What is the concentration (in M) of sodium ions in the product solution?

Solutions

Sodium carbonate, Na_2CO_3, is an ionic compound and, thus, a strong electrolyte that dissociates completely upon dissolution in water:

$$Na_2CO_3(s) \xrightarrow{H_2O} 2\ Na^+(aq) + CO_3^{-2}(aq)$$

Thus, for every one mole of Na_2CO_3 there are two moles of Na^+, and 3.0000×10^{-2} M $Na_2CO_3(aq)$ is equivalent to 6.0000×10^{-2} M $Na^+(aq)$.

Our problem is essentially a dilution of $V_i = 1.0000\times10^3$ L of $C_{Na^+,i} = 6.0000\times10^{-2}$ M with $V_{Ca(NO_3),i} = 2.0000\times10^3$ L to form a solution with a total volume of $V_f = V_i + V_{Ca(NO_3),i} = 3.0000\times10^3$ L (where we assume there are no changes in volume upon mixing). Because the solution of $Ca(NO_3)_2$ introduces no additional sodium ions, we find:

$$n_{Na^+,i} = n_{Na^+,f}$$

$$C_{Na^+,i}V_i = C_{Na^+,f}V_f$$

$$C_{Na^+,f} = \frac{C_{Na^+,i}V_i}{V_f} = \frac{(6.0000\times10^{-2}\ M)(1.0000\times10^3\ L)}{(3.0000\times10^3\ L)} = 2.0000\times10^{-2}\ M$$

Example 6.6: Comparing the Conducting for Different Aqueous Solutions

Determine which of the following aqueous solutions is the worst conductor of electricity: 1 M potassium iodide, 1 M 1-propanol, 1 M butanoic acid, or 1 M sodium sulfate.

Solutions

At equal concentrations solutions of strong electrolytes conduct electricity better than solutions of weak electrolytes, which in turn conduct electricity better than solutions of nonelectrolytes. Potassium iodide and sodium sulfate are ionic compounds and strong electrolytes, butanoic acid is an organic acid and a weak electrolyte, but 1-propanol is nonelectrolyte because it is an organic compound but not an organic acid or base. The solution of 1 M 1-isopropanol will have the lowest conductivity.

6.4 Solubility of Gases in Liquids

For dilute (*i.e.*, relatively unconcentrated) solutions of gases dissolved in liquids, the concentration of dissolved gas—termed its solubility, S_{gas} (in M)—is proportional to the partial pressure of the gas in the vapor phase in contact with solution according to Henry's law:

$$S_{gas} = k_H P_{gas} \qquad [6.7]$$

where k_H is the Henry's law constant, has units of M/atm, and is a function of temperature. Henry's law predicts that the solubility of a gas increases with increasing partial pressure for the gas. Values for Henry's law constants are available in the *CRC Handbook of Chemistry and Physics* (available at http://www.hbcpnetbase.com/) and through the NIST Chemistry Webbook (http://webbook.nist.gov).

Example 6.7: Carbonation of Water

Seltzer (*i.e.*, carbonated water) can be produced by exposing drinking water to a partial pressure of 5.0 atm for $CO_2(g)$. The Henry's law constant for carbon dioxide in water is 3.3×10^{-2} M/atm at 25 °C.

a) What is the solubility (in M) of CO_2 in an unopened bottle of seltzer at 25 °C?

b) When a bottle of seltzer is opened and allowed to sit exposed to air at 25 °C for an extended period, what is the solubility (in M) of CO_2? The mole fraction of CO_2 in air is approximately 3×10^{-4}.

Solutions

a) Apply Henry's law, Equation [6.7], to determine the solubility of CO_2 in water exposed to $P_{CO_2} = 5.0\ \text{atm}$:

$$S_{CO_2} = k_H P_{CO_2} = (3.3 \times 10^{-2}\ \text{M/atm})(5.0\ \text{atm}) = 1.6 \times 10^{-1}\ \text{M}$$

b) Now consider that the ratio of the solubility of a gas at two different partial pressures but a common temperature can be evaluated without including a value of a Henry's law constant as:

$$\frac{S_{gas,2}}{S_{gas,1}} = \frac{k_H P_{gas,2}}{k_H P_{gas,1}} = \frac{P_{gas,2}}{P_{gas,1}}$$

such that we can determine the solubility of CO_2 when its partial pressure is reduced at 25 °C as:

$$\frac{S_{CO_2,2}}{S_{CO_2,1}} = \frac{P_{CO_2,2}}{P_{CO_2,1}}$$

$$S_{CO_2,2} = \frac{P_{CO_2,2} S_{CO_2,1}}{P_{CO_2,1}} = \frac{x_{CO_2,2} P_{Total,2} S_{CO_2,1}}{P_{CO_2,1}} = \frac{(3 \times 10^{-4})(1\ \text{atm})(1.6 \times 10^{-1}\ \text{M})}{(5.0\ \text{atm})} = 1 \times 10^{-5}\ \text{M}$$

The concentration of dissolved CO_2 drops significantly upon prolonged exposure to air, which is consistent with the common observation that an opened bottle of seltzer goes flat.

Henry's law can be applied to understand the phenomenon of the "bends" associated with relatively rapid decompression in scuba diving.

6.5 Colligative Properties of Dilute Solutions

Molarity is not the only basis for specifying the concentration of a dissolved species. In particular, although molarity is often a convenient basis because of the ease in specifying the volume of a solution, the volume of a solution typically is not identical to the volume of the solvent—even considering that the solution is mostly solvent!—due to non-idealities in mixing, and the volume of any solution does depend on temperature. An alternative definition for concentration is useful to address these effects. The molality of a solute can be defined as the number of moles of solute per kg of solvent:

$$b_{solute} = \frac{n_{solute}}{m_{solvent}} \qquad [6.8]$$

where the unit 1 mol/kg is defined as 1 molal (m). Here "m" is a unit, not a mass. Note that because the mass of solvent does not vary with temperature, molality is a metric for concentration that is temperature independent (unlike molarity). The molality and molarity of a solute, however, can be related if the density of the solution is known.

Example 6.8:

Determine the molarity (in M) of a 2.50 m solution of NaCl(*aq*), assuming the density of the solution is 1.202 g/mL.

Solutions

Assume a 1-kg basis for the solvent (*i.e.*, assume we have 1.00 kg of water as the solvent), such that the number of moles of the solute sodium chloride is:

$$n_{NaCl} = b_{NaCl} m_{H_2O} = (2.50 \text{ mol NaCl/kg } H_2O)(1.00 \text{ kg } H_2O) = 2.50 \text{ mol NaCl}$$

The volume of the solution can be determined from its mass and density, with the mass of the solution the sum of the masses of the solvent and solute and the mass of the solute determined based on the number of moles and molar mass of sodium chloride:

$$V_{solution} = \frac{m_{solution}}{d_{solution}} = \frac{m_{H_2O} + m_{NaCl}}{d_{solution}} = \frac{m_{H_2O} + n_{NaCl} M_{NaCl}}{d_{solution}}$$

$$= \frac{\left(1.00 \text{ kg} \times \frac{10^3 \text{ g}}{1 \text{ kg}}\right) + (2.50 \text{ mol NaCl})(58.45 \text{ g/mol NaCl})}{(1.202 \text{ g/mL})}$$

$$V_{solution} = 9.54 \times 10^2 \text{ mL}$$

such that:

$$C_{NaCl} = \frac{n_{NaCl}}{V_{solution}} = \frac{(2.50 \text{ mol NaCl})}{\left(9.54 \times 10^2 \text{ mL} \times \frac{1 \text{ L}}{10^3 \text{ mL}}\right)} = 2.62 \text{ M NaCl}$$

Note that our answer does not depend on the mass of solvent we chose as our basis, as you can verify by assuming a different mass of water (*e.g.*, 10.0 kg) as your basis. We typically find that the numerical value for concentration expressed in terms of molarity is similar but not identical to the numerical value for concentration expressed in terms of molality.

Some of the properties of a dilute solution—termed colligative properties (from the Greek word for "collective")—depend only on the molality of solute particles, not on what these particles are. We frequently encounter dilute solutions for which the solute is non-volatile (*i.e.*, the solute exists only in the solid and liquid phases) and the solute and solvent freeze as separate solid phases. For such solutions the presence of a dilute solute shifts the two-phase coexistence curves on a *P*-*T* phase diagram (Figure 6.2). As a consequence the freezing point of the dilute solution is lower than the freezing point of the pure solvent (*i.e.*, $T_{m,soln} < T_{m,pure}$), a phenomenon known as freezing point depression, and the boiling point of the dilute solution is higher than the boiling point of the pure solvent (*i.e.*, $T_{b,soln} > T_{b,pure}$), a phenomenon known as boiling point elevation.

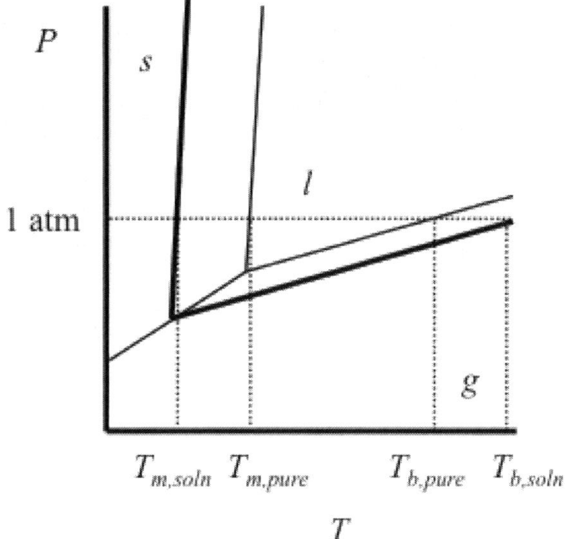

Figure 6.2. Effect of introducing a dilute solute on the *P-T* phase diagram for a solvent. Thin lines depict coexistence curves for the pure solvent; thick lines depict coexistence curves for a dilute solution. The presence of a dilute solute lowers the freezing point and raises the boiling point for the solution relative to the pure solvent.

Both the freezing point depression and the boiling point elevation depend on properties of the solvent as well as whether the solute dissociates in the solvent. We characterize the degree of dissociation using the van't Hoff factor, *i*, which represents how many moles of particles are formed per mole of solute. For example, strong and weak electrolytes dissociate upon dissolution in water, and under some circumstances individual solute particles appear to associate (*e.g.*, due to relatively strong net attractive IM forces). We find that $i > 1$ when the solute experiences net dissociation in solution, $i < 1$ when the solute experiences net association in solution, and $i = 1$ when the solute experiences neither net dissociation nor net association in solution.

Example 6.9: Predicting the van't Hoff Factor

Assuming that strong electrolytes dissociate completely into individual ions and nonelectrolytes neither dissociate nor associate upon dissolution in water, predict the value for the van't Hoff factor for sodium chloride, magnesium chloride, and ethanol in water.

Solutions

Sodium chloride and magnesium chloride are ionic compounds and dissolve in water as strong electrolytes as:

$$NaCl(s) \xrightarrow{H_2O} Na^+(aq) + Cl^-(aq)$$

$$MgCl_2(s) \xrightarrow{H_2O} Mg^{2+}(aq) + 2\ Cl^-(aq)$$

such that we predict $i = 2$ for sodium chloride and $i = 3$ for magnesium chloride (*i.e.*, for every mole of $MgCl_2$, three moles of ions are formed upon dissolution). Ethanol is a nonelectrolyte in water, for which $i = 1$.

The depression of the freezing point for a dilute solution to T_m from $T_{m,pure}$ for the pure solvent can be described quantitatively using the cryoscopic equation for freezing point depression:

$$\Delta T_f \equiv T_m - T_{m,pure} = -iK_f b_{solute} \quad [6.9]$$

where K_f is the cryoscopic constant. Analogously, the elevation of the boiling point for a dilute solution to T_b from $T_{b,pure}$ for the pure solvent can be described quantitatively using the ebullioscopic equation for boiling point elevation:

$$\Delta T_b \equiv T_b - T_{b,pure} = iK_b b_{solute} \quad [6.10]$$

where K_b is the ebullioscopic constant. Both the cryoscopic and ebullioscopic constants are positively-valued properties only of the solvent and do not depend on what the solute is nor how much solute is present, provided the solution is dilute, the solute is non-volatile, and the solute and solvent freeze as separate phases. Values for cryoscopic and ebullioscopic constants are available in the *CRC Handbook of Chemistry and Physics* (available at http://www.hbcpnetbase.com/) for a variety of solvents. Note that the cryoscopic constant is greater than the ebullioscopic constant for a given solvent.

Example 6.10: Determining the Boiling Point Elevation

What is the boiling point elevation when 10.0 g $MgCl_2$ is added to 1.00 kg of water (K_f = 0.515 °C/m)?

Solution

First, determine the molality of $MgCl_2$:

$$b_{solute} = \frac{10.0 \text{ g } MgCl_2}{1.00 \text{ kg } H_2O} \times \frac{1 \text{ mol } MgCl_2}{95.21 \text{ g } MgCl_2} = 0.105 \text{ m } MgCl_2$$

Now apply the ebullioscopic equation, Equation [6.10], with $i = 3$ (from Example 6.9):

$$\Delta T_b = iK_b b_{solute} = (3)(0.515\ ^\circ\text{C/m})(0.105 \text{ m}) = 0.162\ ^\circ\text{C}$$

Example 6.11: Determining the Molar Mass of an Unknown Based on the Cryoscopic Effect

Colligative properties offer simple yet powerful methods for estimating the molar mass of a pure substance. The addition of 5.00 g of an unknown compound lowers the freezing point of 250. g of naphthalene (K_f = 6.899 K-kg/mol) by 0.780 K. What is the molar mass (M, in g/mol) of the unknown compound if it neither dissociates nor associates upon dissolution into naphthalene?

Solutions

Because the solute does not dissociate into ions, $i = 1$. Rearrange the cryoscopic equation, Equation [6.9], to express the molality of the solute in terms of the freezing point depression, cryoscopic constant, and van't Hoff factor, recognizing that the molality of the solute also can be expressed as moles of solute per kg of solvent and the number of moles of solute is the ratio of the mass of the solute and its molar mass:

$$b_{solute} = -\frac{\Delta T_f}{iK_f} = \frac{n_{solute}}{m_{napthalene}} = \frac{\left(\frac{m_{solute}}{M_{solute}}\right)}{m_{napthalene}} = \frac{m_{solute}}{m_{napthalene}M_{solute}}$$

Rearrange to solve for the molar mass of the solute:

$$M_{solute} = -\frac{im_{solute}K_f}{m_{napthalene}\Delta T_f} = -\frac{(1)(5.00\text{ g})(6.94\text{ K - kg/mol})}{\left(250.\text{ g}\times\frac{1\text{ kg}}{10^3\text{ g}}\right)(-0.780\text{ K})} = 1.78\times10^2\text{ g/mol}$$

Osmotic pressure is a third colligative property. Figure 6.3 depicts what happens when a solution of solvent A and solute B is placed on one side of a container separated into two compartments by a semi-permeable barrier that permits the passage of only solvent A. We expect over time that some of the solvent will pass through the semi-permeable membrane to form pure solvent in one compartment in equilibrium with a more-concentrated solution of solute in the other compartment, a process known as osmosis. When equilibrium is reached, osmosis results in an osmotic pressure of π (where π is a variable, not the number 3.14.15 …), the pressure that must be applied to a solution to prevent further influx of pure solvent. This osmotic pressure is related to the number of moles of solute, the van't Hoff factor, the volume of the solution, and temperature by the van't Hoff equation for osmotic pressure:

$$\pi = \frac{in_{solute}RT}{V_{solution}} \qquad [6.11]$$

Note that it is only coincidental that Equation [6.11] has a form similar to the ideal gas law. Osmosis has practical applications in dialysis to concentrate solutions, osmometry to measure molar mass, and transport in biological systems.

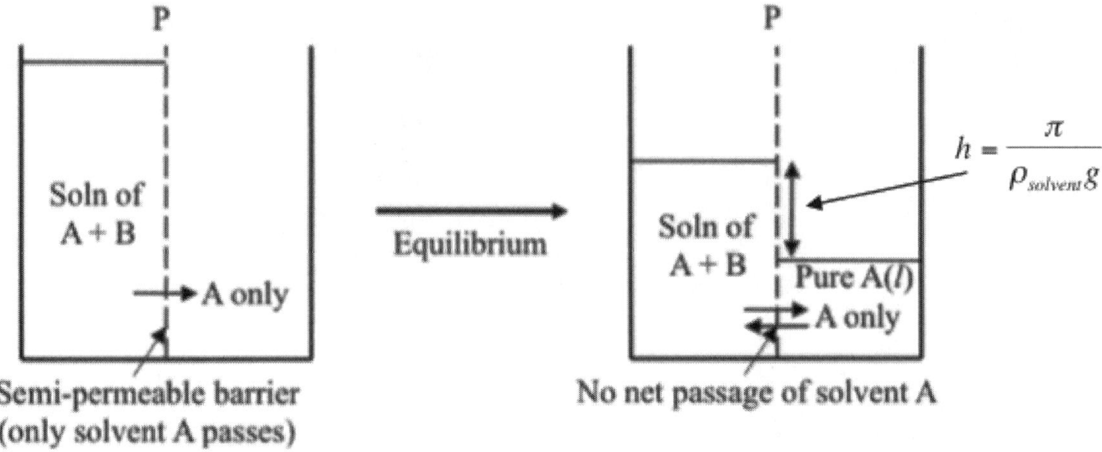

Figure 6.3. Osmosis occurs when equilibrium is reached between two solutions separated by a semi-permeable barrier that permits the passage of only solvent A and not solute B.

Example 6.12: Osmometry

Osmometry—dialysis based on osmosis—is an alternative technique to measure molar mass frequently used for macromolecules. An osmotic pressure of 5.56×10^{-3} atm was measured at 298 K for a 1.00 L solution prepared with 3.22 g of a polymer. What is the molar mass (in kg/mol) of the polymer? Assume the polymer does not dissociate upon dissolution.

Solutions

Rearrange the van't Hoff equation for osmotic pressure, Equation [6.11], to express the number of moles of the solute (the polymer) in terms of the osmotic pressure, the van't Hoff factor (i = 1), the volume of the solution, the temperature, and the gas constant, recognizing that the number of moles of solute is also the ratio of the mass of the solute and its molar mass:

$$\pi = \frac{i n_{solute} RT}{V_{solution}} = \frac{i\left(\frac{m_{solute}}{M_{solute}}\right)RT}{V_{solution}} = \frac{i m_{solute} RT}{V_{solution} M_{solute}}$$

Rearrange to solve for the molar mass of the solute:

$$M_{solute} = \frac{i m_{solute} RT}{\pi V_{solution}} = \frac{(1)(3.22\text{ g})(0.08206\text{ atm} - \text{L/mol} - \text{K})}{(5.56\times10^{-3}\text{ atm})(1.00\text{ L})} \times \frac{1\text{ kg}}{10^3\text{ g}} = 14.2\text{ kg/mol}$$

7:Chemical Reactions

Learning Objectives

By the end of this chapter you should be able to:

1) Balance a chemical equation given the identities of reactants and products.
2) Predict the outcome of simple precipitation reactions.
3) Given the mass of a reactant or product in a chemical reaction, use a balanced chemical equation to calculate the masses of reactants consumed and products formed.
4) Given a set of initial masses of reactants and a balanced chemical equation, identify the limiting reactant and calculate the masses of reactants and products after the reaction has gone to completion.
5) Determine the percentage yield of a reaction from its calculated theoretical yield and its measured actual yield.
6) Find the empirical or molecular formula of a substance from an analysis of its combustion products.

7.1 Balancing Chemical Equations

The first half of this text has focused on the structure of matter and how this structure determines and explains the physical properties we observe macroscopically. Over the remainder of this book we explore the effects of changes in chemical composition, corresponding to chemical properties. We now turn our immediate attention to chemical reactions, processes in which one or more reactants is/are converted to one or more products. How do we know if and when a reaction has occurred? Reactions are often accompanied by a change in color, temperature, state/phase of matter, or an odor. Sometimes none of these changes are obvious, but a reaction has occurred nevertheless.

We define a chemical equation as a representation for a chemical reaction providing the relative number of moles of each reactant that are converted to a corresponding relative number of moles of each product. For both reactant(s) and product(s) these "relative number of moles" are called stoichiometric coefficients; the word "stoichiometric" comes from the Greek for "measure elements." Stoichiometric coefficients are unitless quantities, and ratios of stoichiometric coefficients represent ratios of the number of moles of one species relative to the number of moles of a different species in the balanced chemical equation.

We can write a proper chemical equation using a four-step method:

Step 1: Identify the reactant(s) and product(s).

Step 2: Label each reactant and product with its state of matter—(*g*) for gas, (*l*) for liquid, (*s*) for solid, and (*aq*) for aqueus. For a homogeneous reaction all of the reactants and products are in the same phase. In contrast, for a heterogeneous reaction the reactants and products are present in two or more different phases.

Step 3: Balance the number of atoms for each element on each side of the equation to obtain a balanced chemical equation. This step ensures that the law of conservation of matter is satisfied. General useful practices include balancing one element at a time, balancing elements appearing only once on each side first, and balancing free elements last. When a polyatomic ion appears on both the reactant and product sides of the reaction as a group, it is often useful to try treating this ion as a group. Note that new reactants and/or products should not be introduced at this step! Fractions are allowed for stoichiometric coefficients, although typically integers are preferred.

Step 4: Identify any special conditions for the reaction with a label above the arrow between reactants and products. Commonly encountered conditions include the application of heat (denoted with "Δ"), the requirement of a specific temperature and/or pressure, the requirement of light (often at a specific wavelength), and/or the necessity of a catalyst for the reaction to occur faster.

The following set of examples proceeds from easier to harder problems in balancing.

Example 7.1: The Complete Combustion of Ethanol

Write the balanced chemical equation for the complete combustion of liquid ethanol with oxygen gas to produce carbon dioxide vapor and liquid water upon application of a spark. Note that complete combustion involves the "burning" of a compound with excess molecular oxygen, such that all of the carbon is converted into carbon dioxide. In contrast, incomplete combustion is the reaction of a compound with insufficient oxygen, such that not all of the carbon is converted into carbon dioxide (*e.g.*, some of the carbon is converted to carbon monoxide, soot (solid carbon), *etc.*).

Solutions

Identify liquid ethanol and gaseous molecular oxygen as the reactants and gaseous carbon dioxide and liquid water as the products: $C_2H_5OH(l) + O_2(g) \rightarrow CO_2(g) + H_2O(l)$. Balance the free element oxygen last. We can start balancing with either carbon or hydrogen—the order will not affect the difficulty. Balancing carbon, we find $C_2H_5OH(l) + O_2(g) \rightarrow 2\ CO_2(g) + H_2O(l)$. Subsequently balancing hydrogen we find $C_2H_5OH(l) + O_2(g) \rightarrow 2\ CO_2(g) + 3\ H_2O(l)$. Now balance the free element oxygen as $C_2H_5OH(l) + 3\ O_2(g) \rightarrow 2\ CO_2(g) + 3\ H_2O(l)$. Finally, combustion reactions commonly require the application of heat:

$$C_2H_5OH(l) + 3\ O_2(g) \xrightarrow{\Delta} 2\ CO_2(g) + 3\ H_2O(l) \qquad \text{Rxn [7.E1]}$$

Example 7.2: The Reaction of Metallic Sodium with Water

What is the balanced chemical equation for the reaction (at room temperature) of sodium metal with water to produce aqueous sodium hydroxide and hydrogen gas?

Solutions

First, write the unbalanced equation for the reaction, expecting metallic sodium to be a solid and water a liquid at room temperature: $Na(s) + H_2O(l) \rightarrow NaOH(aq) + H_2(g)$. Balance the free elements sodium and hydrogen last. We sequentially verify that oxygen and sodium are balanced. Balance hydrogen to obtain $Na(s) + H_2O(l) \rightarrow NaOH(aq) + \frac{1}{2} H_2(g)$. Multiply by two to express all the stoichiometric coefficients as integers.

$$2\ Na(s) + 2\ H_2O(l) \rightarrow 2\ NaOH(aq) + H_2(g) \qquad \text{Rxn [7.E2]}$$

Note that there do not appear to be any special conditions for this reaction, which produces impressive explosions!

Example 7.3: Synthesis of Fe_3O_4 for a Ferrofluid

Ferrofluids are suspensions of microscopic solid magnetic particles dispersed in a liquid. Because ferrofluids act like fluids but can be controlled by the application of a magnetic field, these suspensions have been incorporated into a number of commercial processes (*e.g.*, as seals in rotating shafts in heavy machinery, lubricants and sealants in computer hard drives, dampers for background noise in speakers, and inks and media for magnetic writing). Balance the reaction for the synthesis of iron (II/III) oxide, $FeCl_2(aq) + FeCl_3(aq) + NH_3(aq) + H_2O(l) \rightarrow Fe_3O_4(s) + NH_4Cl(aq)$.

Solutions

Balance the elements nitrogen and oxygen first because these elements appear only once on each side. Verify that nitrogen is balanced, and balance oxygen:

$$FeCl_2(aq) + FeCl_3(aq) + NH_3(aq) + 4\ H_2O(l) \rightarrow Fe_3O_4(s) + NH_4Cl(aq)$$

Next balance the elements iron, chlorine, and hydrogen that appear more than once on a side sequentially. Starting with hydrogen, recognize that NH_4^+ has one extra hydrogen compared with NH_3 and that 4 H_2O provides eight extra hydrogen on the reactant side, such that 8 NH_3 and 4 H_2O on the reactant side requires 8 NH_4Cl to balance hydrogen and nitrogen:

$$FeCl_2(aq) + FeCl_3(aq) + 8\ NH_3(aq) + 4\ H_2O(l) \rightarrow Fe_3O_4(s) + 8\ NH_4Cl(aq)$$

Balancing chlorine requires 1 $FeCl_2$ and 2 $FeCl_3$ as reactants for 8 NH_4Cl as products:

$$FeCl_2(aq) + 2\ FeCl_3(aq) + 8\ NH_3(aq) + 4\ H_2O(l) \rightarrow Fe_3O_4(s) + 8\ NH_4Cl(aq) \qquad \text{Rxn [7.E3]}$$

We verify that iron is balanced.

Example 7.4: Balancing a Reaction Involving Polyatomic Ions as Reactants and Products

Write the balanced reaction for the conversion of calcium phosphate powder to aqueous calcium sulfate upon addition of aqueous sulfuric acid, a reaction in which the second product is aqueous phosphoric acid.

Solutions

The unbalanced reaction is $Ca_3(PO_4)_2(s) + H_2SO_4(aq) \rightarrow CaSO_4(aq) + H_3PO_4(aq)$. To balance this reaction consider that there are no free elements but that two polyatomic ions, phosphate (PO_4^{3-}) and sulfate (SO_4^{2-}), appear as both reactants and products, suggesting that we would benefit by treating these polyatomic ions as "groups" in balancing (*i.e.*, we will not count individual numbers of P, S, and O atoms in reactant and products but rather numbers of phosphate and sulfate ions as reactants and products), particularly given that phosphorus, sulfur, and oxygen do not otherwise appear in the unbalanced reaction. Also, note that calcium, hydrogen, and the pair of polyatomic ions each only appear once as reactant and product, suggesting that there is no preferred species to balance first.

Start by balancing the number of Ca^{2+} ions:

$$Ca_3(PO_4)_2(s) + H_2SO_4(aq) \rightarrow 3\ CaSO_4(aq) + H_3PO_4(aq)$$

We now balance the number of sulfate ions:

$$Ca_3(PO_4)_2(s) + 3\ H_2SO_4(aq) \rightarrow 3\ CaSO_4(aq) + H_3PO_4(aq)$$

We then balance the number of protons, H^+:

$$Ca_3(PO_4)_2(s) + 3\ H_2SO_4(aq) \rightarrow 3\ CaSO_4(aq) + 2\ H_3PO_4(aq) \qquad \text{Rxn [7.E4]}$$

We verify that the number of phosphate ions is balanced, such that our overall reaction is balanced.

Chemical reactions are frequently categorized based on either the mechanism by which the reaction occurs or the types of reactants and products. We have seen above one important category of reactions—combustion reactions. Consider a pair of other categories of reaction, encountered in the synthesis of organic polymers. Addition reactions occur when a growing polymer chain extends by addition of an entire monomer. The most common form of addition reactions are free-radical reactions, used in the synthesis of polymers whose monomers contain C-C double and triple bonds. In contrast, condensation reactions occur when an extension of a growing polymer chain by addition of a monomer results in the formation of an additional small molecule. A classic example is the "nylon rope trick," the reaction of the diamine 1,6-diaminohexane and 1,10-decanedioyl dichloride (commonly called sebacoyl dichloride) to form nylon 6,10 and hydrochloric acid:

$$a\ H_2N(CH_2)_6NH_2 + a\ ClOC(CH_2)_8COCl \rightarrow H[HN(CH_2)_6NHOC(CH_2)_8CO]_nCl + b\ HCl \qquad \text{Rxn [7.1]}$$

Here a is the stoichiometric coefficient for both reactants, $n = aN_A$ is the number of repeat units in the polymer, $b = 2a - N_A^{-1} \cong 2a$, and we have assumed that only one long polymer chain is formed (in actuality, a collection of polymer chains with a range of lengths are formed). This reaction can occur as an interfacial polymerization with the relatively water-soluble dichloride in an aqueous phase and the relatively water-insoluble diamine in a non-polar solvent, as demonstrated in a video available at http://www.youtube.com/watch?v=y479OXBzCBQ.

7.2 Precipitation Reactions

We frequently encounter reactions involving dissolved salts. Yet not all ionic compounds dissolve appreciably in water. For example, silver perchlorate ($AgClO_4$) and lithium carbonate (Li_2CO_3) are relatively soluble in water, but silver ions are paired with chloride ions (*i.e.*, AgCl) or magnesium ions paired with carbonate ions (*i.e.*, $MgCO_3$) are relatively insoluble in water, forming precipitates. For now we consider this issue qualitatively; in Chapter 10 we return to this problem for a quantitative perspective.

Table 7.1 presents a set of solubility rules that can be applied to pairs of ions to predict qualitatively whether these pairs are soluble or insoluble (*i.e.*, form precipitates).

Note that two types of ion pairs are always soluble: ion pairs with NH_4^+ or Group 1A cations (Li^+, Na^+, K^+, …), and ion pairs with NO_3^-, ClO_4^-, and CH_3COO^-.

Table 7.1. Qualitative Solubility Rules

Situation	Ions Obeying Rule	Exceptions to Rule
Always soluble cations	NH_4^+ and Group 1A cations	None
Always soluble anions	NO_3^-, ClO_4^-, and CH_3COO^-	None
Usually soluble anions	Cl^-, Br^-, and I^-	Insoluble with Pb^{2+}, Hg_2^{2+}, and Ag^+
	SO_4^{2-}	Insoluble with Pb^{2+}, Hg_2^{2+}, Sr^{2+}, and Ba^{2+}
Insoluble anions (except soluble with NH_4^+ and Group 1A cations)	OH^-	Slightly soluble with Ca^{2+}, Sr^{2+}, and Ba^{2+}
	S^{2-}	Soluble with Group 2A cations
	CO_3^{2-} and PO_4^{3-}	None

Consider now what happens when we mix two solutions with two different soluble ionic compounds. Two possibilities exist: nothing appears (*i.e.*, no reaction occurs), because all of the resulting possible ion pairs are soluble; or a precipitate appears (*i.e.*, a precipitation reaction occurs), because one or two ion pairs in the resulting solution are insoluble. We can apply the solubility rules to determine which of these two possibilities applies for a given situation.

Guided Example 7.1: Mixing Solutions of Sodium Phosphate and Calcium Nitrate

What happens when an aqueous solution of Na_3PO_4 and an aqueous solution $Ca(NO_3)_2$ are mixed?

$$Na_3PO_4(aq) + Ca(NO_3)_2(aq) \rightarrow ?$$

We address this problem with the following systematic strategy:

Step 1: Rewrite the reactants as pairs of soluble cations and anions. We do consider stoichiometry in this step.

$$\{Na^+(aq) + PO_4^{3-}(aq)\} + \{Ca^{2+}(aq) + NO_3^-(aq)\} \rightarrow$$

Step 2: Write the products as pairs of soluble cations and anions, switching ion pairs.

$$\{Na^+(aq) + PO_4^{3-}(aq)\} + \{Ca^{2+}(aq) + NO_3^-(aq)\} \rightarrow$$
$$\{Na^+(aq) + NO_3^-(aq)\} + \{Ca^{2+}(aq) + PO_4^{3-}(aq)\}$$

Step 3: Identify if any of the product ion pairs are insoluble. If all of the product ion pairs are soluble, no reaction occurs, and we write "No reaction" and are done with the problem. But if one or more product ion pairs are insoluble, a precipitation reaction occurs. For this example we identify that all ion pairs with Na^+ are soluble but that Ca^{2+} paired with PO_4^{3-} is insoluble, such that a precipitation reaction does occur!

Step 4: Write any insoluble ion pair as its ionic solid, balancing the charge of cations and ions to form a neutral compound. For this example we need to balance the charges of Ca^{2+} and PO_4^{3-} as $Ca_3(PO_4)_2$.

$$\{Na^+(aq) + PO_4^{3-}(aq)\} + \{Ca^{2+}(aq) + NO_3^-(aq)\} \rightarrow \{Na^+(aq) + NO_3^-(aq)\} + Ca_3(PO_4)_2(s)$$

Step 5: Now cancel out spectator ions—soluble ions that appear on both sides of the reaction—because these ions do *not* participate in the net reaction.

$$\{\cancel{Na^+(aq)} + PO_4^{3-}(aq)\} + \{Ca^{2+}(aq) + \cancel{NO_3^-(aq)}\} \rightarrow \{\cancel{Na^+(aq)} + \cancel{NO_3^-(aq)}\} + Ca_3(PO_4)_2(s)$$

Step 6: Lastly, balance the stoichiometry for the net ionic equation.

$$3\ Ca^{2+}(aq) + 2\ PO_4^{3-}(aq) \rightarrow Ca_3(PO_4)_2(s) \qquad \text{Rxn [7.2]}$$

Example 7.5: Predicting What Happens When Solutions of Strong Electrolytes Are Mixed

Identify what reaction occurs when the following pairs of solutions are mixed and, if precipitation occurs, write the net ionic equation.

a) $Na_2CO_3(aq)$ is mixed with $CaCl_2(aq)$

b) $Pb(NO_3)_2(aq)$ is mixed with $KI(aq)$

c) $CuCl_2(aq)$ is mixed with $MnSO_4(aq)$

Solutions

a) Start by rewriting the reactants as pairs of soluble cations and anions, ignoring the stoichiometry:

$$\{Na^+(aq) + CO_3^{2-}(aq)\} + \{Ca^{2+}(aq) + Cl^-(aq)\} \rightarrow$$

Write the products as pairs of soluble cations and anions, switching ion pairs:

$$\{Na^+(aq) + CO_3^{2-}(aq)\} + \{Ca^{2+}(aq) + Cl^-(aq)\} \rightarrow \{Na^+(aq) + Cl^-(aq)\} + \{Ca^{2+}(aq) + CO_3^{2-}(aq)\}$$

Recognize that Na^+ paired with Cl^- is soluble but Ca^{2+} paired with CO_3^{2-} is insoluble. We then write the insoluble ion pair as $CaCO_3(s)$ to find:

$$\{Na^+(aq) + CO_3^{2-}(aq)\} + \{Ca^{2+}(aq) + Cl^-(aq)\} \rightarrow \{Na^+(aq) + Cl^-(aq)\} + CaCO_3(s)$$

Cancel out spectator ions:

$$\{\cancel{Na^+(aq)} + CO_3^{2-}(aq)\} + \{Ca^{2+}(aq) + \cancel{Cl^-(aq)}\} \rightarrow \{\cancel{Na^+(aq)} + \cancel{Cl^-(aq)}\} + CaCO_3(s)$$

and balance the stoichiometry to obtain the net ionic equation:

$$Ca^{2+}(aq) + CO_3^{2-}(aq) \rightarrow CaCO_3(s) \qquad \text{Rxn [7.E5a]}$$

Note that this reaction occurs in the precipitation of "hard" water (*i.e.*, water with appreciable amounts of dissolved Ca^{+2} and Mg^{+2}).

b) Start by rewriting the reactants as pairs of soluble cations and anions, ignoring the stoichiometry:

$$\{Pb^{2+}(aq) + NO_3^-(aq)\} + \{K^+(aq) + I^-(aq)\} \rightarrow$$

Write the products as pairs of soluble cations and anions, switching ion pairs:

$$\{Pb^{2+}(aq) + NO_3^-(aq)\} + \{K^+(aq) + I^-(aq)\} \rightarrow \{Pb^{2+}(aq) + I^-(aq)\} + \{K^+(aq) + NO_3^-(aq)\}$$

Identify that K^+ paired with NO_3^- is soluble but Pb^{2+} paired with I^- is insoluble. We then write the insoluble ion pair as $PbI_2(s)$ to find:

$$\{Pb^{2+}(aq) + NO_3^-(aq)\} + \{K^+(aq) + I^-(aq)\} \rightarrow PbI_2(s) + \{K^+(aq) + NO_3^-(aq)\}$$

Cancel out spectator ions:

$$\{Pb^{2+}(aq) + \cancel{NO_3^-(aq)}\} + \{\cancel{K^+(aq)} + I^-(aq)\} \rightarrow PbI_2(s) + \{\cancel{K^+(aq)} + \cancel{NO_3^-(aq)}\}$$

and balance the stoichiometry to obtain the net ionic equation:

$$Pb^{2+}(aq) + 2\ I^-(aq) \rightarrow PbI_2(s) \qquad \text{Rxn [7.5b]}$$

Figure 7.1. $PbI_2(s)$ precipitates when $Pb(NO_3)_2(aq)$ is mixed with $KI(aq)$.

c) Start by rewriting the reactants as pairs of soluble cations and anions, ignoring the stoichiometry:

$$\{Cu^{2+}(aq) + Cl^{-}(aq)\} + \{Mn^{2+}(aq) + SO_4^{2-}(aq)\} \rightarrow$$

Write the products as pairs of soluble cations and anions, switching ion pairs:

$$\{Cu^{2+}(aq) + Cl^{-}(aq)\} + \{Mn^{2+}(aq) + SO_4^{2-}(aq)\} \rightarrow$$
$$\{Cu^{2+}(aq) + SO_4^{2-}(aq)\} + \{Mn^{2+}(aq) + Cl^{-}(aq)\}$$

Identify that both Cu^{2+} paired with SO_4^{2-} and Mn^{2+} paired with Cl^- are soluble. Because there are no insoluble ion pairs, no precipitation reaction occurs.

7.3 Stoichiometry, Amounts of Reactants Consumed, and Amounts of Products Formed

Stoichiometry—the use of stoichiometric coefficients—provides quantitative relationships between the number of moles of reactant(s) consumed and number of moles of product(s) produced. For example, consider two of the reactions we discussed in Section 7.1. The balanced chemical equation for the complete combustion of ethanol, Rxn [7.E1], tells us that one mole of ethanol reacts with three moles of molecular oxygen to produce two moles of carbon dioxide and one mole of water. Mathematically, these relationships can be expressed as:

$$1 \text{ mol } C_2H_5OH = 3 \text{ mol } O_2 = 2 \text{ mol } CO_2 = 3 \text{ mol } H_2O \qquad [7.1]$$

Similarly, the balanced chemical equation for the reaction of metallic sodium with water, Rxn [7.E2], tells us that two moles of sodium metal reacts with two moles of water to produce two moles of sodium hydroxide and one mole of molecular hydrogen. Mathematically, these relationships can be expressed as:

$$2 \text{ mol Na} = 2 \text{ mol } H_2O = 2 \text{ mol NaOH} = 1 \text{ mol } H_2 \qquad [7.2]$$

Recognize that we can determine the amounts of reactant(s) consumed and product(s) formed using the number of moles of each species initially present and their stoichiometric coefficients to convert between moles of reactant(s) consumed and moles of product(s) formed. In conjunction we can also use the molar masses of the reactant(s) and product(s) to convert between numbers of moles and mass. In practice, it is more informative for solids and liquids to report changes in mass (*e.g.*, in units of g or kg) than changes in the number of mole. Similarly, for gases we prefer to communicate changes in volume (*e.g.*, in units of L) than changes in moles.

Example 7.6: The Disposal of Surplus Metallic Sodium

In 1947 the federal government dumped drums containing 30,000 lbs (1.36×10^7 g) of metallic sodium into Lake Lenore, an alkaline lake in Washington state; a video of this event is available at http://www.youtube.com/watch?v=HY7mTCMvpEM. The sodium was consumed by reaction with water according to Rxn [7.E2].

a) How many grams of sodium hydroxide are produced?

b) How many grams of water are consumed?

c) What is the volume (in L) of hydrogen gas produced at STP (273.15 K, 1 bar)?

d) What is the increase in the concentration of sodium hydroxide (in M) in the lake after the reaction has occurred if the lake has a volume of 2.41×10^{10} L?

Solutions

a) Determine the number of moles of metallic sodium consumed by converting the mass of Na(*s*) to moles of Na(*s*) using the molar mass of sodium, 23.99 g/mol:

$$1.36\times10^7 \text{ g Na}\times\frac{1 \text{ mol Na}}{22.99 \text{ g Na}} = 5.92\times10^5 \text{ mol Na consumed}$$

Use stoichiometry to determine the number of moles of aqueous sodium hydroxide produced based on the number of moles of Na(*s*) consumed:

$$5.92\times10^5 \text{ mol Na}\times\frac{2 \text{ mol NaOH}}{2 \text{ mol Na}} = 5.92\times10^5 \text{ mol NaOH produced}$$

Determine the mass of NaOH(*aq*) produced from the number of moles of NaOH(*aq*) produced using the molar mass of sodium hydroxide:

$$5.92\times10^5 \text{ mol NaOH}\times\frac{40.00 \text{ g NaOH}}{1 \text{ mol NaOH}} = 2.37\times10^7 \text{ g NaOH produced}$$

b) Determine the mass of $H_2O(l)$ consumed based on the number of moles of sodium consumed, the stoichiometry, and the molar mass of water:

$$5.92\times10^5 \text{ mol Na}\times\frac{2 \text{ mol H}_2\text{O}}{2 \text{ mol Na}}\times\frac{18.02 \text{ g H}_2\text{O}}{1 \text{ mol H}_2\text{O}} = 1.07\times10^7 \text{ g H}_2\text{O consumed}$$

Assuming water has a density of ~1 g/mL, this mass of water corresponds to ~10^4 L, which is much smaller than the amount of water in the lake.

c) Determine the number of moles of $H_2(g)$ produced based on the number of moles of sodium consumed and the stoichiometry:

$$5.92\times10^5 \text{ mol Na}\times\frac{1 \text{ mol H}_2}{2 \text{ mol Na}} = 2.96\times10^5 \text{ mol H}_2 \text{ produced}$$

Determine the volume of hydrogen gas produced by applying the ideal gas law, Equation [1.6]:

$$V_{H_2} = \frac{n_{H_2}RT}{P} = \frac{(2.96\times 10^5 \text{ mol})(8.206\times 10^{-2} \text{ atm - L/mol - K})(273.15 \text{ K})}{\left(10^5 \text{ Pa}\times\frac{1.00 \text{ atm}}{1.013\times 10^5 \text{ Pa}}\right)} = 6.72\times 10^6 \text{ L}$$

d) Determine the change in concentration of sodium hydroxide based on the ratio of the number of moles of NaOH(*aq*) produced to the volume of the lake, neglecting the change in volume of the lake due to the consumption of water:

$$\Delta C_{NaOH} = \frac{\Delta n_{NaOH}}{V_{solution}} = \frac{(+5.92\times 10^5 \text{ mol NaOH})}{(2.41\times 10^{10} \text{ L})} = 2.46\times 10^{-5} \text{ M NaOH}$$

This change in concentration of 24.6 μM NaOH is relatively small.

We can generalize our observations by recognizing that the magnitude of the ratio of the change in the number of moles for a species participating in a reaction (regardless of whether the species is a reactant or product) to the stoichiometric coefficient for the species is a constant. For example, in Example 7.6 we find:

$$\frac{5.92\times 10^5 \text{ mol Na consumed}}{2 \text{ mol Na}} = \frac{5.92\times 10^5 \text{ mol Na consumed}}{2 \text{ mol } H_2O} = \frac{5.92\times 10^5 \text{ mol NaOH produced}}{2 \text{ mol NaOH}} = \frac{2.96\times 10^5 \text{ mol } H_2 \text{ produced}}{1 \text{ mol } H_2} \quad [7.3a]$$

$$\left|\frac{\Delta n_{Na}}{\upsilon_{Na}}\right| = \left|\frac{\Delta n_{H_2O}}{\upsilon_{H_2O}}\right| = \left|\frac{\Delta n_{NaOH}}{\upsilon_{NaOH}}\right| = \left|\frac{\Delta n_{H_2}}{\upsilon_{H_2}}\right| \quad [7.3b]$$

where Δn_i for a species is negative if species *i* is a reactant and positive if species *i* is a product. Thus, for the following generic reaction:

$$\upsilon_A \text{ A} + \upsilon_B \text{ B} + \ldots \rightarrow \upsilon_C \text{ C} + \upsilon_D \text{ D} + \ldots \quad \text{Rxn [7.3]}$$

where υ_i (υ is the Greek letter "upsilon"), the stoichiometric coefficient for species *i*, is positively valued for products but negatively valued for reactants, we find:

$$\frac{\Delta n_A}{\upsilon_A} = \frac{\Delta n_B}{\upsilon_B} = \frac{\Delta n_C}{\upsilon_C} = \frac{\Delta n_D}{\upsilon_D} = \ldots \quad [7.4]$$

7.4 Limiting Reactants, Stoichiometry, and Amounts of Reactants and Products After a Reaction Has Gone to Completion

When a reaction requires more than one reactant, two possibilities exist. One possibility is that the reactants are in stoichiometric proportion, such that all of the reactants are completely and simultaneously consumed (*i.e.*, none remain). For example, the reaction of iron (III) oxide with metallic aluminum to produce aluminum oxide and metallic iron, known as the thermite reaction, is used for welding (Figure 7.2) and in some ammunitions:

$$Fe_2O_3(s) + 2\,Al(s) \xrightarrow{KClO_3,H_2SO_4} Al_2O_3(s) + 2\,Fe(s) \qquad \text{Rxn [7.4]}$$

In this reaction Fe_2O_3 and Al are in stoichiometric proportion if we have exactly one mole of Fe_2O_3 for every two moles of Al. Do you think this condition is typical?

Figure 7.2. Use of the thermite reaction to weld railroad ties.

More commonly, reactants are not in stoichiometric proportion, such that exactly one of the reactants is limiting (and is the limiting reactant) and the other reactant(s) is/are in excess. For this second possibility all of the limiting reactant is consumed, with some of each reactant in excess remaining (*i.e.*, we have leftovers for reactants in excess). The amount of each product formed depends on the amount of the limiting reactant and the stoichiometry for the reaction. Likewise, the amount of each reactant in excess remaining depends on the amount of each reactant in excess present initially, as well as the amount of the limiting reactant and the stoichiometry for the reaction.

We seek a general approach for determining if we have a limiting reactant and, if so, which reactant is limiting. Further, our approach must enable us to determine the amounts of products formed given a limiting reactant as well as the amounts of any reactants in excess remaining. Two strategies are available for us, each leading to identical answers: we can assume that each reactant is limiting and calculate the amount of one product formed, with the actual limiting reactant forming the least amount of the product, or as an alternative strategy we can compare ratios of moles among reactants.

Example 7.7: Quantitative Evaluation of a Thermite Reaction

Consider the reaction of 500.0 g of iron ore (Fe_2O_3) and 200.0 g of metallic aluminum according to Rxn [7.4].

a) Which reactant, if either, is limiting?

b) What are the masses (in g) of metallic iron and aluminum oxide produced?

c) What is the mass (in g) of the reactant in excess remaining?

For any problem a useful first approach is to think about the problem qualitatively based on the available information. For example, do you expect to find a limiting reactant, and, if so, which reactant do you think is limiting?

Solutions

a) We illustrate for this example the first strategy for determining the limiting reactant introduced above. First, assume that Fe_2O_3 is the limiting reactant (*i.e.*, Al is in excess) and determine the amount of one product—*e.g.*, Fe—produced:

$$500.0 \text{ g } Fe_2O_3 \times \frac{1 \text{ mol } Fe_2O_3}{159.70 \text{ g } Fe_2O_3} \times \frac{2 \text{ mol Fe}}{1 \text{ mol } Fe_2O_3} \times \frac{55.85 \text{ g Fe}}{1 \text{ mol Fe}} = 349.7 \text{ g Fe produced}$$

Now, assume that the other reactant, Al, is limiting (*i.e.*, Fe_2O_3 is in excess) and determine the amount of Fe produced:

$$200.0 \text{ g Al} \times \frac{1 \text{ mol Al}}{26.98 \text{ g Al}} \times \frac{2 \text{ mol Fe}}{2 \text{ mol Al}} \times \frac{55.85 \text{ g Fe}}{1 \text{ mol Fe}} = 414.0 \text{ g Fe produced}$$

Because less metallic iron is produced when we assume that Fe_2O_3 is limiting, Fe_2O_3 is the limiting reactant, and Al is in excess.

b) Based on the above analysis, we expect that 349.7 g of Fe is produced. With Fe_2O_3 as the limiting reactant, we determine the mass of Al_2O_3 produced as:

$$500.0 \text{ g } Fe_2O_3 \times \frac{1 \text{ mol } Fe_2O_3}{159.70 \text{ g } Fe_2O_3} \times \frac{1 \text{ mol } Al_3O_3}{1 \text{ mol } Fe_2O_3} \times \frac{101.96 \text{ g } Al_3O_3}{1 \text{ mol } Al_3O_3} = 319.2 \text{ g } Al_3O_3 \text{ produced}$$

c) We determine the mass of excess aluminum remaining by first determining the mass of Al consumed as:

$$500.0 \text{ g } Fe_2O_3 \times \frac{1 \text{ mol } Fe_2O_3}{159.70 \text{ g } Fe_2O_3} \times \frac{2 \text{ mol Al}}{1 \text{ mol } Fe_2O_3} \times \frac{26.98 \text{ g Al}}{1 \text{ mol Al}} = 168.9 \text{ g Al consumed}$$

and then determining the mass of Al remaining as:

$$200.0 \text{ g Al initially} - 168.9 \text{ g Al consumed} = 31.1 \text{ g Al remaining}$$

Note that with only one reactant in excess (*i.e.*, a limiting reactant and a total of two reactants), it would have been easier to apply the law of conservation of matter if all we needed to know was the amount of Al remaining:

$$\begin{Bmatrix}\text{Initial amount}\\ \text{of reactants}\end{Bmatrix} - \begin{Bmatrix}\text{Amount of}\\ \text{products produced}\end{Bmatrix} = \begin{Bmatrix}\text{Amount of}\\ \text{reactants in excess remaining}\end{Bmatrix}$$

$$[(500.0 \text{ g } Fe_2O_3) + (200.0 \text{ g Al})] - [(349.7 \text{ g Fe}) + (319.2 \text{ g } Al_3O_3)] = 31.1 \text{ g Al remaining}$$

Example 7.8: Synthesis of Nylon 6,10

In an experiment 1.0659 g of 1,6-diaminohexane dissolved in water and 1.0325 g of 1,10-decanedioyl dichloride dissolved in hexane are allowed to react by interfacial polymerization according to Rxn [7.1].

a) Is there a limiting reactant and, if so, which reactant is limiting and which is in excess?

b) What is the mass of nylon 6,10 produced (in g), assuming only one long chain is formed? How much of the reactant in excess is left (in g)?

Solutions

a) Determine the limiting reactant by first determining the number of moles each of $H_2N(CH_2)_6NH_2$ and $ClOC(CH_2)_8COCl$ available:

$$1.0659 \text{ g } H_2N(CH_2)_6NH_2 \times \frac{1 \text{ mol } H_2N(CH_2)_6NH_2}{116.24 \text{ g } H_2N(CH_2)_6NH_2}$$
$$= 9.1698 \times 10^{-3} \text{ mol } H_2N(CH_2)_6NH_2$$

$$1.0325 \text{ g } ClOC(CH_2)_8COCl \times \frac{1 \text{ mol } ClOC(CH_2)_8COCl}{239.16 \text{ g } ClOC(CH_2)_8COCl}$$
$$= 4.3172 \times 10^{-3} \text{ mol } ClOC(CH_2)_8COCl$$

The stoichiometry for the reaction is *a:a* (*i.e.*, 1:1), and the fewer available moles of the dichloride means that we will theoretically consume all of the dichloride prior to consuming all of the diamine Thus, $ClOC(CH_2)_8COCl$ is the limiting reactant, and $H_2N(CH_2)_6NH_2$ is the reactant in excess.

b) Determine the mass of nylon formed as product by first determining the mass of diamine consumed (based on consuming all of the mass of the limiting dichloride), the mass of diamine remaining, and the mass of HCl produced:

$$1.0325 \text{ g ClOC(CH}_2)_8\text{COCl} \times \frac{1 \text{ mol ClOC(CH}_2)_8\text{COCl}}{239.16 \text{ g ClOC(CH}_2)_8\text{COCl}} \times \frac{1 \text{ mol H}_2\text{N(CH}_2)_6\text{NH}_2}{1 \text{ mol ClOC(CH}_2)_8\text{COCl}}$$
$$\times \frac{116.24 \text{ g H}_2\text{N(CH}_2)_6\text{NH}_2}{1 \text{ mol H}_2\text{N(CH}_2)_6\text{NH}_2} = 0.5018 \text{ g H}_2\text{N(CH}_2)_6\text{NH}_2 \text{ consumed}$$
$$1.0659 \text{ g H}_2\text{N(CH}_2)_6\text{NH}_2 \text{ initially} - 0.5018 \text{ g H}_2\text{N(CH}_2)_6\text{NH}_2 \text{ consumed}$$
$$= 0.5641 \text{ g H}_2\text{N(CH}_2)_6\text{NH}_2 \text{ remaining}$$

$$1.0325 \text{ g ClOC(CH}_2)_8\text{COCl} \times \frac{1 \text{ mol ClOC(CH}_2)_8\text{COCl}}{239.16 \text{ g ClOC(CH}_2)_8\text{COCl}} \times \frac{2 \text{ mol HCl}}{1 \text{ mol ClOC(CH}_2)_8\text{COCl}}$$
$$\times \frac{36.45 \text{ g HCl}}{1 \text{ mol HCl}} = 0.3148 \text{ g HCl produced}$$

The mass of nylon 6,10 produced then can be determined by applying th law of conservation of mass, as the initial masses of diamine and dichloride must equal the combined masses of diamine remaining and hydrochloric acid and nylon produced:

$$m_{diamine,i} + m_{dichloride,i} = m_{diamine,f} + m_{\text{HCl}} + m_{nylon}$$
$$m_{nylon} = (m_{diamine,i} + m_{dichloride,i}) - (m_{diamine,f} + m_{\text{HCl}})$$
$$= [(1.0659 \text{ g}) + (1.0325 \text{ g})] - [(0.5641 \text{ g}) + (0.3418 \text{ g})]$$
$$m_{nylon} = 1.2195 \text{ g}$$

7.5 Theoretical and Actual Yields

What we have determined when we calculated amounts of products produced are actually theoretical yields. In practice, we frequently find the actual yield—the measured amounts of products—is less than the theoretical yield (and never more than the theoretical yield). We define the percent yield as:

$$\% - \text{yield} \equiv \frac{\{\text{Actual yield}\}}{\{\text{Theoretical yield}\}} \times 100\% \qquad [7.5]$$

Example 7.9: The Interfacial Polymerization of Nylon 6,10 Revisited

The interfacial polymerization of nylon 6,10 by reaction of 1.0659 g of aqueous 1,6-diaminohexane dissolved in water and 1.0325 g of 1,10-decanedioyl dichloride dissolved in hexane, as introduced in Example 7.8, resulted in the formation of 0.8558 g of nylon 6,10 thread (after washing with acetone and drying overnight). What was the %-yield?

Solutions

Determine the %-yield from the theoretical yield of 1.2195 g determined in Example 7.8, the reported actual yield of 0.8558 g, and Equation [7.5] as:

$$\% - \text{yield} = \frac{(0.8558 \text{ g nylon})}{(1.2195 \text{ g nylon})} \times 100\% = 70.18\%$$

Example 7.10: The Thermite Reaction Revisited

If the reaction of 500.0 g of iron ore with 200.0 g of metallic aluminum in Example 7.7 has a percent yield of 56.35%, what is the actual yield of metallic iron (in g)?

Solutions

Rearrange Equation [7.5] to solve for the actual yield with a %-yield = 56.35% and the theoretical yield of 414.0 g Fe(*s*) determined in Example 7.7:

$$\{\text{Actual yield}\} = \frac{\% - \text{yield} \times \{\text{Theoretical yield}\}}{100\%} = \frac{(56.35\%)(414.0 \text{ g Fe})}{100\%} = 233.3 \text{ g Fe}$$

Observations of amounts of products less than predicted by stoichiometry are indicative of a variety of phenomena, such as competing/side reactions that produce by-products, inadequate mixing, slow kinetics, and reversible reactions that do not go to completion. We will examine the kinetics and reversibility of reactions in more detail in Chapters 9 and 10, respectively.

7.5 Combustion Analysis

We can determine the formula for compounds by applying the concepts introduced in this chapter in a method known as combustion analysis based on the complete combustion of a compound with excess oxygen, such that some oxygen is left over after the reaction has occurred. For hydrocarbons and organic compounds containing carbon, hydrogen, and oxygen, complete combustion produces carbon dioxide and water alone; complete combustion of compounds containing other elements produces a more complex product mixture.

Consider how the process of complete combustion can be used to determine the formula for hydrocarbons and organic molecules containing only carbon, hydrogen, and oxygen, as depicted in Figure 7.3. Our sample sits in a pan in the blast furnace and has a generic molecular formula $C_xH_yO_z$. Excess oxygen is fed into the furnace. The carbon dioxide and water produced can be removed selectively using scrubbers (typically absorbents and adsorbents). We assume no other products are formed in this complete combustion and that no sample remains (*i.e.*, the percent yield is 100%). As long as excess oxygen is supplied, the water vapor contains all the hydrogen in the original sample and the carbon dioxide all the carbon in the original sample. Note that if not enough oxygen is fed to the furnace (*e.g.*, oxygen is not in excess), incomplete combustion occurs and some of the carbon in the compound will be converted to carbon monoxide and other carbon-bearing species rather than carbon dioxide.

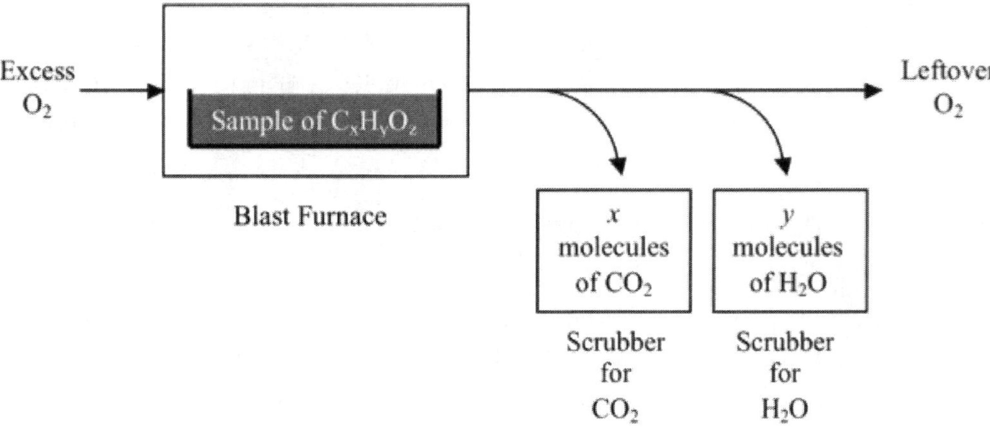

Figure 7.3. Schematic of a combustion analysis to determine the composition of an organic compound.

The overall chemical reaction taking place in the furnace can be written as:

$$C_xH_yO_z(s) + a\,O_2(g) \rightarrow x\,CO_2(g) + (y/2)\,H_2O(g) + (a + z/2 - x - y/4)\,O_2(g) \qquad \text{Rxn [7.5]}$$

The amounts of carbon dioxide and water produced are typically measured as masses. From these masses the empirical formula can be determined by first determining the percent composition and then following the strategy we introduced in Section 3.5.

Example 7.11: Molecular Formula for a Formulation of Biodiesel

Rapeseed methyl ester is a vegetable-based oil used as biodiesel for automobiles. Formulations of this biodiesel fuel tend to be poorly characterized. In an experiment 33.04 g of this compound, known to contain only carbon, hydrogen, and oxygen, is burned with excess oxygen to yield 91.99 g of CO_2 and 37.66 g of H_2O.

a) What is the empirical formula for this compound?

b) If the molar mass of rapeseed oil methyl ester is 284.54 g/mol, what is the molecular formula for this compound?

Solutions

a) First, determine the individual masses of carbon, hydrogen, and oxygen in the original sample from the measured masses of combustion products, their molar masses, and the original mass of the sample. Determine the mass of carbon in the original sample from the mass of carbon dioxide as:

$$m_C = 91.99\text{ g CO}_2 \times \frac{1\text{ mol CO}_2}{44.01\text{ g CO}_2} \times \frac{1\text{ mol C}}{1\text{ mol CO}_2} \times \frac{12.01\text{ g C}}{1\text{ mol C}} = 25.10\text{ g C}$$

Next, determine the mass of hydrogen in the original sample from the mass of water, recognizing that there are two moles of hydrogen for each mole of water:

$$m_H = 37.66\text{ g H}_2\text{O} \times \frac{1\text{ mol H}_2\text{O}}{18.02\text{ g H}_2\text{O}} \times \frac{2\text{ mol H}}{1\text{mol H}_2\text{O}} \times \frac{1.01\text{ g H}}{1\text{ mol H}} = 4.22\text{ g H}$$

Lastly, determine the mass of oxygen in the original sample by subtracting the masses of carbon and hydrogen from the mass of the sample:

$$m_O = m_{sample} - (m_C + m_H) = (33.04\text{ g}) - [(25.10\text{ g C}) + (4.22\text{ g H})] = 3.72\text{ g O}$$

Now determine the number of moles of each element in the sample as:

$$25.10\text{ g C} \times \frac{1\text{ mol C}}{12.01\text{ g C}} = 2.090\text{ mol C}$$

$$4.22\text{ g H} \times \frac{1\text{ mol H}}{1.01\text{ g H}} = 4.18\text{ mol H}$$

$$3.72\text{ g O} \times \frac{1\text{ mol O}}{16.00\text{ g O}} = 0.232\text{ mol O}$$

As our tentative formula of $C_{2.090}H_{4.18}O_{0.232}$ is nonsensical, we divide by the smallest subscript (the "0.232" for oxygen) to yield $C_{9.00}H_{18.0}O$, which in whole numbers is the empirical formula $C_9H_{18}O$.

b) To determine the molecular formula we first need to determine the empirical mass per mole:

$$1 \text{ mol } C_9H_{18}O = 9 \text{ mol C} + 18 \text{ mol H} + 1 \text{ mol O} = 9(12.01 \text{ g}) + 18(1.01 \text{ g}) +(16.00 \text{ g})$$

$$1 \text{ mol } C_9H_{18}O = 142.27 \text{ g } C_9H_{18}O$$

Determine the molecular formula as the integer multiple (*n*) of the empirical formula:

$$\frac{284.54 \text{ g/mol } C_{9n}H_{18n}O_n}{142.27 \text{ g/mol } C_9H_{18}O} = n = 2$$

The molecular formula is twice the empirical formula: $C_{18}H_{36}O_2$.

8: Thermochemistry

Learning Objectives

By the end of this chapter you should be able to:

1) Provide a physical interpretation for the concept of *P-V* work and calculate the work done on/by a chemical system when its volume changes.
2) Identify a physical interpretation for the concept of heat and relate the amount of heat transferred to or from a given quantity of a substance to the change in temperature the substance experiences and the heat capacity for the substance.
3) Apply the first law of thermodynamics to relate the change in energy experienced by a system to the flow of heat and work between the system and its surroundings.
4) Find the final temperature reached when multiple substances at different initial temperatures are placed in thermal contact.
5) Perform thermochemical calculations for processes involving phase changes.
6) Determine the heat evolved or absorbed and change in temperature for systems in which a reaction occurs based on the reaction energy and enthalpy.
7) Use Hess's law to relate changes in enthalpy for sets of related reactions.
8) Calculate standard-state changes in enthalpy for a reaction based on standard molar enthalpies of formation for reactants and products.

8.1 Introduction to Thermodynamics

Our discussion of stoichiometry and its use in relating amounts of reactant consumed and products produced provides a foundation for considering how to address changes in energy associated with reactions. This subject matter, known as thermochemistry, is part of the larger field of thermodynamics, describing the behavior of matter based on transformations between different forms of energy. Much of the foundation for thermodynamics was developed in the nineteenth century and the birth of the industrial age, a period of time during which the conversion of energy to power transportation and manufacturing systems came to prominence. Today an understanding of thermodynamics is important in explaining and designing a wide range of phenomena encountered in science and engineering, including the properties of fuels and design of engines, HVAC (heating, ventilation, and air conditioning) and refrigeration systems, the behavior of materials in response to changes in temperature, weather forecasting and climate modeling, and the physiology of living organisms.

From the perspective of thermodynamics we define the universe as all matter and space, the system as the part of the universe containing the matter and space whose behavior we wish to study, and the surroundings as all of the universe that is outside the system (Figure 81). A system is separated from its surroundings by a boundary that may be real or imaginary. We distinguish between boundaries based on whether they are rigid or flexible, insulated or diathermal (un-insulated), and permeable or impermeable.

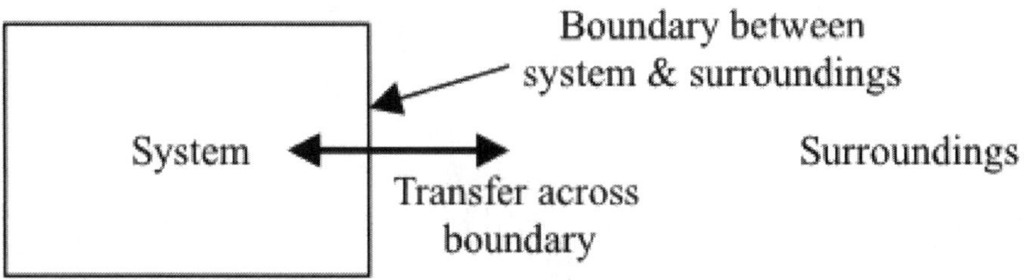

Figure 8.1. Conceptual schematic of a system and its surroundings

A system and its surroundings interact through the transfer of matter and energy across the boundary. We classify systems based on whether matter and/or energy can be transferred: in an open system both matter and energy can be transferred, in a closed system only energy can be transferred, and in an isolated system neither matter nor energy can be transferred. Note that the transfer of matter without energy is impossible because all matter contains energy. Most of our focus will be on closed and isolated systems, as a detailed discussion of open systems involving flow is typically deferred to more advanced treatments in chemical and mechanical engineering.

For closed and open systems we focus on energy transformations that occur between different states of the system that are in equilibrium with the surroundings. We identify the state of a system based on its thermodynamic properties, including pressure, temperature, volume, and composition. We remind ourselves of the distinction introduced in Chapter 1 between extensive properties that depend on the size of the system (*e.g.*, mass, volume, and number of moles) and intensive properties that are independent of the size of the system (*e.g.*, density, pressure, and temperature). Whenever an extensive property has corresponding intensive properties on a mass (specific) basis (*i.e.*, per unit mass) or mole basis (per unit mole), we will distinguish between these properties using the following conventions: the extensive property will be noted with an upper-case letter (*e.g.*, V for volume), the intensive property on a molar basis by the corresponding lower-case letter (*e.g.*, $v \equiv \frac{V}{n}$ for molar volume), and the intensive property on a mass basis by the corresponding lower-case letter with a circumflex (a "hat") (*e.g.*, $\hat{v} \equiv \frac{V}{m} = \rho^{-1}$ for specific volume).

Equilibrium occurs when the thermodynamic properties describing the state of a system do not change with time and there is no net driving force for change. We commonly encounter several different types of forms for equilibrium. For example, we introduced concepts for phase equilibrium in Section 5.3. Mechanical equilibrium occurs when a system is at the same pressure as its surroundings. Thermal equilibrium occurs when a system is at the same temperature as its surroundings and can also be applied to pairs of isolated systems (*i.e.*, at the same temperature) the zeroth law of thermodynamics states that if a pair of systems, A and B, are in thermal equilibrium (*i.e.*, $T_A = T_B$) and system B is in thermal equilibrium with a third system C (*i.e.*, $T_B = T_C$), then systems A and C are also in thermal equilibrium (*i.e.*, $T_A = T_C$). The zeroth law is the underlying basis for all scales and measurements for temperature. We will defer further consideration of chemical equilibrium, which occurs when the concentration of each species participating in the reaction does not change with time, to Chapters 10 and 11.

From a chemical perspective energy can manifest itself in the forms of macroscopic kinetic energy (the energy associated with the macroscopic motion and denoted E_k), macroscopic potential energy (the energy associated with the macroscopic position in an external field, such as a gravitational field, and denoted E_p), and internal energy. Internal energy (denoted U) is the sum of the kinetic energy of atomic/molecular motion and chemical potential energy (*i.e.*, the energy stored chemically in matter). Chemical kinetic energy is associated with the translation, vibration, and rotation of individual atoms and molecules. Chemical potential energy is energy stored in covalent and ionic bonds, IM interactions, and excited electrons. Each form of energy is a state property that depends only on what the state is, not the history of the system.

If we neglect nuclear reactions (*i.e.*, mass and energy are not transmutable) and the universe consists only of the system and its surroundings, then the total energy of the universe, $E_{uni} = E_{sys} + E_{surr}$, must satisfy the law of conservation of energy, $\Delta E_{uni} = 0$, which with $\Delta E_{uni} = \Delta E_{sys} + \Delta E_{surr}$ means that:

$$\Delta E_{sys} = -\Delta E_{surr} \qquad [8.1]$$

Thus, the only way for a system to change its energy is by exchange of energy with the surroundings, and we can focus on tracking changes in energy solely from the perspective of the system. In particular, we focus be on changes in energy, not absolute energies. Only two mechanisms have ever been identified for the exchange of energy between a system and its surroundings: work (w), the transfer of energy due to forces causing motion; and heat (q), the transfer of energy due to differences in temperature.

8.2 Work

From a thermodynamic perspective work is the flow of energy between a system and its surroundings that can be used to change the height of a mass in the surroundings. Work can occur in the form of pressure-volume (P-V) work associated with the expansion or compression of bulk matter, interfacial work associated with the expansion or compression of an interface between two phases (*e.g.*, the expansion or collapse of a soap bubble), line tension work associated with the stretching or linear compression of bulk matter (*e.g.*, a rubber band or strand of nucleic acid), electrical work associated with movement of charges through an electric field (*e.g.*, in the charging or discharge of a battery), and shaft work associated with the rotation of a shaft or propeller. At present we focus solely on P-V work, deferring further discussion of shaft work to Section 8.8 and electrical work to Chapter 12.

The classic model for considering P-V work is based on defining our system as the contents inside an ideal frictionless piston/cylinder assembly (Figure 8.2). The volume of the system changes from its initial value of V_i to a final value of V_f when mechanical equilibrium is disturbed, either because the pressure in the surroundings changes or the pressure in the system changes (*e.g.*, due to a chemical reaction that increases or decreases the number of moles of gas). For example, airbags activate by the rapid decomposition upon heating of sodium azide, NaN_3, to produce metallic sodium and nitrogen gas, followed by a pair of reactions to consume the sodium:

$2\ NaN_3(s) \xrightarrow{\Delta} 2\ Na(s) + 3\ N_2(g)$ Rxn [8.1a]
$10\ Na(s) + 2\ KNO_3(s) \rightarrow K_2O(s) + 5\ Na_2O(s) + N_2(g)$ Rxn [8.1b]
$K_2O(s) + Na_2O(s) + 2\ SiO_2(s) \rightarrow$ silicate glass Rxn [8.1c]

We find $V_f > V_i$ because the number of moles of gas increases due to Rxn [81a] and [8.1b]. In another example, carbon dioxide can be removed from a gas mixture in a sealed bag by reacting it with solid calcium oxide to form solid calcium carbonate:

$CaO(s) + CO_2(g) \rightarrow CaCO_3(s)$ Rxn [8.2]

For this process $V_f < V_i$ because the number of moles of gas decreases. We identify that expansions correspond to $V_f > V_i$, compressions to $V_f < V_i$, with the expectation that in the final state the system is at mechanical equilibrium (*i.e.*, $P_{sys,f} = P_{surr}$).

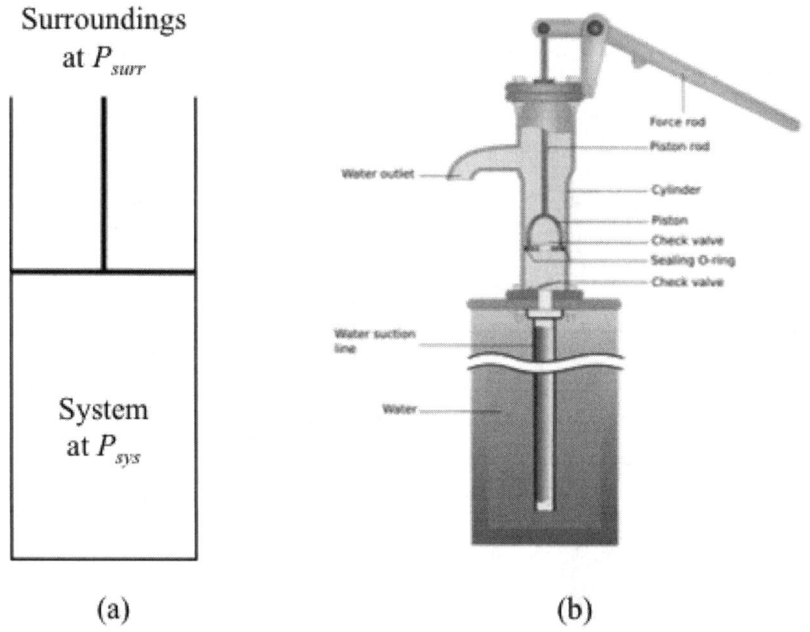

Figure 8.2. Piston/cylinder assemblies. (a) Conceptual schematic; (b) implementation as a water pump.

Using calculus we calculate *P*–*V* work quantitatively as:

$$w = -\int_{V_i}^{V_f} P_{surr}\,dV \qquad [8.2]$$

The negative sign arises because the amount of work and change in internal energy is considered from the perspective of system. In particular, $w < 0$ represents when the system expands and loses energy because the system does work to push the surroundings out of the way, such that energy is transferred via work from the system to the surroundings. In contrast, $w > 0$ is when the system is compressed and gains energy because the particles of matter comprising the system are pushed closer together, such that energy is transferred via work from the surroundings to the system.

Example 8.1: Work Done in Deployment of an Airbag

How much work (in kJ) is done when an airbag with 65.0 g of sodium azide and a stoichiometric amount of potassium nitrate deploys suddenly at sea level to reach a final temperature of 25. °C? Assume a constant atmospheric pressure of 1.0 atm.

Solutions

To determine the work done we first need to determine the pressure in the surroundings and the initial and final volumes of the airbag. Defining the system as the contents of the airbag, we identify the surroundings are at a constant pressure of P_{surr} = 1.0 atm. We assume the initial volume of the airbag is negligible relative to its final, expanded volume (*i.e.*, $V_i \cong 0$ L) because the solid reactants take up negligible volume relative to the gaseous product. Nitrogen gas is produced by two reactions: Rxn [8.1a] and [8.1b]. Determine first the number of moles of azide consumed because we will need this quantity to calculate the number of moles of $N_2(g)$ and Na(*s*) produced by Rxn [8.1a]:

$$65.0 \text{ g NaN}_3 \times \frac{1 \text{ mol NaN}_3}{65.0 \text{ g NaN}_3} = 1.00 \text{ mol NaN}_3$$

$$1.00 \text{ mol NaN}_3 \times \frac{3 \text{ mol N}_2}{2 \text{ mol NaN}_3} = 1.50 \text{ mol N}_2$$

$$1.00 \text{ mol NaN}_3 \times \frac{2 \text{ mol Na}}{2 \text{ mol NaN}_3} = 1.00 \text{ mol Na}$$

With a stoichiometric amount of KNO_3 (*i.e.*, the airbag was loaded with NaN_3 and KNO_3 in stoichiometric proportion), the number of moles of $N_2(g)$ produced by Rxn [8.1b] as:

$$1.00 \text{ mol Na} \times \frac{1 \text{ mol N}_2}{10 \text{ mol Na}} = 0.100 \text{ mol N}_2$$

such that deployment of the airbag produces a total of 1.60 mol N_2. Apply the rearranged ideal gas law to determine the final volume corresponding to the volume of $N_2(g)$ at the final temperature and pressure of 298. K and 1.0 atm, respectively.

$$V_f \cong V_{N_2} = \frac{n_{N_2}RT}{P} = \frac{(1.60 \text{ mol})(8.206 \times 10^{-2} \text{ atm} - \text{L/mol} - \text{K})(298.\text{ K})}{(1.0 \text{ atm})} = 39.\text{ L}$$

The integral for work, Equation [8.2], is then evaluated with a constant P_{surr} as:

$$w = -\int_{V_i}^{V_f} P_{surr} dV = -P_{surr}\int_{V_i}^{V_f} dV = -P_{surr}(V_f - V_i) = -(1.0 \text{ atm})[(39.\text{ L}) - (0 \text{ L})] = 39.\text{ atm} - \text{L}$$

Two strategies are available for converting our value for work to units of joules: the more labor-intensive strategy of converting atm to Pa and L to m^3 to express the product atm-L in the SI units of 1 J = 1 Pa-m^3, or the more convenient strategy of using the gas constant as a conversion factor of R = 8.314 J/mol-K = 8.206×10^{-2} atm-L/mol-K. Implementing the second strategy we find:

$$w = -39.\text{ atm} - \text{L} \times \frac{8.314 \text{ J/mol} - \text{K}}{8.206 \times 10^{-2} \text{ atm} - \text{L/mol} - \text{K}} = -4.0 \times 10^3 \text{ J} \times \frac{1 \text{ kJ}}{10^3 \text{ J}} = -4.0 \text{ kJ}$$

8.3 Heat

From a thermodynamic perspective heat is the flow of energy between a system and its surroundings that is not work. Heat transfer can occur by conduction when energy is transferred by collisions between faster- and slower-moving atoms and/or molecules, by convection when energy is transferred by the bulk flow of fluid, and by radiation when energy is transferred by emission and/or absorption of light. Note that we never refer to the amount of heat an object has—heat is a quantity of transfer only—and our focus is on the amount of heat transferred and its consequences, not the rate of heat transfer. We apply the convention that the heat transferred $q > 0$ if the system gains energy from the surroundings and $q < 0$ if the system loses energy to the surroundings.

Heat transfer can occur either as sensible heat associated with changes in temperature or as latent heat associated with changes in phase at constant temperature. Sensible heat is evaluated in terms of the ability of the system to hold energy transferred in the form of heat, which we represent in terms of a heat capacity, C, as:

$$C = \frac{dq}{dT} \quad [8.2a]$$

$$q = \int_{T_i}^{T_f} C dT \quad [8.2b]$$

Heat capacity is an extensive property that can be expressed as an intensive property either as a specific heat capacity ($\hat{c} \equiv \frac{C}{m}$) or molar heat capacity($c \equiv \frac{C}{n}$). We observe that different substances have different heat capacities (available at the NIST Chemistry Webbook, http://webbook.nist.gov) and that heat capacities are always positively valued.

Heat capacities are functions of temperature (except for noble gases behaving ideally) and depend on both the phase of the substance and whether the substance is held at constant volume or constant pressure, although this distinction is very small for condensed phases. To simplify our descriptions we will adopt the model that the heat capacity for a substance is simply a constant:

$$q = C(T_f - T_i) \qquad [8.2c]$$

Note that if there are multiple substances in the system, the overall heat transferred to the system is the sum of the heat transferred to each individual substance:

$$q_{total} = q_1 + q_2 + \ldots = \sum q_i \qquad [8.3]$$

Example 8.2: Heat Needed to Increase the Temperature of Liquid Water by 1 °C

A calorie (abbreviated cal) is a metric unit defined as the amount of heat needed to raise the temperature of 1 g of $H_2O(l)$, with a specific heat capacity of 4.184 J/g-°C, from 25 °C to 26 °C. How many joules are there in a calorie?

Solutions

Determine the amount of heat in joules equal to one calorie using Equation [8.2c] with $C = m\hat{c}$:

$$q = m\hat{c}(T_f - T_i) = (1.000\text{ g})(4.184\text{ J/g}-{}^\circ\text{C})[(26.000\,^\circ\text{C}) - (25.000\,^\circ\text{C})] = 4.184\text{ J} = 1\text{ cal}$$

Note that a "calorie" used in nutritional labeling actually is a kcal! The English unit for heat (and energy in general) is the British thermal unit (BTU), defined as the amount of heat needed to raise the temperature of 1 lb_m of water at its maximum density (39 °F) by 1 °F.

Example 8.3: Heat Transferred to the Tiles of a Spacecraft Reentering the Atmosphere

Recently retired space shuttles had 24,000 1.5-kg ceramic tiles made of silica aerogel with a specific heat capacity of 0.84 kJ/kg-K. How much heat (in GJ) is transferred to these tiles upon re-entry, which is accompanied by a temperature change from 150 °C to 1300 °C?

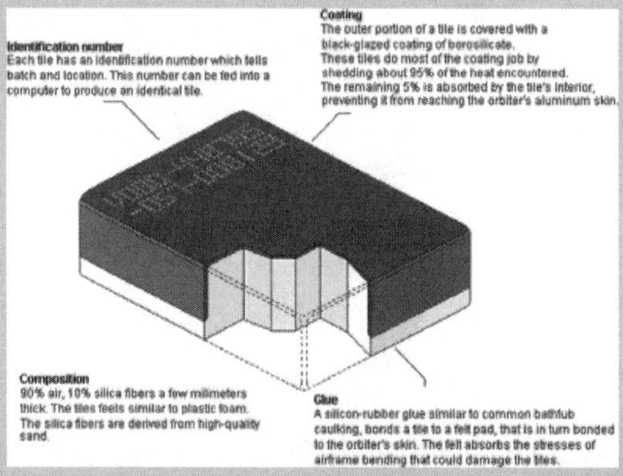

Solution

Apply Equation [8.2c] with $m = 24,000 \text{ tiles} \times 1.5 \text{ kg/tile} = 3.6 \times 10^4 \text{ kg}$ and $\hat{c} = 0.84 \text{ kJ/kg-°K}$ and recognizing that $\Delta T = +1450$ °C = +1450 K:

$$q = m\hat{c}(T_f - T_i) = (3.6 \times 10^4 \text{ kg})(0.84 \text{ kJ/kg-K})(+1450 \text{ K}) = 4.4 \times 10^7 \text{ kJ} \times \frac{1 \text{ GJ}}{10^6 \text{ kJ}} = 44. \text{ GJ}$$

8.4 The First Law of Thermodynamics for Closed and Isolated Systems

Based on the law of conservation of energy, the concept that energy can manifest as macroscopic kinetic energy, macroscopic potential energy, internal energy, and work and heat are the only forms for transfer of energy, we obtain the first law of thermodynamics for a closed or isolated system:

$$\Delta U + \Delta E_k + \Delta E_p = q + w \qquad [8.4]$$

The first law tells us that no machine can exist that can create energy out of nothing.

Example 8.4: Standard Versus Regenerative Braking

A driver makes a panic stop to a standstill from 56 miles/hr (~25. m/s) in an 8.0-metric ton (8.0×10^3-kg) car.

a) What is the maximum temperature rise (in °C) reached by a standard braking system consisting of a set of four 30.-kg steel disk brakes ($\hat{c}_{disc} = 4.6 \times 10^2$ J/kg - °C), each disc with a pair of 2.0-kg pads ($\hat{c}_{pad} = 8.40 \times 10^2$ J/kg - °C)?

b) If the car was a hybrid with a regenerative braking system, what is the maximum amount of energy (in J) that could be stored in a set of on-board batteries?

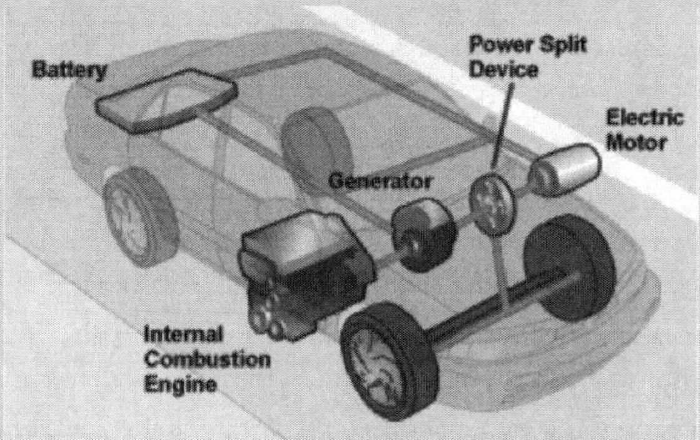

Solutions

For both parts a and b we assume no changes in macroscopic potential energy upon braking (*i.e.*, $\Delta E_p = 0$), no *P-V* work (*i.e.*, *w* = 0), and apply Equation (8.4) with:

$$\Delta E_k = E_{k,f} - E_{k,i} = \frac{1}{2}m_{car}|\vec{V}_f|^2 - \frac{1}{2}m_{car}|\vec{V}_i|^2 = -\frac{m_{car}|\vec{V}_i|^2}{2} = -\frac{(8.0 \times 10^3 \text{ kg})(25.\text{ m/s})^2}{2}$$

$$\Delta E_k = -2.5 \times 10^6 \text{ J}$$

a) In a standard braking system the decrease in macroscopic kinetic energy is not stored (*i.e.*, $\Delta U = 0$) but rather is transferred from the car as heat to the brakes with $\Delta E_k = q_{car}$, $-q_{car} = q_{brakes} = C_{brakes}\Delta T$, and $C_{brakes} = 4(m_{disk}\hat{c}_{disk} + 2m_{pad}\hat{c}_{pad})$, such that:

$$-\Delta E_k = 4(m_{disk}\hat{c}_{disk} + 2m_{pad}\hat{c}_{pad})\Delta T$$

$$\Delta T = -\frac{\Delta E_k}{4(m_{disk}\hat{c}_{disk} + 2m_{pad}\hat{c}_{pad})}$$

$$= \frac{(-2.5 \times 10^6 \text{ J})}{4[(30.\text{ kg})(4.60 \times 10^2 \text{ J/kg - °C}) + 2(2.0\text{ kg})(8.40 \times 10^2 \text{ J/kg - °C})]}$$

$$\Delta T = 36.\text{ °C}$$

b) In a regenerative braking system, ideally no energy is lost from the car as heat to the brakes (*i.e.*, *q* = 0), but rather the decrease in macroscopic kinetic energy is stored in a set of batteries to increase the car's internal energy as $\Delta U + \Delta E_k = 0$, such that the maximum amount of energy stored chemically in the batteries is:

$$\Delta U = -\Delta E_k = -(-2.5 \times 10^6 \text{ J}) = 2.5 \times 10^6 \text{ J}$$

In a hybrid vehicle this energy can be "recovered" by powering an electric motor to restart motion.

We commonly encounter processes with specific types of attributes. For example, a process may occur for a stationary system (*i.e.*, $\Delta E_k = \Delta E_p = 0$), such that the first law is expressed as:

$$\Delta U = q + w \qquad [8.5]$$

Some processes are adiabatic, involving no heat transfer (*i.e.*, $q = 0$, such as is appropriate if the boundary between the system and surroundings is insulated); others are isothermal at a constant temperature (such as is appropriate if the boundary between the system and surroundings is diathermal without resistance to heat transfer. Frequently we encounter processes that occur as a sequence of steps. Cycles are sequences of steps in which the system starts and ends in the same state, such that for a stationary system $\Delta U_{cycle} = U_f - U_i = 0$. We commonly encounter cycles in the operation of engines and HVAC and refrigeration systems.

Example 8.5: Thermodynamic Cycle for an Engine

An engine operates with the following cycle of four steps:

Step 1: The burning of fuel results in the transfer of 200 kJ of heat to the engine. No work is performed during this step.

Step 2: The engine performs work on the surroundings adiabatically.

Step 3: The engine transfers 150 kJ of heat to the surroundings. No work is performed during this step.

Step 4: 50 kJ of work is done on the engine adiabatically.

a) What is the work (in kJ) associated with Step 2?

b) What is the net work (in kJ) for the cycle?

Solutions

a) Based on the engine as the system, identify the heat transfer and work associated with each step in the cycle:

Step	Heat (q, in kJ)	Work (w, in kJ)
1	+200 (transfer to engine)	0 (no work)
2	0 (adiabatic)	$w_2 < 0$ (work done on surroundings)
3	-150 (transfer to surroundings)	0 (no work)
4	0 (adiabatic)	+50 (work done on engine)

Now apply the first law to the cycle:

$$\Delta U_{cycle} = \Delta U_1 + \Delta U_2 + \Delta U_3 + \Delta U_4 = (q_1 + w_1) + (q_2 + w_2) + (q_3 + w_3) + (q_4 + w_4)$$

$$\Delta U_{cycle} = q_1 + q_3 + w_2 + w_4 = 0$$

$$w_2 = - q_1 - q_3 - w_4 = -(+200 \text{ kJ}) - (-150 \text{ kJ}) - (+50 \text{ kJ}) = -100 \text{ kJ}$$

b) The net work is the sum of the work for each step:

$$w_{net} = w_1 + w_2 + w_3 + w_4 = w_2 + w_4 = (0 \text{ kJ}) + (-100 \text{ kJ}) + (0 \text{ kJ}) + (+50 \text{ kJ}) = -50 \text{ kJ}$$

A useful engine performs a negative amount of net work (*i.e.*, the engine does net work on the surroundings). You might wonder why we cannot obtain more net work from this engine by eliminating Step 3, performing an additional 150 kJ of work on the surroundings instead of "venting" this amount of heat to the surroundings. Explaining why Step 3 is necessary requires introduction of additional concepts from thermodynamics and will be examined in Chapter 11.

How the work is performed—in particular, how the pressure in the surroundings is varied—affects the evaluation of the integral for P-V work and, consequently, the amount of work for a process. For a reversible process in which there are infinitesimal driving forces, the system can be returned to its original state with no net effects on the surroundings. A reversible process can be reversed at any point in the process by making an infinitesimal change in the opposite direction. Such processes occur with no appreciable gradients in temperature, pressure, or velocity, with the system in mechanical and thermal equilibrium with the surroundings (*i.e.*, $P_{sys} = P_{surr}$ and $T_{sys} = T_{surr}$) throughout the process. In contrast, an irreversible process is any process that is not reversible (*i.e.*, the system is not in equilibrium with its surroundings throughout the process). Most processes we encounter are described better as irreversible than as reversible. For example, internal combustion engines generate more power and torque when operated irreversibly than reversibly, and batteries generate no current when operated reversibly. Reversibility, however, provides us limits on the best performance possible for a given process: a reversible expansion does more work than any corresponding irreversible expansion, and a reversible compression requires less work than corresponding irreversible compression.

8.5 Heat Transfer Among Substances at Different Initial Temperatures in Thermal Contact

Consider the application of the first law to an isolated stationary system (*i.e.*, a system experiencing no energy exchange with the surroundings, such that $q = w\,0$, and $\Delta E_k = \Delta E_p = 0$, such that $\Delta U = 0$. For no *P–V* work to occur, either the surroundings are a vacuum (*i.e.*, $P_{surr} = 0$) and/or the system undergoes a process that is isochoric (*i.e.*, the system is rigid with constant volume. In general, we represent the change in internal energy for a process with $\Delta V = 0$ as $\Delta U = q_V$, where the subscript "V" indicates that any heat transfer occurs at constant volume.

If our isolated stationary system contains two substances at different initial temperatures (Figure 8.3), the absence of heat transfer with the surroundings means that the system is insulated, such that the sum of the heat transferred within the system is:

$$q_{total} = q_1 + q_2 = C_1\Delta T_1 + C_2\Delta T_2 = C_1(T_f - T_{i,1}) + C_2(T_f - T_{i,2}) = 0 \qquad [8.6]$$

where $\Delta T_1 = T_{f,1} - T_{i,1}$ and $\Delta T_2 = T_{f,2} - T_{i,2}$. What happens if the substances within the system are at different initial temperatures (*i.e.*, $T_{i,1} \neq T_{i,}$)? We expect the substances to reach a common final temperature (*i.e.*, $T_{f,1} = T_{f,2} = T_f$)!

Example 8.6: Heat Transfer Between a Copper Rod and Water in an Insulated Container

A 7.3-g copper rod ($\hat{c}_{Cu} = 0.384$ J/g – °C) at a temperature of 18.8 °C is placed in 10.0 g of water at 54.0 °C in a rigid, insulated container. What is the final temperature (in °C)?

Solutions

Assume that heat transfer occurs only between the copper rod and the water within the container. We expect that the common final temperature will be between the pair of initial temperatures (*i.e.*, 18.8 °C < T_f < 54.0 °C) and anticipate that it will be closer to 54.0 °C than 18.8 °C because the heat capacity (the product of the mass and specific heat capacity) for the water is greater than the heat capacity for the copper rod. Apply Equation [8.6] with $q_{total} = 0$:

$$m_{Cu}\hat{c}_{Cu}(T_f - T_{i,Cu}) + m_w\hat{c}_w(T_f - T_{i,w}) = 0$$

$$m_{Cu}\hat{c}_{Cu}T_f + m_w\hat{c}_wT_f = m_{Cu}\hat{c}_{Cu}T_{i,Cu} + m_w\hat{c}_wT_{i,w}$$

$$(m_{Cu}\hat{c}_{Cu} + m_w\hat{c}_w)T_f = m_{Cu}\hat{c}_{Cu}T_{i,Cu} + m_w\hat{c}_wT_{i,w}$$

$$T_f = \frac{m_{Cu}\hat{c}_{Cu}T_{i,Cu} + m_w\hat{c}_wT_{i,w}}{m_{Cu}\hat{c}_{Cu} + m_w\hat{c}_w}$$

$$= \frac{(7.5\text{ g})(0.384\text{ J/g}-^\circ\text{C})(18.8\,^\circ\text{C}) + (10.0\text{ g})(4.184\text{ J/g}-^\circ\text{C})(54.0\,^\circ\text{C})}{(7.5\text{ g})(0.384\text{ J/g}-^\circ\text{C}) + (10.0\text{ g})(4.184\text{ J/g}-^\circ\text{C})}$$

$$T_f = 52.\,^\circ\text{C}$$

Our answer is consistent with our initial expectations.

We remind ourselves that we can say nothing about the rate of heat transfer (*i.e.*, we know nothing about how long it takes to reach the final temperature). Our consideration of thermochemistry tells us only about the final state reached and not the dynamics to reach that final state. Students majoring in chemical and mechanical engineering will have the opportunity to consider rates of heat transfer later in their studies.

Example 8.7: Using Hot Rocks to Warm Air

A relatively simple way to harness solar energy for home heating is based on using flat plate collectors with a black aluminum back plate to heat a flow of air that is then used to warm a bin of rocks during the day, with a recirculating flow of air from the interior of the house warmed at night by passing over the warm rocks. What is the volume (in m^3) of air (with specific heat capacity of 1.00 J/g-°C and density of 1.20×10^3 g/m^3) that can be heated from 0.0 °C to 20.0 °C by passing through 100.0 kg of hot crushed rock (with specific heat capacity of 0.800 J/g-°C) initially at 110.0 °C?

Solutions

Apply Equation [8.6] to the air/rock system with q_{total} = 0, assuming that the rock and air reach a common final temperature of 20.0 °C, and rearrange to solve for the mass of air:

$$q_{total} = m_{air}\hat{c}_{air}(T_f - T_{air,i}) + m_{rock}\hat{c}_{rock}(T_f - T_{rock,i}) = 0$$

$$m_{air}\hat{c}_{air}(T_f - T_{air,i}) = -\,m_{rock}\hat{c}_{rock}(T_f - T_{rock,i})$$

$$m_{air} = -\frac{m_{rock}\hat{c}_{rock}(T_f - T_{rock,i})}{\hat{c}_{air}(T_f - T_{air,i})} = -\frac{\left(100.0\text{ kg}\times\frac{10^3\text{ g}}{1\text{ kg}}\right)(0.800\text{ J/g-}^\circ\text{C})[(20.0\,^\circ\text{C})-(110.0\,^\circ\text{C})]}{(1.00\text{ J/g-}^\circ\text{C})[(20.0\,^\circ\text{C})-(0.0\,^\circ\text{C})]}$$

$$m_{air} = 3.60\times10^5\text{ g}$$

Determine the corresponding volume using the density:

$$\rho_{air} = \frac{m_{air}}{V_{air}}$$

$$V_{air} = \frac{m_{air}}{\rho_{air}} = \frac{(3.60\times10^5\text{ g})}{(1.20\times10^3\text{ g/m}^3)} = 3.00\times10^2\text{ m}^3$$

This volume corresponds to the a medium-sized room.

As introduced in Section 8.3, sometimes heat is transferred as latent heat associated with a change in phase. In Section 5.3 we observed that pairs of phases coexist at equilibrium at a common temperature and pressure, such that phase transitions occur without a change in either temperature or pressure. Any process that occurs at constant pressure is termed isobaric and corresponds to a closed system with a flexible boundary with the surroundings and $P = P_{sys} = P_{surr}$. Application of the first law for a stationary system then yields:

$$\Delta U = q_P + w = q_P - \int_{V_i}^{V_f} P_{surr} dV = q_P - P\Delta V$$

$$\Delta U + P\Delta V = \Delta U + \Delta(PV) = \Delta(U + PV)$$

$$\Delta H = q_P \qquad [8.7]$$

where the subscript "P" indicates that any heat transfer occurs at constant pressure and we have defined enthalpy as:

$$H \equiv U + PV \qquad [8.8]$$

Enthalpy is just another way to measure energy, useful for isobaric processes.

The sign for the latent heat for a phase change depends on whether the transition is endothermic and occurs by absorption of heat (*e.g.*, fusion, vaporization, and sublimation), or exothermic and occurs by release of heat (*e.g.*, freezing, condensation, and deposition). Note that endothermic phase changes correspond to transitions for which heat is needed to overcome attractive IM forces, and that exothermic phase changes correspond to transitions for which increases in attractive IM forces result in the release of heat. Data for latent heats for pure substances are typically tabulated in terms of molar changes in enthalpy at a specified temperature, with the magnitude for the latent heat dependent on the relative strength of IM forces overcome. For example, for water we find $\Delta h_{vap} > \Delta h_{fus}$, with $\Delta h_{fus} = 6.01$ kJ/mol and $\Delta h_{vap} = 40.6$ kJ/mol, because more energy is required to almost entirely overcome attractive IM forces in the liquid phase and move water molecules far apart into the vapor phase than to partially overcome attractive IM forces in the solid phase and move water molecules into the liquid phase. Note also that enthalpy (like internal energy) is a state property. As a consequence, we can relate latent heats for fusion and freezing based on the sequence of steps $s \to l \to s$: $\Delta h_{s \to l \to s} = \Delta h_{fus} + \Delta h_{freezing} = 0$ means $\Delta h_{fus} = -\Delta h_{freezing}$. Similarly, we find $\Delta h_{cond} = -\Delta h_{vap}$ and $\Delta h_{deposition} = -\Delta h_{sublimation}$. Data for changes in enthalpy for phase transitions for different substances is available at the NIST Chemistry Webbook (http://webbook.nist.gov).

Example 8.8: Evaluating a Process with Both Sensible and Latent Heats

Determine the heat (in kJ) necessary to transform 10.0 g of $H_2O(s)$ at 0.00 °C to $H_2O(g)$ at 100.00 °C in an open container.

Solutions

Because the process occurs in an open container, the heat transfer is isobaric, and we seek $q_P = \Delta H$. Further, because enthalpy is a state property, we are free to evaluate changes in enthalpy with whatever path is most convenient. We choose the path (s, $T_i = T_m = 0.00$ °C) → (l, 0.00 °C) → (l, $T_b = 100.00$ °C) → (g, $T_b = T_f = 100.0$ °C), depicted below as a heating curve providing a graphical representation of relationships between temperature and heat.

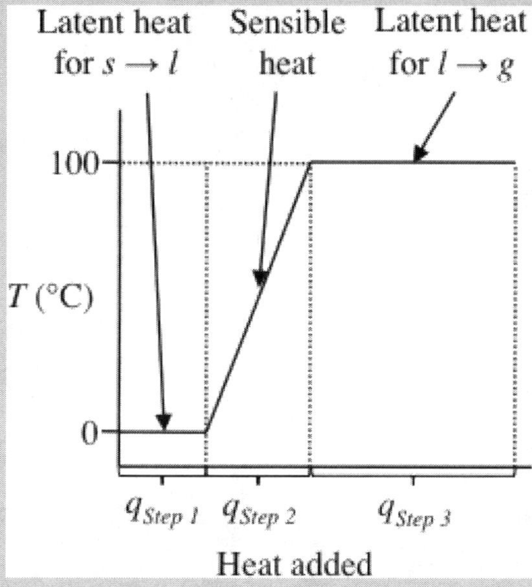

The heat needed is the sum of the changes in enthalpy (*i.e.*, heat transferred at constant pressure) for each step, where we make sure to convert J to kJ:

$$q = q_{Step1} + q_{Step2} + q_{Step3} = n\Delta h_{fus} + m\hat{c}(T_b - T_m) + n\Delta h_{vap}$$
$$= m\hat{c}(T_b - T_m) + \left(\frac{m}{M}\right)\Delta h_{fus} + \left(\frac{m}{M}\right)\Delta h_{vap} = m\left[\hat{c}(T_b - T_m) + \frac{\Delta h_{fus} + \Delta h_{vap}}{M}\right]$$
$$= (10.0\ \text{g})\left[\left(4.184\ \text{J/g}-{}^\circ\text{C} \times \frac{1\ \text{kJ}}{10^3\ \text{J}}\right)[(100.00\ {}^\circ\text{C}) - (0.00\ {}^\circ\text{C})] + \frac{(6.01\ \text{kJ/mol}) + (40.6\ \text{kJ/mol})}{(18.02\ \text{g/mol})}\right]$$
$$q = 30.0\ \text{kJ}$$

Under some circumstances heat transfer between materials in thermal contact that have different initial temperatures results in the materials reaching a common final temperature with one or more of the materials experiencing a phase change, as illustrated in the example below.

Example 8.9: Heating an Iron Radiator with Steam

What is the final temperature reached when a 26.0-kg cast iron radiator ($\hat{c}_{Fe} = 0.449$ J/g − °C) at 10.0 °C is heated with 200.0 g of steam (*i.e.*, water vapor) at 100.0 °C? Assume that the air around the radiator is a perfect insulator, which is a reasonable first approximation.

Solutions

With the air around the radiator acting as a perfect insulator, we find the sum of the heat transferred from the perspective of the water, q_{water}, and from the perspective of the iron radiator, q_{Fe}, must be zero:

$$q_{total} = q_{water} + q_{Fe} = 0$$

We have to make one of two assumptions: either all of the steam initially at 100 °C condenses to liquid water at 100 °C and the condensed water then cools to a common final temperature 10.0 °C < T_f < 100.0 °C reached by the warming of the radiator, or only some of the steam condenses to liquid water and the radiator warms to a common final temperature of 100 °C. If we make the wrong assumption, we obtain an answer that is inconsistent with our assumption and/or the problem.

Based on assuming that the final temperature is less than T_b = 100 °C, we find that the heat transferred from the perspective of the water has contributions from the latent heat of condensation and from the sensible heat of cooling (converting units as necessary):

$$n_{water}\Delta h_{cond} + m_{water}\hat{c}_{H_2O(l)}(T_f - T_b) + m_{Fe}\hat{c}_{Fe}(T_f - T_{i,Fe}) = 0$$

$$-n_{water}\Delta h_{cond} + m_{water}\hat{c}_{H_2O(l)}T_b + m_{Fe}\hat{c}_{Fe}T_{i,Fe} = (m_{water}\hat{c}_{H_2O(l)} + m_{Fe}\hat{c}_{Fe})T_f$$

$$T_f = \frac{\left(\frac{m_{water}}{M}\right)\Delta h_{vap} + m_{water}\hat{c}_{H_2O(l)}T_b + m_{Fe}\hat{c}_{Fe}T_{i,Fe}}{m_{water}\hat{c}_{H_2O(l)} + m_{Fe}\hat{c}_{Fe}} = \frac{m_{water}\left(\frac{\Delta h_{vap}}{M} + \hat{c}_{H_2O(l)}T_b\right) + m_{Fe}\hat{c}_{Fe}T_{i,Fe}}{m_{water}\hat{c}_{H_2O(l)} + m_{Fe}\hat{c}_{Fe}}$$

$$T_f = \frac{\left[\begin{array}{l}(200.0\text{ g})\left[\frac{\left(40.6\text{ kJ/mol}\times\frac{10^3\text{ J}}{1\text{ kJ}}\right)}{(18.02\text{ g/mol})} + (4.184\text{ J/g}-{}^\circ\text{C})(100.0\,{}^\circ\text{C})\right] \\ +\left(26.0\text{ kg}\times\frac{10^3\text{ g}}{1\text{ kg}}\right)(0.449\text{ J/g}-{}^\circ\text{C})(10.0\,{}^\circ\text{C})\end{array}\right]}{\left[(200.0\text{ g})(4.184\text{ J/g}-{}^\circ\text{C}) + \left(26.0\text{ kg}\times\frac{10^3\text{ g}}{1\text{ kg}}\right)(0.449\text{ J/g}-{}^\circ\text{C})\right]} = 52.0\,{}^\circ\text{C}$$

Our assumption that the final temperature is less than the boiling point of water is validated.

What would have happened if we had made the other assumption, namely that the final temperature was 100 °C? For this assumption the amount of heat necessary to warm the radiator to final temperature would be:

$$q_{Fe} = m_{Fe}\hat{c}_{Fe}(T_b - T_{i,Fe}) = \left(26.0\text{ kg}\times\frac{10^3\text{ g}}{1\text{ kg}}\right)(0.449\text{ J/g}-{}^\circ\text{C})[(100.0\,{}^\circ\text{C}) - (10.0\,{}^\circ\text{C})]$$

$$q_{Fe} = 1.05\times10^6\text{ J}$$

Determine how much heat would be released if all of the steam condensed:

$$q_{cond,max} = n_{water}\Delta h_{cond} = -\frac{m_{water}\Delta h_{vap}}{M} = -\frac{(200.0\text{ g})\left(40.6\text{ kJ/mol}\times\frac{10^3\text{ J}}{1\text{ kJ}}\right)}{(18.02\text{ g/mol})} = -4.51\times10^5\text{ J}$$

We find that we do not have sufficient steam to heat the radiator to 100 °C! Note that if we perform our calculations correctly, we always find only one possible assumption can be correct.

8.6 Thermochemistry of Reactions

We also can apply the first law of thermodynamics to systems in which a chemical reaction occurs. Because the types, numbers, and/or arrangements of specific bonds typically differ between reactants and products, the products typically have a different internal energy than the reactants (*i.e.*, $U_{products} \neq U_{reactants}$) and $\Delta U_{rxn} \equiv U_{products} - U_{reactants} \neq 0$. Two types of situations commonly arise: endothermic reactions (Figure 8.4a) in which $U_{products} > U_{reactants}$, $\Delta U_{rxn} > 0$, and energy is absorbed; and exothermic reactions (Figure 8.4b) in which $U_{products} < U_{reactants}$, $\Delta U_{rxn} < 0$, and energy is released. Endothermic reactions are used in cold packs, exothermic reactions in hot pacs. Combustion reactions as well as the thermite reaction are other examples of exothermic reactions.

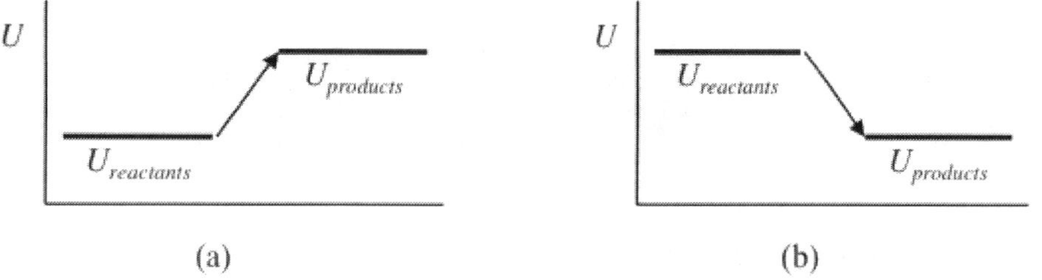

Figure 8.4. Energy changes for (a) an endothermic reaction and (b) an exothermic reaction.

Because the energetics of a reaction also depend on the conditions under which the reaction occurs (*e.g.*, temperature and pressure) and the state of matter for the participating species, reaction energetics are commonly referenced with respect to standard state conditions, specified with a subscript "o" and corresponding to a reference pressure of standard pressure, $P^{\circ} \equiv 1$ bar, and requiring each reactant and product to be in its pure form and reference state (*i.e.*, its most stable state) at 1 bar and the temperature of interest. For example, we expect O_2 to be a gas, water to be a liquid, and solid carbon to be in the form of graphite in their reference states at 1 bar and 25.0 °C. Note that no reference temperature is specified but that data are commonly presented with reference to "room temperature," 298.5 K. Finally, we need to recognize that values for changes in energy for reactions are reported frequently based on the stoichiometry of the reaction, with an implied set of units of "per mole reaction."

Example 8.10: Complete Combustion of Biodiesel

The complete combustion of rapeseed methyl ester, a form of biodiesel introduced in Example 7.11, occurs through the following balanced chemical equation:

$$C_{18}H_{36}O_2(l) + 26\ O_2(g) \xrightarrow{\Delta} 18\ CO_2(g) + 18\ H_2O(l) \qquad \text{Rxn [8.E1]}$$

This reaction is strongly exothermic—signifying that $C_{18}H_{36}O_2$ is potentially a good fuel—with a reported heat of reaction at constant volume of -1.17×10^4 kJ Note that because this reaction is a combustion process, we sometimes use the phrase the heat of combustion at constant volume is -1.17×10^4 k. This value implies that 1.17×10^4 kJ of energy is released per mole of $C_{18}H_{36}O_2$ reacting with 26 mol O_2 to produce 18 mol each of CO_2 and H_2O.

a) How much energy (in kJ) is released when 2.00 mol of $C_{18}H_{36}O_2$ is burned?

b) How much energy (in kJ) is released when 1.00 g of $C_{18}H_{36}O_2$ is burned?

Solutions

a) When 2.00 mol $C_{18}H_{36}O_2$ is consumed:

$$2.00 \text{ mol } C_{18}H_{36}O_2 \times 1.17 \times 10^4 \text{ kJ/mol } C_{18}H_{36}O_2 = 2.34 \times 10^4 \text{ kJ of energy released}$$

b) When 1.00 g $C_{18}H_{36}O_2$ is consumed, based on a molar mass of 284.54 g/mol for $C_{18}H_{36}O_2$:

$$1.00 \text{ g } C_{18}H_{36}O_2 \times \frac{1 \text{ mol } C_{18}H_{36}O_2}{284.54 \text{ g } C_{18}H_{36}O_2} \times 1.17 \times 10^4 \text{ kJ/mol } C_{18}H_{36}O_2 = 41.1 \text{ kJ of energy released}$$

How we generalize these observations depends on whether the volume or pressure is constant and whether we are interested in molar or specific changes in energy. In particular, we prefer to evaluate changes in energy occurring at constant volume in terms of changes in internal energy and at constant pressure in terms of changes in enthalpy, as introduced in Section 8.5. With n_{rxn} and m_{rxn} as the number of moles and mass of a specific reactant, respectively, we identify for constant-volume systems (*i.e.*, the reaction occurs inside a rigid, sealed vessel):

$$q_V = q_{rxn} + q_{sys} = q_{rxn} + C_{sys}\Delta T_{sys} \qquad [8.9a]$$

$$q_{rxn} = n_{rxn}\Delta u^{\circ}_{rxn} = m_{rxn}\Delta \hat{u}^{\circ}_{rxn} \qquad [8.9b]$$

and for constant-pressure systems (*i.e.*, the reaction occurs inside a vessel with flexible walls or a vessel open to the atmosphere):

$$q_P = q_{rxn} + q_{sys} = q_{rxn} + C_{sys}\Delta T_{sys} \qquad [8.10a]$$

$$q_{rxn} = n_{rxn}\Delta h^{\circ}_{rxn} = m_{rxn}\Delta \hat{h}^{\circ}_{rxn} \qquad [8.10b]$$

For example, we found in Example 8.10 that for the complete combustion of rapeseed methyl ester $\Delta u^{\circ}_{rxn} = -1.17 \times 10^{4}$ kJ/mol $C_{18}H_{36}O_2$ and that $\Delta\hat{u}^{\circ}_{rxn}$ = -41.1kJ/g $C_{18}H_{36}O_2$. The absolute value of $\Delta\hat{u}^{\circ}_{rxn}$ is also the fuel value for rapeseed methyl ester. Note that typically the difference between reaction energetics at constant volume versus at constant pressure is negligible.

A bomb calorimeter (Figure 8.5) is the classic model for a constant-volume system in which reaction energetics are studied. Such calorimeters are typically operated as insulated systems (*i.e.*, $q_V = 0$). A variety of constant-pressure calorimeters are encountered in practice, including the ultra-simple coffee-cup calorimeters used in many high-school labs and consisting of nested Styrofoam cups with lidding. We also typically assume these systems are insulated (*i.e.*, $q_P = 0$).

(a)

(b)

Figure 8.5. Parr Instruments, Inc. bomb calorimeters. (a) Assembled calorimeter; (b) steel "bomb" in which a combustion reaction is conducted after pressurizing with ~30 atm of oxygen.

Example 8.11: Calibrating a Bomb Calorimeter

Bomb calorimeters are calibrated with benzoic acid prior to use to determine their heat capacity. What is the heat capacity (in kJ/°C) of an insulated bomb calorimeter for which the complete combustion of 0.9776 g of benzoic acid ($\Delta\hat{u}^{\circ}_{rxn} = -26.434$ kJ/g) results in a temperature increase from 21.5 °C to 23.8 °C?

Solutions

Apply Equation [8.9] using a mass basis:

$$q_V = 0 = q_{rxn} + C_{sys}\Delta T_{sys} = m_{rxn}\Delta\hat{u}^{\circ}_{rxn} + C_{sys}(T_f - T_i)$$

and rearrange to solve for the heat capacity for the system (*i.e.*, the calorimeter):

$$C_{sys}(T_f - T_i) = -m_{rxn}\Delta\hat{u}^{\circ}_{rxn}$$

$$C_{sys} = -\frac{m_{rxn}\Delta\hat{u}^{\circ}_{rxn}}{T_f - T_i} = -\frac{(0.9776\ \text{g})(-26.434\ \text{kJ/g})}{(23.8\ ^{\circ}\text{C}) - (21.5\ ^{\circ}\text{C})} = 11.3\ \text{kJ}/^{\circ}\text{C}$$

Example 8.12: Evaluating the Energy Content of Kerosene

The complete combustion of *n*-dodecane ($C_{12}H_{26}$), one of the major components of kerosene, occurs by the following balanced chemical equation:

$$C_{12}H_{26}(l) + \frac{37}{2}O_2(g) \xrightarrow{\Delta} 12\,CO_2(g) + 13\,H_2O(l) \qquad \text{Rxn [8.E2]}$$

What is the heat of combustion at constant volume, Δu°_{rxn} (in kJ/mol $C_{12}H_{26}$), if a temperature increase of 4.2 °C is observed in the bomb calorimeter from Example 8.11 when 1.00 g of *n*-dodecane is consumed?

Solutions

Apply Equation [8.9] using a mole basis:

$$q_V = 0 = q_{rxn} + C_{sys}\Delta T_{sys} = n_{rxn}\Delta u^{\circ}_{rxn} + C_{sys}(T_f - T_i)$$

and rearrange to solve for the heat of combustion at constant volume with a molar mass of 170.38 g/mol for *n*-dodecane:

$$-n_{rxn}\Delta u^{\circ}_{rxn} = C_{sys}\Delta T_{sys}$$

$$\Delta u^{\circ}_{rxn} = -\frac{C_{sys}\Delta T_{sys}}{n_{rxn}} = -\frac{C_{sys}\Delta T_{sys}}{\left(\frac{m_{rxn}}{M_{C_{12}H_{26}}}\right)} = -\frac{M_{C_{12}H_{26}}C_{sys}\Delta T_{sys}}{m_{rxn}}$$

$$\Delta u^{\circ}_{rxn} = -\frac{(170.38\text{ g/mol})(11.3\text{ kJ/°C})(4.2\text{ °C})}{(1.00\text{ g})} = -8.1\times10^{3}\text{ kJ/mol}$$

Example 8.13: Heating Water Using Biodiesel

The complete combustion of 2.00 mol of rapeseed methyl ester in an insulated, unsealed (*i.e.*, open) furnace is used to heat 100. kg of water at 1 atm. If the water is initially at 20.0 °C, what is the final temperature reached if all of the heat released by combustion is absorbed by the water?

Solutions

The furnace is operated as a constant-pressure system. We assume that $\Delta h^{\circ}_{rxn} \cong \Delta u^{\circ}_{rxn} = -1.17\times10^{4}$ kJ/mol $C_{18}H_{36}O_2$ for the reaction and that the water does not boil, such that if all of the heat released by the reaction is absorbed by the water, the heat capacity of the system is equivalent to the heat capacity of the water. Applying Equation [8.10] using a mole basis and rearranging we find:

$$q_P = 0 = q_{rxn} + C_{sys}\Delta T_{sys} = n_{rxn}\Delta h^{\circ}_{rxn} + C_{sys}(T_f - T_i) = n_{rxn}\Delta h^{\circ}_{rxn} + m_w\hat{c}_w(T_f - T_i)$$

$$m_w\hat{c}_w(T_f - T_i) = -n_{rxn}\Delta h^{\circ}_{rxn}$$

$$T_f - T_i = -\frac{n_{rxn}\Delta h^{\circ}_{rxn}}{m_w\hat{c}_w}$$

$$T_f = T_i - \frac{n_{rxn}\Delta h^{\circ}_{rxn}}{m_w\hat{c}_w} = (20.0\ ^{\circ}\text{C}) - \frac{(2.00\ \text{mol})\left(-1.17\times10^4\ \text{kJ/mol}\times\frac{10^3\ \text{J}}{1\ \text{kJ}}\right)}{(100.\times10^3\ \text{g})(4.184\ \text{J/g}-^{\circ}\text{C})} = 75.9\ ^{\circ}\text{C}$$

Our assumption that the water does not boil is justified.

Example 8.14: Boiling Water Using Gasoline

A gasoline-powered stove is used to boil water initially at 100 °C and a constant pressure of 1 atm. Assuming that gasoline can be modeled as pure octane (C_8H_{18}, *l*) with a heat of combustion of $\Delta h^{\circ}_{rxn} = -5.47\times10^3$ kJ/mol and that all of the heat evolved by the combustion of octane goes to boil the water, how much water (in g) can be boiled by burning 10.0 g of gasoline?

Solutions

The stove is operated as a constant-pressure system. We assume that the water is merely to be transformed to vapor at its normal boiling point. Apply Equation [8.10] using a mole basis for the reaction with a molar mass of 114.0 g/mol for octane and $q_{sys} = n_w\Delta h_{vap} = \frac{m_w\Delta h_{vap}}{M_w}$, rearranging to find:

$$q_P = 0 = \left(\frac{m_{rxn}}{M_{C_8H_{18}}}\right)\Delta h^{\circ}_{rxn} + \frac{m_w\Delta h_{vap}}{M_w}$$

$$\frac{m_w\Delta h_{vap}}{M_w} = -\frac{m_{rxn}\Delta h^{\circ}_{rxn}}{M_{C_8H_{18}}}$$

$$m_w = -\frac{M_w m_{rxn}\Delta h^{\circ}_{rxn}}{M_{C_8H_{18}}\Delta h_{vap}} = -\frac{(18.02\ \text{g/mol})(10.0\ \text{g})(-5.47\times10^3\ \text{kJ/mol})}{(114.0\ \text{g/mol})(40.6\ \text{kJ/mol})} = 2.13\times10^2\ \text{g}$$

8.7 Strategies to Evaluate Reaction Enthalpies

An understanding of how exothermic or endothermic a reaction is can allow us to evaluate how to use the reaction and its potential consequences (*e.g.*, Does the temperature increase or decrease when the reaction is conducted? How useful is the reaction for releasing energy to power an engine?) In Section 8.6 we identified how we can determine reaction energetics from experimental observations, but this approach has the drawback that for each new reaction we encounter we need to conduct at least one experiment We seek alternative strategies that can provide predictions based on previously-conducted experiments.

One such strategy takes advantage of the fact that changes in internal energy and enthalpy are independent of path. For example, consider the reaction for the preparation of metallurgical-grade silicon for semiconductors from quartz sand (SiO_2):

$$SiO_2(s) + 2\,Mg(s) \rightarrow Si(s) + 2\,MgO(s) \quad \Delta h^{\circ}_{rxn} = ? \qquad \text{Rxn [8.3]}$$

Thermochemical data for the following pair of related reactions is available:

$$Si(s) + O_2(g) \rightarrow SiO_2(s) \quad \Delta h^{\circ}_{Rxn\,(1)} = -\,911 \text{ kJ/mol} \qquad \text{[8.4a]}$$

$$Mg(s) + \tfrac{1}{2}\,O_2(g) \rightarrow MgO(s) \quad \Delta h^{\circ}_{Rxn\,(2)} = -\,602 \text{ kJ/mol} \qquad \text{Rxn [8.4b]}$$

Because internal energy and enthalpy are state properties, changes in internal energy or enthalpy for a reaction have the following three properties:

Property 1: The change in internal energy/enthalpy for a reaction depends on the stoichiometry for the reaction. Thus:

$$2\,Mg(s) + O_2(g) \rightarrow 2\,MgO(s) \quad \Delta h^{\circ}_{2\times Rxn\,(2)} = 2(-602 \text{ kJ/mol}) = -\,1204 \text{ kJ/mol}$$

Property 2: The change in internal energy/enthalpy for the reverse of a reaction is equal to the *negative* of the change in internal energy/enthalpy for the reaction. Thus:

$$SiO_2(s) \rightarrow Si(s) + O_2(g) \quad -\Delta h^{\circ}_{Rxn\,(1)} = -(-911 \text{ kJ/mol}) = 911 \text{ kJ/mol}$$

Property 3: If an overall reaction can be written as a sum of real or hypothetical steps, the net change in enthalpy for the overall reaction is the sum of the change in enthalpy for each step. Thus:

$$\begin{array}{llll} & 2\,Mg(s) + O_2(g) \rightarrow 2\,MgO(s) & & \Delta h^{\circ}_{2\times Rxn\,(2)} = -\,1204 \text{ kJ/mol} \\ + & SiO_2(s) \rightarrow Si(s) + O_2(g) & & -\Delta h^{\circ}_{Rxn\,(1)} = 911 \text{ kJ/mol} \\ \hline & SiO_2(s) + 2\,Mg(s) \rightarrow Si(s) + 2\,MgO(s) & & \Delta h^{\circ}_{rxn} = (-1204 \text{ kJ/mol}) + (911 \text{ kJ/mol}) \\ & & & \Delta h^{\circ}_{rxn} = -\,293 \text{ kJ/mol} \end{array}$$

Property 3 is known as Hess's law and essentially tells us that we can determine the change in energy associated with a targeted reaction (or process) if we can express that targeted reaction (or process) as a linear combination of a set of reactions (and/or processes) with known changes in energy. Hess's law is stated explicitly in terms of changes in enthalpy but can also be applied to determine changes in internal energy.

We can apply Hess's law to predict a bond dissociation energy, the energy that must be added to break one mole of bonds in the vapor phase (see Section 4.4). Estimates of bond dissociation energies are made on gas-phase species because, if the system is sufficiently dilute, we can neglect IM energetics.

Example 8.15: Predicting the Dissociation Energy of an O-H Bond

Determine Δh°_{rxn} (in kJ/mol) at 25 °C for the reaction OH(g)→ O(g) + H(g) using the following data at 25 °C:

$\frac{1}{2} H_2(g) + \frac{1}{2} O_2(g) \rightarrow OH(g)$ $\quad \Delta h^\circ_{Rxn(1)} = 38.95$ kJ/mol $\quad$ Rxn [8.E3a]

$H_2(g) \rightarrow 2\,H(g)$ $\quad \Delta h^\circ_{Rxn(2)} = 435.994$ kJ/mol $\quad$ Rxn [8.E3b]

$O_2(g) \rightarrow 2\,O(g)$ $\quad \Delta h^\circ_{Rxn(3)} = 498.34$ kJ/mol $\quad$ Rxn [8.E3c]

Note that it is impractical to run the targeted reaction but that Rxns [8.E3a], [8.E3b], and [8.E3c] can be examined in separate experiments with relatively high precision.

Solutions

We seek a linear combination of Rxns [8.E3a], [8.E3b], and [8.E3c] yielding the targeted reaction. Consider that our targeted reactant, OH, appears only in Rxn [8.E3a] and as a product, and that our targeted products, O and H, appear only in Rxn [8.E3c] and [8.E3b], respectively, and as products in twice the targeted amounts, such that:

$OH(g) \rightarrow \frac{1}{2} H_2(g) + \frac{1}{2} O_2(g)$ $\quad -\Delta h^\circ_{Rxn(1)}$

$\frac{1}{2}\,[O_2(g) \rightarrow 2\,O(g)]$ $\quad +\frac{1}{2}(\Delta h^\circ_{Rxn(3)})$

$\frac{1}{2}\,[H_2(g) \rightarrow 2\,H(g)]$ $\quad +\frac{1}{2}(\Delta h^\circ_{Rxn(2)})$

$OH(g) + \cancel{\frac{1}{2} O_2(g)} + \cancel{\frac{1}{2} H_2(g)} \rightarrow \cancel{\frac{1}{2} H_2(g)} + \cancel{\frac{1}{2} O_2(g)} + O(g) + H(g)$ $\quad -\Delta h^\circ_{Rxn(1)}$

$OH(g) \rightarrow O(g) + H(g)$ $\quad +\frac{1}{2}(\Delta h^\circ_{Rxn(3)})$

$\quad +\frac{1}{2}(\Delta h^\circ_{Rxn(2)})$

$$\Delta h^\circ_{rxn} = -(38.95 \text{ kJ/mol}) + \frac{1}{2}(498.34 \text{ kJ/mol}) + \frac{1}{2}(435.994 \text{ kJ/mol}) = 428.22 \text{ kJ/mol}$$

Example 8.16: Predicting the Reaction Enthalpy for the Production of Substitute Natural Gas

Substitute natural gas (SNG) is a gaseous mixture containing $CH_4(g)$ that can be used as a fuel. One reaction for the production of SNG is:

$$4\ CO(g) + 8\ H_2(g) \rightarrow 3\ CH_4(g) + CO_2(g) + 2\ H_2O(l) \qquad \text{Rxn [8.E4]}$$

Use the following set of data at 298 K to determine Δh° (in kJ) at this temperature for Rxn [8.E4]:

$$C(s,\text{graphite}) + \tfrac{1}{2}\ O_2(g) \rightarrow CO(g) \qquad \Delta h^\circ = -110.5\ \text{kJ/mol} \qquad \text{Rxn [8.E5a]}$$

$$C(s,\text{graphite}) + 2\ H_2(g) \rightarrow CH_4(g) \qquad \Delta h^\circ = -74.81\ \text{kJ/mol} \qquad \text{Rxn [8.E5b]}$$

$$CH_4(g) + 2\ O_2(g) \rightarrow CO_2(g) + 2\ H_2O(l) \qquad \Delta h^\circ = -890.3\ \text{kJ/mol} \qquad \text{Rxn [8.E5c]}$$

Solutions

The following combination of the above reactions yields Rxn [8.E4]:

$$CH_4(g) + 2\ O_2(g) \rightarrow CO_2(g) + 2\ H_2O(l) \qquad \Delta h^\circ_{\text{Rxn [E8.5c]}}$$

$$4\ C(s,\ \text{graphite}) + 8\ H_2(g) \rightarrow 4\ CH_4(g) \qquad 4\Delta h^\circ_{\text{Rxn [E8.5b]}}$$

$$4\ CO(g) \rightarrow 4\ C(s,\ \text{graphite}) + 2\ O_2(g) \qquad -4\Delta h^\circ_{\text{Rxn [E8.5a]}}$$

$$4\ CO(g) + 8\ H_2(g) \rightarrow 3\ CH_4(g) + CO_2(g) + 2\ H_2O(l) \qquad \Delta h^\circ_{\text{Rxn [E8.5c]}} + 4\Delta h^\circ_{\text{Rxn [E8.5b]}} - 4\Delta h^\circ_{\text{Rxn [E8.5a]}}$$

The targeted reaction then has an enthalpy change of:

$$\Delta h^\circ_{\text{Rxn [E8.4]}} = \Delta h^\circ_{\text{Rxn [E8.5c]}} + 4\Delta h^\circ_{\text{Rxn [E8.5b]}} - 4\Delta h^\circ_{\text{Rxn [E8.5a]}}$$
$$= (-890.3\ \text{kJ/mol}) - 4(-74.81\ \text{kJ/mol}) - 4(-110.5\ \text{kJ/mol})$$
$$\Delta h^\circ_{\text{Rxn [E8.4]}} = -747.5\ \text{kJ/mol}$$

Application of Hess's law as we have considered it thus far is limited and tedious—we would need to acquire data for specific sets of reactions. We would prefer to have a strategy that is based less on data for specific reactions and more on data for specific substances. Consider reformulating Hess's law for the generic reaction Rxn [7.3] conducted at 1 bar:

$$\upsilon_A\,A + \upsilon_B\,B + \ldots \rightarrow \upsilon_C\,C + \upsilon_D\,D + \ldots \qquad \text{Rxn [7.3]}$$

where we remind ourselves from our discussion in Section 7.3 that υ_i, the stoichiometric coefficient for species *i*, is positively-valued for products but negatively-valued for reactants. Because enthalpy is a state property and changes in enthalpy are independent of path, we can express the molar change in enthalpy for Rxn [7.3] at standard state using the hypothetical two-step path depicted in Figure 8.6 as:

$$\Delta h^{\circ}_{rxn} = \Delta h^{\circ}_{Step\ 1} + \Delta h^{\circ}_{Step\ 2} \qquad [8.11]$$

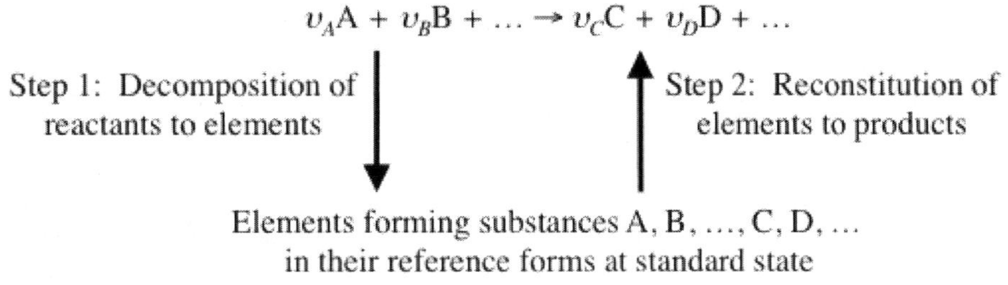

$\upsilon_A A + \upsilon_B B + \ldots \rightarrow \upsilon_C C + \upsilon_D D + \ldots$

Step 1: Decomposition of reactants to elements

Step 2: Reconstitution of elements to products

Elements forming substances A, B, ..., C, D, ... in their reference forms at standard state

Figure 8.6. Two-step path for applying Hess's law to evaluate reaction enthalpy using standard molar enthalpies of formation.

We define the <u>standard molar enthalpy of formation</u> of substance *i*, $\Delta h^{\circ}_{f,i}$ (expressed in kJ/mol), as the change in enthalpy for the reaction to form one mole of substance *i* at standard state from its constitutive pure elements in their reference form at standard state (based on this definition $\Delta h^{\circ}_{f,i} = 0$ kJ/mol for a pure element in its reference form). Data for standard molar enthalpies of formation are tabulated at 298.15 K and are available in the *CRC Handbook of Chemistry and Physics* (http://www.hbcpnetbase.com/) and through the NIST Chemistry Webbook (http://webbook.nist.gov). We observe that substances with positively-valued standard molar enthalpies of formation have bonds that are more complex than for their constituent elements in their reference forms, and that substances with negatively-valued standard molar enthalpies of formation have bonds that are less complex than for their constituent elements in their reference forms.

Applying the definition of standard molar enthalpies of formation we find:

$$\Delta h^{\circ}_{Step\ 1} = \upsilon_A \Delta h^{\circ}_{f,A} + \upsilon_B \Delta h^{\circ}_{f,B} + \ldots = \sum_{Reactants} \upsilon_i \Delta h^{\circ}_{f,i} \qquad [8.12a]$$

$$\Delta h^{\circ}_{Step\ 2} = \upsilon_C \Delta h^{\circ}_{f,C} + \upsilon_D \Delta h^{\circ}_{f,D} + \ldots = \sum_{Products} \upsilon_i \Delta h^{\circ}_{f,i} \qquad [8.12b]$$

Substitute of Equations [8.12a] and [8.12b] into Equation [8.11] yields:

$$\Delta h^{\circ}_{rxn} = \sum \upsilon_i \Delta h^{\circ}_{f,i} \qquad [8.13]$$

where the summation is over both reactants and products. Application of this expression is relatively straightforward—provided data for standard molar enthalpies of formation. Typically it is reasonable to assume the reaction enthalpy is relatively independent of temperature and pressure; only for large deviations in temperature from room temperature does the difference in heat capacity between reactants and products contribute significantly to Δh°_{rxn}.

Example 8.17: Predicting the Reaction Enthalpy for the Complete Combustion of Methane

The complete combustion of methane, the principal component of natural gas, occurs by the following reaction (with standard molar enthalpies of formation for each species at 298.15 K):

$$CH_4(g) \quad + \quad 2 \quad O_2(g) \xrightarrow{\Delta} CO_2(g) \quad + \quad 2 \quad H_2O(g) \qquad \text{Rxn [8.E6]}$$

Δh_f°	-74.6	0	-393.5	-285.8

(in kJ/mol)

How much heat (in kJ) is released per mole of methane consumed at 298.15 K?

Solution

Apply Hess's law formulated in terms of standard molar enthalpies of formation, Equation [8.13]:

$$\begin{aligned}\Delta h_{rxn}^{\circ} &= \upsilon_{CH_4}\Delta h_{f,CH_4(g)}^{\circ} + \upsilon_{O_2}\Delta h_{f,O_2(g)}^{\circ} + \upsilon_{CO_2}\Delta h_{f,CO_2(g)}^{\circ} + \upsilon_{H_2O}\Delta h_{f,H_2O(l)}^{\circ} \\ &= (-1)(-74.6 \text{ kJ/mol}) + (-2)(0 \text{ kJ/mol}) + (1)(-393.5 \text{ kJ/mol}) + (2)(-285.8 \text{ kJ/mol}) \\ \Delta h_{rxn}^{\circ} &= -890.5 \text{ kJ/mol}\end{aligned}$$

The reaction releases 890.5 kJ of heat per mole of CH_4 consumed.

Example 8.18: Predicting the Standard Molar Enthalpy of Formation for Liquid Ethanol

The complete combustion of liquid ethanol, an additive introducing oxygen into some blends of gasoline such that it burns cleaner (*e.g.*, "E85" fuel is 85% ethanol and 15% gasoline), was introduced in Section 7.1:

$$C_2H_5OH(l) + 3\ O_2(g) \xrightarrow{\Delta} 2\ CO_2(g) + 3\ H_2O(l) \qquad \text{Rxn [7.E1]}$$

If $\Delta h_{rxn}^{\circ}(298.15\text{ K}) = -1368.8$ kJ/mol for this reaction, use data for standard molar enthalpies of formation from Example 8.17 for $O_2(g)$, $CO_2(g)$, and $H_2O(l)$ to determine the standard molar enthalpy of formation of liquid ethanol (in kJ/mol).

Solutions

Apply Hess's law formulated in terms of standard molar enthalpies of formation, Equation [8.13], and rearrange to solve for the standard molar enthalpy of formation for liquid ethanol:

$$\Delta h^{\circ}_{rxn} = \upsilon_{C_2H_5OH}\Delta h^{\circ}_{f,C_2H_5OH(g)} + \upsilon_{O_2}\Delta h^{\circ}_{f,O_2(g)} + \upsilon_{CO_2}\Delta h^{\circ}_{f,CO_2(g)} + \upsilon_{H_2O}\Delta h^{\circ}_{f,H_2O(l)}$$

$$\upsilon_{C_2H_5OH}\Delta h^{\circ}_{f,C_2H_5OH(g)} = \Delta h^{\circ}_{rxn} - \upsilon_{CO_2}\Delta h^{\circ}_{f,CO_2(g)}\upsilon_{H_2O}\Delta h^{\circ}_{f,H_2O(l)}$$

$$\Delta h^{\circ}_{f,C_2H_5OH(g)} = \frac{\Delta h^{\circ}_{rxn} - \upsilon_{O_2}\Delta h^{\circ}_{f,O_2(g)} - \upsilon_{CO_2}\Delta h^{\circ}_{f,CO_2(g)}\upsilon_{H_2O}\Delta h^{\circ}_{f,H_2O(l)}}{\upsilon_{C_2H_5OH}}$$

$$= \frac{\begin{bmatrix}(-1368.8\text{ kJ/mol})-(-3)(0\text{ kJ/mol})-(2)(-393.5\text{ kJ/mol})\\ -(3)(-285.8\text{ kJ/mol})\end{bmatrix}}{(-1)}$$

$$\Delta h^{\circ}_{f,C_2H_5OH(g)} = -275.6\text{ kJ/mol}$$

Table 8.1 compares heats of combustion and fuel values for fuels we have considered in this chapter.

Table 8.1 Heats of Combustion and Fuel Values for Selected Fuels

Fuel	Δh°_{comb} (kJ/mol)	M (g/mol)	Fuel Value ($\lvert\Delta\hat{h}^{\circ}_{comb}\rvert$, in kJ/g)	State at 25 °C and 1 atm
Natural gas (CH_4)	-8.90×10^2	16	55.5	*g*
Ethanol (C_2H_5OH)	-1.37×10^3	46	29.8	*l*
Gasoline (C_8H_{18})	-5.47×10^3	114	47.9	*l*
Kerosene ($C_{12}H_{26}$)	-8.1×10^3	170	48	*l*
Biodiesel ($C_{18}H_{36}O_2$)	-1.17×10^4	284	41.1	*l*

8.8 The First Law of Thermodynamics for Open Systems

Many important processes involve not closed but rather open systems in which the flow of matter is important. For open systems with inflows and outflows of matter the concept of equilibrium is inapplicable; instead a corresponding concept of steady state is used—a state that does not change with time. For example, consider the generic schematic for an open system with one inlet stream and one outlet stream depicted in Figure 8.7, where $\dot{m}_{in}$ and $\dot{m}_{out}$ are the mass flow rate for the inlet and outlet, respectively. At steady state the law of conservation of matter—applied as a material balance—yields:

$$\frac{dm_{sys}}{dt} = \dot{m}_{in} - \dot{m}_{out} = 0 \qquad [8.14]$$

We find that the rate of flow of mass into the system matches the rate of flow of mass out of the system (*i.e.*, $\dot{m}_{in} = \dot{m}_{out}$). For this system the first law of thermodynamics can be expressed for steady state as:

$$\frac{d(U_{sys} + E_{k,sys} + E_{p,sys})}{dt} = \dot{m}_{in}\left(\hat{h}_{in} + \hat{e}_{k,in} + \hat{e}_{p,in}\right) - \dot{m}_{out}\left(\hat{h}_{out} + \hat{e}_{k,out} + \hat{e}_{p,out}\right) + \dot{Q} + \dot{W}_s = 0 \qquad [8.15]$$

where $\hat{h}_{in}$ and $\hat{h}_{out}$ are the specific enthalpies of the inlet and outlet, respectively, $\hat{e}_{k,in}$ and $\hat{e}_{k,out}$ are the specific macroscopic kinetic energies of the inlet and outlet, respectively, $\hat{e}_{p,in}$ and $\hat{e}_{p,out}$ are the specific macroscopic potential energies of the inlet and outlet, respectively, $\dot{Q}$ is the rate of heat transfer, and $\dot{W}_s$ is the rate of shaft work. The terms with specific enthalpy capture the combination of specific internal energy and flow work associated with each stream.

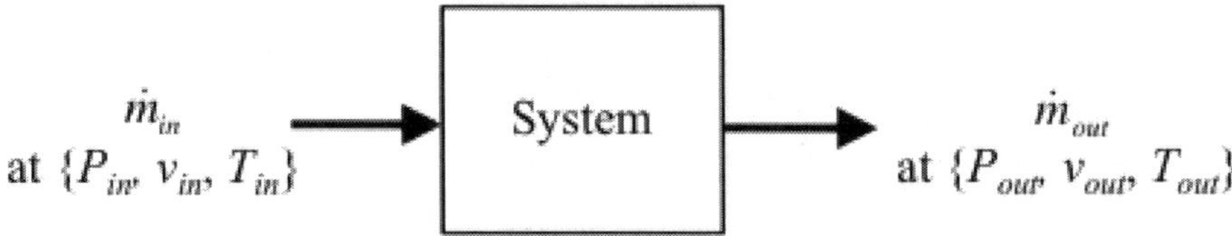

Figure 8.7. Schematic of an open system with a single inlet and outlet.

We can frequently neglect specific terms describing rates of specific mechanisms of energy transfer. Which terms can be neglected depends on the specific process operation, as outlined in Table 8.2. Further discussion of these devices is beyond our present scope, but interested readers are referred to an excellent text by M. D. Koretsky (*Engineering and Chemical Thermodynamics*, Hoboken, NJ: John Wiley and Sons, 2004).

Tab 8.2. Examples of Process Equipment for Application of Equations [8.14] and [8.15]

<table>
<tr><th>Process Equipment</th><th>Utility</th><th>Terms Neglected in Energy Balance</th></tr>
<tr><td>Turbines</td><td>Generate shaft power</td><td rowspan="3">Heat transfer</td></tr>
<tr><td>Compressors</td><td>Increase pressure of gases based on shaft work</td></tr>
<tr><td>Pumps</td><td>Increase macroscopic potential energy of exit streams based on shaft work</td></tr>
<tr><td>Heat exchangers</td><td>Exchange thermal energy between flow streams</td><td>Macroscopic kinetic energy
Macroscopic potential energy
Shaft work</td></tr>
<tr><td>Nozzles</td><td>Constrict flow to increase macroscopic kinetic energy</td><td rowspan="2">Macroscopic potential energy
Heat transfer
Shaft work</td></tr>
<tr><td>Diffusers</td><td>Dilate flow to decrease macroscopic kinetic energy</td></tr>
</table>

9:Chemical Kinetics

Learning Objectives

By the end of this chapter you should be able to:

1) Relate the rate(s) of disappearance of reactant(s) and rate(s) of appearance of product(s) for a reaction.

2) Deduce a rate law for a reaction by evaluating experimental measurements of initial rates as a function of initial concentrations.

3) Deduce a rate law for a reaction by evaluating experimental measurements of concentration versus time, and use an integrated rate law to predict how the concentrations evolve over time.

4) Use a reaction profile to relate the forward and reverse activation energies for a reaction.

5) Apply the Arrhenius model to explain how temperature and activation energy affect a rate constant.

6) Appreciate how complex rate laws arise as a consequence of multistep reaction mechanisms.

7) Describe different types of catalysts, and characterize their effect on reactions based on reaction mechanisms.

9.1 Relating Rates of Consumption of Reactants and Formation of Products

In Chapter 7 we identified that reactions frequently result in smaller amounts of products than predicted by stoichiometry, and that one of the reasons for this behavior is that reactions do not occur instantaneously—reactions occur with finite rates. There are two types of complimentary approaches that are insightful for describing the kinetics of a reaction. Phenomenological representations allow us to identify relationships between rates of consumption of reactants, rates of formation of products, and the concentrations of reactants and products as a function of time. Mechanistic representations allow us to explain how reactions actually occur in terms of atomic rearrangements, identifying how temperature and catalysts affect observed rates.

We begin by constructing a fundamental framework for describing rates of reaction, focusing as a guided example on a reaction that occurs in air polluted by the emission of nitrogen-bearing products from internal combustion engines:

$$2\ N_2O_5(g) \rightarrow 4\ NO_2(g) + O_2(g) \qquad \text{Rxn [9.1]}$$

If the volume is constant and the reaction proceeds only toward the products, we expect that the concentration of the reactant N_2O_5 decreases and the concentrations of the products NO_2 and O_2 increase over time. We define the rate at which the concentration of a species *i*—regardless of whether the species is a reactant or product—changes using differential calculus as:

$$r_i \equiv \lim_{\Delta t \to 0} \frac{\Delta C_i}{\Delta t} = \frac{dC_i}{dt} \qquad [9.1]$$

where the rates are evaluated in units of moles per liter per unit time. Applying this definition to Rxn [9.1] we find $r_{N_2O_5} = \frac{dC_{N_2O_5}}{dt}$, $r_{NO_2} = \frac{dC_{NO_2}}{dt}$, and $r_{O_2} = \frac{dC_{O_2}}{dt}$. We observe that the rates for the reactant N_2O_5 are negative (*i.e.*, $r_{N_2O_5} < 0$) because the concentration of N_2O_5 decreases over time and that the rates for the products NO_2 and O_2 are positive (*i.e.*, $r_{NO_2} > 0$ and $r_{O_2} > 0$) because the concentrations of NO_2 and O_2 increase over time. We also suspect that the rate of consumption of N_2O_5 and rates of formation of NO_2 and O_2 are related: the stoichiometry tells us that two moles of N_2O_5 are consumed for every four moles of NO_2 and one mole of O_2 produced.

Example 9.1: Relating Rates of Individual Species for Rxn [9.1]

If $r_{N_2O_5} = -2.0 \times 10^{-4}$ M/s, determine r_{NO_2} and r_{O_2} (in M/s) according to Rxn [9.1].

Solutions

Apply the stoichiometry of Rxn [9.1] with the convention for the stoichiometric coefficients defined in Section 8.7:

$$-2.0 \times 10^{-4} \frac{\text{mol } N_2O_5}{\text{L - s}} \times \frac{4 \text{ mol } NO_2}{-2 \text{ mol } N_2O_5} = 4.0 \times 10^{-4} \frac{\text{mol } NO_2}{\text{L - s}} \qquad [9.E1a]$$

$$-2.0 \times 10^{-4} \frac{\text{mol } N_2O_5}{\text{L - s}} \times \frac{1 \text{ mol } O_2}{-2 \text{ mol } N_2O_5} = 1.0 \times 10^{-4} \frac{\text{M } O_2}{\text{s}} \qquad [9.E1b]$$

Note that we can rearrange Equations [9.E1a] and [9.E1b] as:

$$\frac{-2.0 \times 10^{-4} \frac{\text{M}}{\text{s}} N_2O_5}{-2 \text{ mol } N_2O_5} = \frac{4.0 \times 10^{-4} \frac{\text{M}}{\text{s}} NO_2}{4 \text{ mol } NO_2} = \frac{1.0 \times 10^{-4} \frac{\text{M}}{\text{s}} O_2}{1 \text{ mol } O_2} \qquad [9.2a]$$

$$\frac{r_{N_2O_5}}{\upsilon_{N_2O_5}} = \frac{r_{NO_2}}{\upsilon_{NO_2}} = \frac{r_{O_2}}{\upsilon_{O_2}} \qquad [9.2b]$$

where u_i is the stoichiometric coefficient for species *i*, with $u_i < 0$ for the reactant N_2O_5 and $u_i > 0$ for the products NO_2 and O_2. Now consider how we can apply our conclusions to the generalized reaction structure introduced in Section 8.7:

$$\upsilon_A \text{A} + \upsilon_B \text{B} + \ldots \rightarrow \upsilon_C \text{C} + \upsilon_D \text{D} + \ldots \qquad \text{Rxn [7.3]}$$

We generalize Equation [9.2b] to provide a generic relationship between rates for reactants and products for Rxn [7.3] as:

$$r_{overall} = \frac{r_A}{\upsilon_A} = \frac{r_B}{\upsilon_B} = \frac{r_C}{\upsilon_C} = \frac{r_D}{\upsilon_D} = \ldots \qquad [9.3]$$

where the "overall" rate of reaction $r_{overall} > 0$.

We evaluate rates experimentally based on measuring the concentration of one or more participating species as a function of time. Our choice of which technique we use to make these measurements depends on the properties of the species we examine. For example, does one species absorb light at a particular wavelength not absorbed by other species? Does this species exhibit fluorescence (*i.e.*, does the species absorb light at one wavelength and emit light at a longer wavelength) that is distinct from other species present? How long does it take to make a single measurement, and is this technique fast enough to capture sufficient details of the dynamics? In particular, we seek a technique that allows us to make measurements over short enough time intervals such that we can approximate the derivative in the definition for the rate as $r_i \cong \frac{\Delta C_i}{\Delta t}$.

Once we have collected a set of data for concentration as a function of time, the preferable strategy for approximating the rate is based on a three-point method. Consider a set of $n + 1$ samples for the concentration of a reactant A, $\{C_{A,0}, C_{A,1}, C_{A,2}, \ldots, C_{A,n-1}, C_{A,n}\}$, measured at the corresponding set of times $\{t_0, t_1, t_2, \ldots, t_{n-1}, t_n\}$. Three members of these matched sets are depicted in Figure 9.1 for $0 < j < n$. We approximate the rate of change in the concentration of A at a time t_j for this set as the average of the slopes of the lines coming into and out of the point $(t_j, C_{A,j})$ using the three points $(t_{j-1}, C_{A,j-1})$, $(t_j, C_{A,j})$, and $(t_{j+1}, C_{A,j+1})$:

$$\left(\frac{dC_A}{dt}\right)_{0<j<n} \cong \frac{1}{2}\left[\left(\frac{C_{A,j} - C_{A,j-1}}{t_j - t_{j-1}}\right) + \left(\frac{C_{A,j+1} - C_{A,j}}{t_{j+1} - t_j}\right)\right] \qquad [9.4a]$$

For the end points $(t_0, C_{A,0})$ and $(t_n, C_{A,n})$ we use a two-point approximation for the slope and rate:

$$\left(\frac{dC_A}{dt}\right)_{j=0} \cong \frac{C_{A,1} - C_{A,0}}{t_1 - t_0} \qquad [9.4b]$$

$$\left(\frac{dC_A}{dt}\right)_{j=n} \cong \frac{C_{A,n} - C_{A,n-1}}{t_n - t_{n-1}} \qquad [9.4c]$$

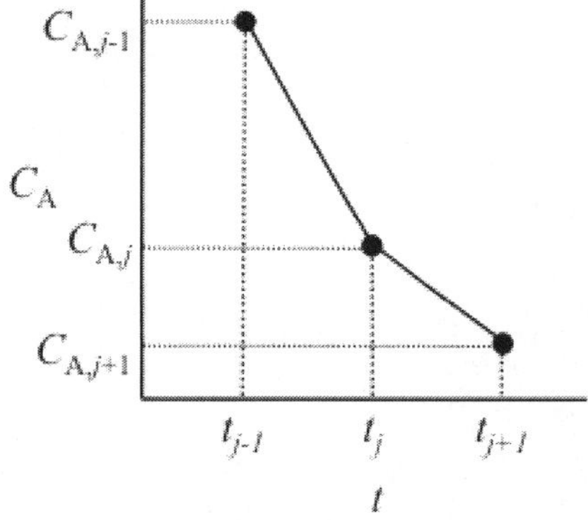

Figure 9.1. Evaluating the rate for a reactant A based on the three-point method.

Frequently, we find it convenient to perform our measurements of concentration using equally spaced time intervals Δt, such that measurements are made at $t_0 = 0$, $t_1 = \Delta t$, $t_2 = 2\Delta t$, $t_3 = 3\Delta t$, ...). For such equally spaced time intervals, Equations [9.4a], [9.4b], and [9.4c] simplify to:

$$\left(\frac{dC_A}{dt}\right)_{0<j<n} \cong \frac{C_{A,j+1} - C_{A,j-1}}{2\Delta t} \qquad [9.5a]$$

$$\left(\frac{dC_A}{dt}\right)_{j=0} \cong \frac{C_{A,1} - C_{A,0}}{\Delta t} \qquad [9.5b]$$

$$\left(\frac{dC_A}{dt}\right)_{j=n} \cong \frac{C_{A,n} - C_{A,n-1}}{\Delta t} \qquad [9.5c]$$

Example 9.2: Kinetics of Decomposition of NO2

NO_2 is a brown-colored gas primarily responsible for the smog associated with polluted air, and it decomposes according to the reaction:

$$2\ NO_2(g) \rightarrow 2\ NO(g) + O_2(g) \qquad \text{Rxn [9.E1]}$$

Determine the rate (in mM/s) at which NO_2 decomposes based on the data below, obtained at 383 °C.

t (s)	$[NO_2]$ (mM)
0.0	100.0
5.0	17.0
10.0	9.0
15.0	6.2
20.0	4.7

Solutions

As the data were obtained with equal time intervals of $\Delta t = 5.0$ s, apply Equations [9.5a], [9.5b], and [9.5c]:

$$r_{NO_2,t=0\,s} = \left(\frac{dC_{NO_2}}{dt}\right)_{t=0\,s} \cong \frac{C_{NO_2,t=5\,s} - C_{NO_2,t=0\,s}}{\Delta t} = \frac{(17.0\text{ mM}) - (100.0\text{ mM})}{(5.0\text{ s})} = -1.7\times10^{1}\text{ mM/s}$$

$$r_{NO_2,t=5\,s} = \left(\frac{dC_{NO_2}}{dt}\right)_{t=5\,s} \cong \frac{C_{NO_2,t=10\,s} - C_{NO_2,t=0\,s}}{2\Delta t} = \frac{(9.0\text{ mM}) - (100.0\text{ mM})}{2(5.0\text{ s})} = -9.1\text{ mM/s}$$

$$r_{NO_2,t=10\,s} = \left(\frac{dC_{NO_2}}{dt}\right)_{t=10\,s} \cong \frac{C_{NO_2,t=15\,s} - C_{NO_2,t=5\,s}}{2\Delta t} = \frac{(6.2\text{ mM}) - (17.0\text{ mM})}{2(5.0\text{ s})} = -1.1\text{ mM/s}$$

$$r_{NO_2,t=15\,s} = \left(\frac{dC_{NO_2}}{dt}\right)_{t=15\,s} \cong \frac{C_{NO_2,t=20\,s} - C_{NO_2,t=10\,s}}{2\Delta t} = \frac{(4.7\text{ mM}) - (9.0\text{ mM})}{2(5.0\text{ s})} = -4.3\times10^{-1}\text{ mM/s}$$

$$r_{NO_2,t=20\,s} = \left(\frac{dC_{NO_2}}{dt}\right)_{t=20\,s} \cong \frac{C_{NO_2,t=20\,s} - C_{NO_2,t=15\,s}}{\Delta t} = \frac{(4.7\text{ mM}) - (6.2\text{ mM})}{(5.0\text{ s})} = -3.0\times10^{-1}\text{ mM/s}$$

We observe the rate is not constant, which can also be seen graphically from a plot of C_{NO_2} *vs. t*.

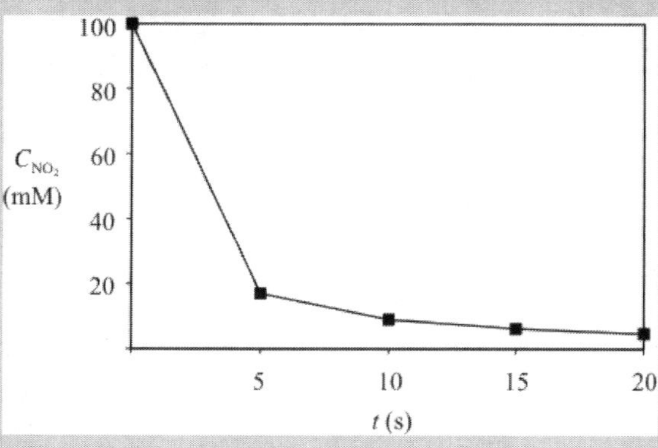

Example 9.3: Kinetics of Decomposition of Hydrogen Peroxide

Consider the following data for the decomposition of hydrogen peroxide:

$H_2O_2(aq) \rightarrow H_2O(l) + ½\, O_2(g)$ Rxn [9.E2]

t (s)	$[H_2O_2]$ (M)
0.	2.32
200.	2.01
400.	1.72
600.	1.49
1200.	0.98
1800.	0.62
3000.	0.25

a) What is the rate of reaction for H_2O_2 (in M/s) at 1800. s?

b) What is the rate of reaction for O_2 (in M/s) at 1800. s?

<u>Solutions</u>

a) As the data were obtained with unequal time intervals, apply Equation [9.4a]:

$$r_{H_2O_2,t=1800\,s} = \left(\frac{dC_{H_2O_2}}{dt}\right)_{t=1800\,s} \cong \frac{1}{2}\left[\left(\frac{C_{H_2O_2,t=1800\,s} - C_{H_2O_2,t=1200\,s}}{t_{t=1800\,s} - t_{t=1200\,s}}\right) + \left(\frac{C_{H_2O_2,t=3000\,s} - C_{H_2O_2,t=1800\,s}}{t_{t=3000\,s} - t_{t=1800\,s}}\right)\right]$$

$$= \frac{1}{2}\left[\left[\frac{(0.62\text{ M}) - (0.98\text{ M})}{(1800.\text{ s}) - (1200.\text{ s})}\right] + \left[\frac{(0.25\text{ M}) - (0.62\text{ M})}{(3000.\text{ s}) - (1800.\text{ s})}\right]\right]$$

$$r_{H_2O_2,t=1800\,s} = -4.54 \times 10^{-4}\text{ M/s}$$

For Rxn [9.E2] further analysis also leads to the conclusion that the rate is not constant, as seen graphically from a plot of $C_{H_2O_2}$ *vs. t.*

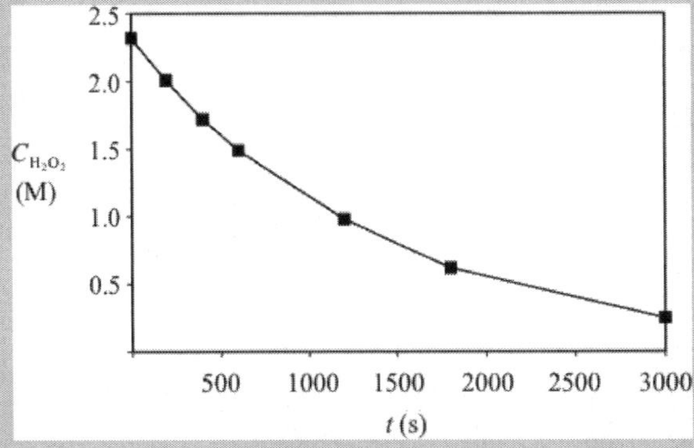

b) Apply Equation [9.3] to relate the rates for H_2O_2 and O_2 with $\upsilon_{H_2O_2} = -1$ and $\upsilon_{O_2} = +\frac{1}{2}$:

$$\frac{r_{H_2O_2,t=1800\,s}}{\upsilon_{H_2O_2}} = \frac{r_{O_2,t=1800\,s}}{\upsilon_{O_2}}$$

$$r_{O_2,t=1800\,s} = \left(\frac{\upsilon_{O_2}}{\upsilon_{H_2O_2}}\right) r_{H_2O_2,t=1800\,s} = \frac{\left(\frac{1}{2}\right)(-4.54 \times 10^{-4}\text{ M/s})}{(-1)} = +2.27 \times 10^{-4}\text{ M/s}$$

9.2 Rate Laws and the Method of Initial Rates

When we consider data for experimental measurements of rates of reaction, we typically observe that the rate of reaction depends on composition and temperature. This functionality is termed the rate law. Rate laws commonly exhibit a power-law form, with the rate proportional to the product of the concentrations of reactants and products, with the concentration of each species raised to a power:

$$r_{overall} = kC_A^{m_A} C_B^{m_B} \ldots C_C^{m_C} C_D^{m_D} \ldots \qquad [9.6]$$

where the proportionality constant k is termed the rate constant and is a positively valued function of temperature and not composition, and the power m_i is defined as the reaction order for species i. The sum of these powers, $m = \sum_i m_i$, is defined as the overall reaction order. Although in general reaction order may be any real number, we frequently encounter power-law rate laws for which only the concentrations of the reactants appear in the rate law, and the reaction order for each reactant is either $m_i = 0$, corresponding to zeroth order, $m_i = 1$, corresponding to first order, or $m_i = 2$, corresponding to second order. Rate laws for which the concentration of one or more products appear with positive reaction orders correspond to auto-catalytic reactions. In contrast, if a product has a negatively-valued reaction order, the product is inhibitory. Note that the units for the rate constant can be determined from the overall reaction order as $[k] = M^{1-m}$-time^{-1}.

One relatively simple strategy for determining reaction orders and the rate constant for a rate law with a power-law form is the method of initial rates, which is based on measuring how the initial rate of reaction at $t = 0$ (denoted with the subscript "o") varies with the initial concentration of reactants. Consider that a log-log plot of an initial rate for a power-law rate law as a function of the initial concentration of a species A, $C_{A,o} \equiv C_A(t = 0)$, should be linear with a slope equal to the reaction order for A:

$$\ln r_{overall,o} = \ln(kC_{A,o}^{m_A} C_{B,o}^{m_B} \ldots) = m_A \ln C_{A,o} + \ln(kC_{B,o}^{m_B} \ldots) \quad [9.7]$$

where $C_{B,o} \equiv C_B(t = 0)$ analogously. Applying this strategy separately to each species that contributes to the rate law allows independent evaluation of the reaction order for each species. The rate constant can then be determined from the intercept ln k of a log-log plot of $\ln(r_{overall,o})$ *vs.* $\ln(C_{A,o}^{m_A} C_{B,o}^{m_B} \ldots)$, where this plot should have a slope of 1. In practice, the method of initial rates is commonly applied by considering ratios of initial rates for pairs of data with fixed initial concentrations for all species but one.

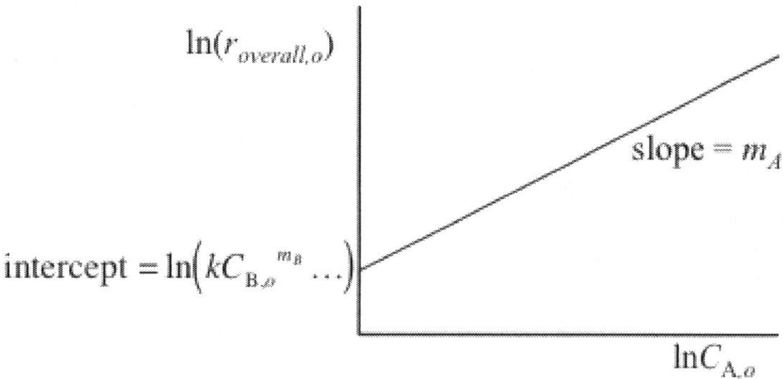

Figure 9.2. Graphical form for Method of Initial Rates.

Example 9.4: Applying the Method of Initial Rates to Matched Pairs of Data

A rate law of the form $(-r_A) = kC_A^{m_A} C_B^{m_B}$ is postulated for the reaction A + B → products and the following data obtained:

Experiment	$C_{A,o}$ (M)	$C_{B,o}$ (M)	$(-r_A)_o$ (M/s)
1	0.200	0.050	2.5×10^{-4}
2	0.400	0.100	2.0×10^{-3}
3	0.200	0.100	1.0×10^{-3}

Determine the reaction orders for A and B, the overall reaction order, and the value and units for the rate constant.

Solutions

The rate law based on initial rates can be formulated as $(-r_A)_o = kC_{A,o}^{m_A}C_{B,o}^{m_B}$. Determine the reaction orders for A and B separately by comparing sets of data for which the concentration of either A or B varies. Start by comparing the data for Experiments 2 and 3, for which the initial concentrations of B are identical:

$$\frac{(-r_A)_{o,2}}{(-r_A)_{o,3}} = \frac{(2.0\times10^{-3}\text{ M/s})}{(1.0\times10^{-3}\text{ M/s})} = 2 = \frac{k(C_{A,o})_2^{m_A}(C_{B,o})_2^{m_B}}{k(C_{A,o})_3^{m_A}(C_{B,o})_3^{m_B}} = \left(\frac{(C_{A,o})_2}{(C_{A,o})_3}\right)^{m_A} = \left[\frac{(0.400\text{ M})}{(0.200\text{ M})}\right]^{m_A} = 2^{m_A}$$

$$m_A = 1$$

Now compare the data for Experiments 1 and 3, for which the initial concentrations of A are identical:

$$\frac{(-r_A)_{o,1}}{(-r_A)_{o,3}} = \frac{(2.5\times10^{-4}\text{ M/s})}{(1.0\times10^{-3}\text{ M/s})} = \frac{1}{4} = \frac{k(C_{A,o})_1(C_{B,o})_1^{m_B}}{k(C_{A,o})_3(C_{B,o})_3^{m_B}} = \left(\frac{(C_{B,o})_1}{(C_{B,o})_3}\right)^{m_B} = \left[\frac{(0.050\text{ M})}{(0.100\text{ M})}\right]^{m_B} = \left(\frac{1}{2}\right)^{m_B}$$

$$m_B = 2$$

Thus, the reaction is first order in A, second order in B, and (with $m = m_A + m_B = 3$) third order overall. The rate constant can then be determined from any of the experiments (and most properly by using data from all three experiments!) as:

$$(-r_A)_{o,i} = k(C_{A,o})_i(C_{B,o})_i^2$$

$$k = \frac{(-r_A)_{o,i}}{(C_{A,o})_i(C_{B,o})_i^2} = \frac{(-r_A)_{o,1}}{(C_{A,o})_1(C_{B,o})_1^2} = \frac{(2.5\times10^{-4}\text{ M/s})}{(0.200\text{ M})(0.050\text{ M})^2} = 0.50\text{ M}^{-2}\text{s}^{-1}$$

9.3 Integrated Rate Laws

Although the method of initial rates can provide insight into reaction orders for a rate law, it does not provide us explicit predictions for how the concentration of a reactant or product changes over time. Further, it is typically not the most reliable approach for quantitatively determining values for rate constants. The integral method based on applying integral calculus to integrate a proposed power-law rate law offers a powerful alternative that addresses both of these issues. We focus on reactions for which the rate depends solely on the concentration of a single reactant, A, such that the rate law is $(-r_A) = kC_A^{m_A}$. The concentration of A evolves over time as:

$$r_A = \frac{dC_A}{dt} = -kC_A^{m_A} \qquad [9.8]$$

with initial concentration $C_{A,o}$. Rearrange the rate law and integrate from initial conditions ($t = 0$, $C_{A,o}$) to (t, C_A):

$$\frac{dC_A}{C_A^{m_A}} = -kdt \qquad [9.9a]$$

$$\int_{C_{A,o}}^{C_A} \frac{dC_A}{C_A^{m_A}} = \int_0^t -kdt = -kt \qquad [9.9b]$$

Further evaluation of the remaining integral in Equation [9.9b] depends on the value for reaction order m_A. If $m_A = 1$, corresponding to a first-order reaction with respect to A, we find:

$$\int_{C_{A,o}}^{C_A} \frac{dC_A}{C_A} = [\ln C_A]_{C_{A,o}}^{C_A} = \ln C_A - \ln C_{A,o} = -kt \qquad [9.10a]$$

$$\ln C_A = \ln C_{A,o} - kt \qquad [9.10b]$$

However, if $m_A \neq 1$, corresponding to reaction that is not first order with respect to A, we find:

$$\int_{C_{A,o}}^{C_A} C_A^{-m_A} dC_A = \left[\frac{C_A^{-m_A+1}}{-m_A+1}\right]_{C_{A,o}}^{C_A} = \frac{C_A^{1-m_A} - C_{A,o}^{1-m_A}}{1-m_A} = -kt \qquad [9.11a]$$

$$C_A^{1-m_A} = C_{A,o}^{1-m_A} + (m_A - 1)kt \qquad [9.11b]$$

Equations [9.10b] and [9.11b] are called integrated rate law and suggest that a plot of $\ln C_A$ *vs.* t should be linear, with a slope equal to the negative of the rate constant, for a reaction that is not first order with respect to A (Figure 9.3a), and that a plot of $C_A^{1-m_A}$ *vs.* t should be linear, with a slope that is proportional to the rate constant, for a reaction that is not first order with respect to A (Figure 9.3b). Linearity is demonstrated best using rigorous statistical techniques.

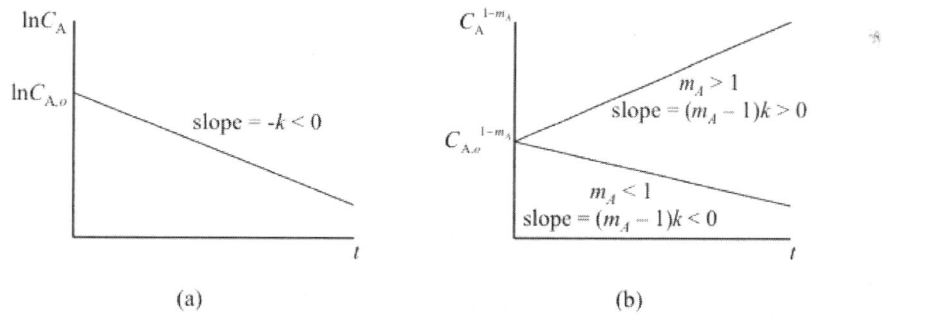

Figure 9.3. Linearized plots for integrated rate laws for power-law rate laws depending only on the concentration of a single reactant A. (a) Reaction is first-order with respect to A; (b) reaction is not first-order with respect to A.

Let's compare the behaviors predicted for the common cases of zeroth-order (*i.e.*, $m_A = 0$), first-order (*i.e.*, $m_A = 1$), and second-order (*i.e.*, $m_A = 2$) reactions. When $m_A = 0$, corresponding to a rate that is independent of the concentration of A (*i.e.*, the rate is constant) and a rate constant with units of moles per liter per unit time, we find upon substitution into Equation [9.11b]:

$$C_A = C_{A,o} - kt \qquad [9.12]$$

such that the concentration of A decreases linearly with time (Figure 9.4a), with the rate constant equal to the negative of the slope. We find that the concentration of the reactant decreases linearly with time, with the reactant disappearing at a finite time, t_{max}:

$$C_A(t = t_{max}) = C_{A,o} - kt_{max} = 0 \quad [9.13a]$$

$$t_{max} = \frac{C_{A,o}}{k} \quad [9.13b]$$

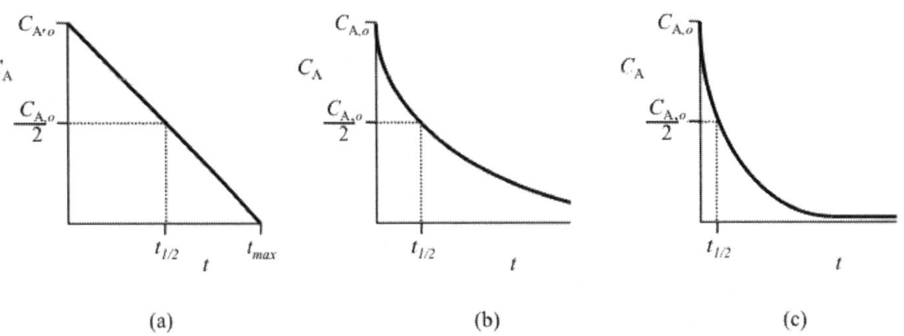

Figure 9.4. Plots of concentration of reactant A vs. time for power-law rate laws depending only on the concentration of a single reactant A. (a) zeroth-order; (b) first-order; (c) second-order.

We also can determine a time—defined as the half-life, $t_{½}$—at which the concentration of the reactant is half its initial value:

$$C_A(t = t_{1/2}) = C_{A,o} - kt_{1/2} = \frac{C_{A,o}}{2} \quad [9.14a]$$

$$t_{1/2} = \frac{C_{A,o}}{2k} \quad [9.14b]$$

The half-life for a zeroth-order reaction is proportional to the initial concentration of the reactant.

When $m_A = 1$, corresponding to a rate that is proportional to the concentration of A (*i.e.*, the rate decreases as the concentration of the reactant A decreases) and a rate constant with units of reciprocal time, we can rearrange Equation [9.10b] as:

$$C_A = C_{A,o}e^{-kt} \quad [9.15]$$

such that a plot of concentration of A versus time is nonlinear (Figure 9.4b) and exhibits exponential decay with time (*i.e.*, the concentrations decreases exponentially with time). Note that it takes an infinite amount of time for the concentration of reactant to decrease to zero—the reactant never completely disappears! We determine the half-life as:

$$C_A(t = t_{1/2}) = \frac{C_{A,o}}{2} = C_{A,o}e^{-kt/} \quad [9.16a]$$

$$\frac{1}{2} = e^{-kt/} \quad [9.16b]$$

$$\ln\left(\frac{1}{2}\right) = -\ln 2 = -kt_{1/2} \quad [9.16c]$$

$$t_{1/2} = \frac{\ln 2}{k} \quad [9.16d]$$

The half-life for a first-order reaction is independent of the initial concentration of the reactant.

When $m_A = 2$, corresponding to a rate that is proportional to the square of the concentration of A (*i.e.*, the rate decreases strongly as the concentration of the reactant A decreases) and a rate constant with units of liter per mole per unit time, we find upon substitution into Equation [9.11b]:

$$\frac{1}{C_A} = \frac{1}{C_{A,o}} + kt \tag{9.17}$$

such that a plot of concentration of A versus time is highly nonlinear (Figure 9.4c). Note that a plot of the reciprocal of the concentration of the reactant increases linearly with time, with a slope equal to the rate constant, and (as we observed for first-order kinetics) it takes an infinite amount of time for the concentration of reactant to decrease to zero. We determine the half-life as:

$$\frac{1}{C_A(t = t_{1/2})} = \frac{1}{\left(\frac{C_{A,o}}{2}\right)} = \frac{2}{C_{A,o}} = \frac{1}{C_{A,o}} + kt_{1/2} \tag{9.18a}$$

$$\frac{1}{C_{A,o}} = kt_{1/2} \tag{9.18b}$$

$$t_{1/2} = \frac{1}{kC_{A,o}} \tag{9.18c}$$

The half-life for a first-order reaction is inversely proportional to the initial concentration of the reactant.

Example 9.5: Identifying the Reaction Order for the Decomposition of Hydrogen Peroxide

The decomposition of hydrogen peroxide, Rxn [9.E2], obeys power-law kinetics that depend only on the concentration of the reactant. We saw in Example 9.3 that a plot of the concentration of hydrogen peroxide versus time was nonlinear. Plots of the data based on Equations [9.10b] and [9.17] are provided below on the top and bottom, respectively. What is the reaction order for hydrogen peroxide in Rxn [9.E2] and the value and units for the rate constant?

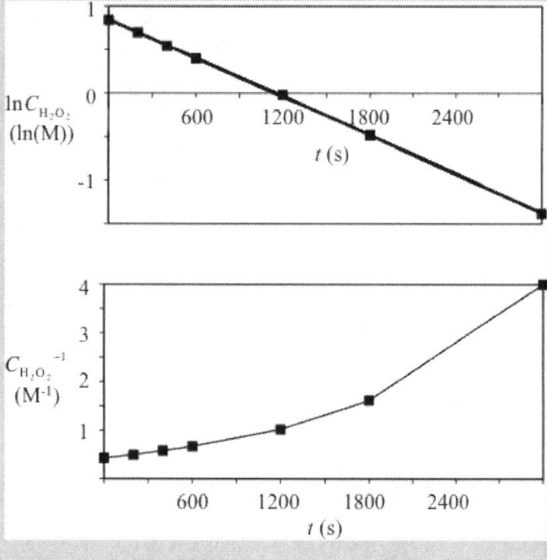

Solutions

We identify that the reaction order for H_2O_2 is not 0-based on the nonlinearity of the plot of $C_{H_2O_2}$ *vs.* t from Example 9.3, and not 1-based on the nonlinearity of the plot of $C_{H_2O_2}^{-1}$ *vs.* t. But the plot of $\ln C_{H_2O_2}$ *vs.* t is linear, such that the reaction order for H_2O_2 is 1. A linear regression of the data for this plot identifies a slope of -7.41×10^{-4} s^{-1}, such that the rate constant is $k = 7.41\times10^{-4}$ s^{-1}.

Example 9.6: Kinetics of Chemical Vapor Deposition of Titanium Nitride

Thin films of titanium nitride (TiN) provide wear-resistant coatings on tools and activated surfaces for growth of copper films for ultra large-scale integration (ULSI) applications for the manufacture of computer chips. These films of TiN are deposited by a technique known as chemical vapor deposition (CVD). Deposition occurs as TiN(g) $\rightarrow$ film at relatively high concentrations of TiN(g) and obeys power-law kinetics with respect to the concentration of TiN(g) with a rate constant of 2.00×10^{-4} M/s. If the initial concentration of TiN(g) is 5.00 mM:

a) What is the half-life (in s) for the reaction?

b) What is the concentration (in mM) of TiN(g) after 10.0 s?

Solutions

a) We identify the overall reaction order as zeroth-order based on the units for the rate constant, such that the half-life depends on the initial concentration of TiN(g) according to Equation [9.14b]:

$$t_{1/2} = \frac{C_{\text{TiN},o}}{2k} = \frac{(5.00 \text{ mM})}{2\left(2.00\times10^{-4} \text{ M/s}\times\frac{1 \text{ mM}}{10^{-3} \text{ M}}\right)} = 12.5 \text{ s}$$

b) Apply Equation [9.12], the integrated rate law for a zeroth-order reaction with $[\text{TiN}]_o = 5.00 \text{ mM} = 5.00\times10^{-3} \text{ M}$:

$$C_{\text{TiN}} = C_{\text{TiN},o} - kt = (5.00 \text{ mM}) - \left(2.00\times10^{-4} \text{ M/s}\times\frac{1 \text{ mM}}{10^{-3} \text{ M}}\right)(10.0 \text{ s}) = 3.00 \text{ mM}$$

Example 9.7: Kinetics of Interfacial Polymerization of Nylon 6,10

The interfacial polymerization of nylon 6,10, introduced as Rxn [7.1] in Chapter 7, is first-order with respect to each monomer and obeys overall second-order kinetics. When the initial concentrations of the dichloride and diamine are identical, the kinetics are identical to second-order kinetics with respect to a single reactant, A, and the average degree of polymerization can be determined as $n_{ave} \equiv \frac{C_{A,o}}{C_A}$. If the initial concentration of monomer (the reactant) is 5.00 M and the rate constant is 1.00×10^2 $M^{-1}min^{-1}$, at what time (in min) is the average degree of polymerization 2.50×10^3?

Solutions

The integrated rate law for second-order kinetics, Equation [9.17], tells us that the concentration of monomer varies with time as $\frac{1}{C_A} = \frac{1}{C_{A,o}} + kt$, such that the average degree of polymerization is:

$$n_{ave} = \frac{C_{A,o}}{C_A} = 1 + kC_{A,o}t$$

Rearranging for t yields:

$$kC_{A,o}t = n_{ave} - 1$$

$$t = \frac{n_{ave} - 1}{kC_{A,o}} = \frac{(2.50\times10^3) - 1}{(1.00\times10^2\ \text{M}^{-1}\text{min}^{-1})(5.00\ \text{M})} = 5.00\ \text{min}$$

9.4 Activation Energy and the Arrhenius Model for the Effects of Temperature on Rate Constants

In general, any reaction can be considered reversible (*i.e.*, occur in both the forward and reverse directions), a concept we referenced in Chapter 7 and will examine in detail in Chapter 10. We represent the energetics for a reaction with a reaction profile, a map of the energy changes that occur during the reaction. For example, Figure 9.5 illustrates the shape of a reaction profile for an exothermic reaction involving no detectable intermediates; a reaction profile for a similar but endothermic reaction would be similar except the product(s) would have higher energy than the reactant(s). The ordinate, termed the reaction coordinate, represents the progress of the reaction and can be conceived as a map of the positions of the atoms exchanged between reactant(s) and products. The transition state, denoted commonly with the symbol ‡, is the highest energy state as the reaction proceeds, occurring as a transient intermediate in which the reactant(s) is/are partially rearranged to product(s).

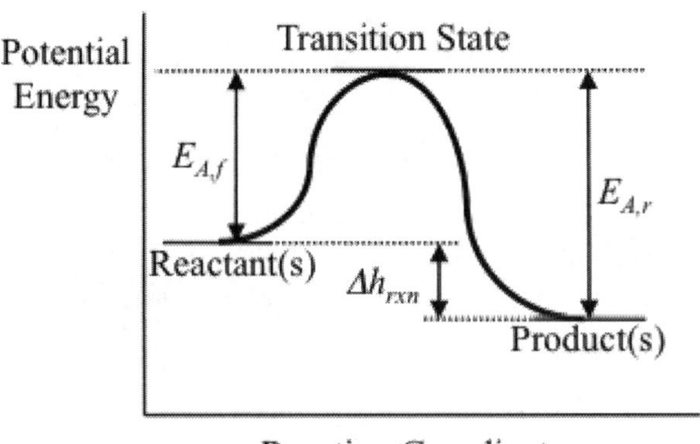

Figure 9.5. Reaction profile for an exothermic reaction with no detectable intermediates.

For every reaction we identify two positively valued activation energies: an activation energy $E_{A,f}$ in the forward reaction, representing the difference between the potential energy of the transition state and the reactant(s), and an activation energy $E_{A,r}$ in the reverse reaction, representing the difference between the potential energy of the transition state and the product(s).

The difference between the forward and reverse activation energies is the molar change in enthalpy for the reaction:

$$\Delta h_{rxn} = E_{A,f} - E_{A,r} \tag{9.19}$$

We can generalize whether a reaction is exothermic or endothermic based on the magnitude of the activation energies for the forward and reverse reactions: for exothermic reactions $E_{A,f} < E_{A,r}$ (the reaction is net "downhill" on a reaction profile), and for endothermic reactions $E_{A,f} > E_{A,r}$ (the reaction is net "uphill" on a reaction profile).

We commonly observe that reactions occur more rapidly as the temperature increases (think of whether milk sours more rapidly in the refrigerator or on a countertop on a warm summer day). The rate constant characterizes the effects of temperature on reaction rates and can be described as the product of the frequency that a reaction event (*e.g.*, the collision of two reactants or the over-stretching of a bond) occurs and the fraction of these events with sufficient energy, $E^\star$, to overcome the activation energy, E_A. Only events with $E^\star \geq E_A$ should result in a reaction. Based on a Boltzmann distribution the probability that $E^\star \geq E_A$ is proportional to $e^{-\frac{E_A}{RT}}$, such that we expect the rate constant to be proportional to $e^{-\frac{E_A}{RT}}$, consistent with Arrhenius' proposal that the rate constant for a forward or reverse reaction depends on the activation energy and absolute temperature as:

$$k = Ae^{-\frac{E_A}{RT}} \tag{9.20}$$

where the parameter A is a constant (termed the Arrhenius constant) that accounts for the stability of the transition state and any orientational effects.

Equation [9.20] can be rearranged by taking the natural log of both sides:

$$\ln k = \ln A - \frac{E_A}{R}\left(\frac{1}{T}\right) \tag{9.21}$$

Equation [9.21] predicts a linear relationship between the natural log of the rate constant and the inverse of temperature, as is frequently observed experimentally (Figure 9.6); however, we typically encounter problems for which we do not need to know the value for the Arrhenius constant but rather need to compare values for the rate constant at two different temperatures. Consider that if $k_1 = k(T_1)$ and $k_2 = k(T_2)$ for a reaction:

$$\ln k_2 - \ln k_1 = \left(\ln A - \frac{E_A}{RT_2}\right) - \left(\ln A - \frac{E_A}{RT_1}\right) = -\frac{E_A}{R}\left(\frac{1}{T_2} - \frac{1}{T_1}\right) \quad [9.22]$$

which can be rearranged as the Arrhenius equation:

$$\ln\left(\frac{k_2}{k_1}\right) = -\frac{E_A}{R}\left(\frac{1}{T_2} - \frac{1}{T_1}\right) \quad [9.23]$$

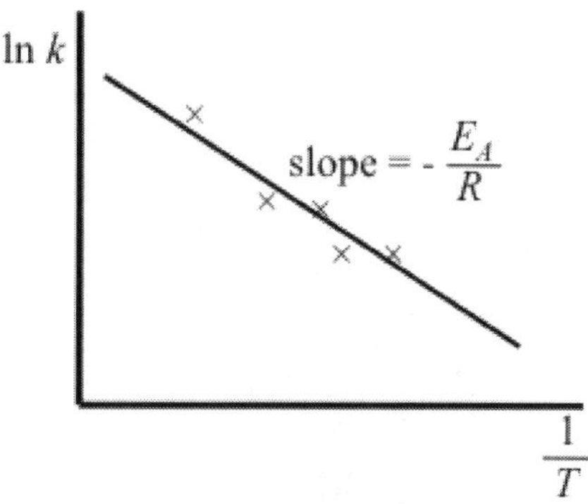

Figure 9.6. Arrhenius plot of lnk vs. T-1. The symbol "x" depicts experimental data.

When we apply the Arrhenius equation we express the activation energy in units of J/mol, use the gas constant expressed as 8.314 J/mol-K, and express temperature in units of K.

Example 9.8: Applying the Arrhenius Model to Interpret Experimental Observations

Rate constants of 2.3×10^{-4} s^{-1} and 4.6×10^{-4} s^{-1} were measured for a reaction at 20.0 °C and 30.0 °C, respectively.

a) What is the overall order for the reaction?

b) What is the activation energy, E_a (in kJ/mol), for the reaction?

c) What is the rate constant at 40.0 °C?

Solutions

a) Because the rate constants have units of reciprocal time, the reaction is first-order overall.

b) Assign T_1 = 20.0 °C = 293.2 K, with its corresponding rate constant $k_1 = 2.3\times10^{-4}$ s^{-1}, and T_2 = 30.0 °C = 303.2 K, with its corresponding rate constant $k_2 = 4.6\times10^{-4}$ s^{-1}. Apply the Arrhenius equation, Equation [9.23], and rearrange to solve for the activation energy, converting temperatures to units of K, and expressing the gas constant in units of J/mol-K:

$$\ln\left(\frac{k_2}{k_1}\right) = -\frac{E_A}{R}\left(\frac{1}{T_2} - \frac{1}{T_1}\right)$$

$$E_A = -\frac{R\ln\left(\frac{k_2}{k_1}\right)}{\frac{1}{T_2} - \frac{1}{T_1}} = -\frac{(8.314\text{ J/mol} - \text{K})\ln\left[\frac{(4.6\times 10^{-5}\text{ s}^{-1})}{(2.3\times 10^{-5}\text{ s}^{-1})}\right]}{\frac{1}{(303.2\text{ K})} - \frac{1}{(293.2\text{ K})}}$$

$$E_A = 5.1\times 10^4\text{ J/mol} = 51.\text{ kJ/mol}$$

Note that it would not have mattered which rate constant was assigned an index of 1 and which was assigned an index of 2 as long as the temperatures were matched correctly.

c) Again apply the Arrhenius equation, but now with $T_3 = 40.0$ °C $= 313.2$ K and using the activation energy determined in part b (expressed in J/mol):

$$\ln\left(\frac{k_3}{k_1}\right) = -\frac{E_A}{R}\left(\frac{1}{T_3} - \frac{1}{T_1}\right)$$

$$\frac{k_3}{k_1} = \exp\left[-\frac{E_A}{R}\left(\frac{1}{T_3} - \frac{1}{T_1}\right)\right]$$

$$k_3 = k_1\exp\left[-\frac{E_A}{R}\left(\frac{1}{T_3} - \frac{1}{T_1}\right)\right]$$

$$= (2.3\times 10^{-4}\text{ s}^{-1})\exp\left[-\frac{(5.1\times 10^4\text{ J/mol})}{(8.314\text{ J/mol} - \text{K})}\left[\frac{1}{(313.2\text{ K})} - \frac{1}{(293.2\text{ K})}\right]\right]$$

$$k_3 = 8.7\times 10^{-4}\text{ s}^{-1}$$

Note that the commonly applied heuristic that the rate of reaction doubles for every 10 °C increase in temperature is not always true!

Example 9.9: Determining the Temperature at Which a Reaction Occurs

Parisi and coworkers found that rate of etching of SiO_2(*s*) with HF(*aq*), used in the manufacture of electronics, had an activation energy of 41.4 kJ/mol. At what temperature (in K) does this reaction occur twice as fast as at 298. K?

Solutions

Identify the unknown temperature as T_2 with $k_2 = 2k_1$ for $T_1 = 298.$ K and $E_A = 41.4$ kJ/mol. Rearrange the Arrhenius equation to solve for T_2, converting the activation energy to J/mol:

$$\frac{1}{T_2}-\frac{1}{T_1}=-\frac{R}{E_A}\ln\left(\frac{k_2}{k_1}\right)$$

$$\frac{1}{T_2}=\frac{1}{T_1}-\frac{R}{E_A}\ln\left(\frac{k_2}{k_1}\right)$$

$$T_2=\left[\frac{1}{T_1}-\frac{R}{E_A}\ln\left(\frac{k_2}{k_1}\right)\right]^{-1}=\left[\frac{1}{(298.\ \mathrm{K})}-\frac{(8.314\ \mathrm{J/mol-K})}{\left(41.4\ \mathrm{kJ/mol}\times\frac{10^3\ \mathrm{J}}{1\ \mathrm{kJ}}\right)}\ln\left[\frac{(2k_1)}{k_1}\right]\right]^{-1}$$

$$T_2=311.\ \mathrm{K}$$

Note that some reactions show a different dependence on temperature. A more general empirical model for the effect of temperature on the rate constant is:

$$k=aT^m e^{-\frac{E_{A,app}}{RT}} \qquad [9.24]$$

where the constant a and the apparent activation energy $E_{A,app}$ are independent of temperature. The power m can be matched to a specific value based on the appropriate theory for the reaction energetics. We can identify the appropriate theory by considering mechanistic bases for the rate constant (as opposed to phenomenological Arrhenius model). Further discussion is beyond the scope of this text.

9.5 Reaction Mechanisms Can Explain Observed Kinetics

Power-law rate laws are relatively easy to apply to describe how concentrations of reactant(s) and product(s) vary temporally. But many important reactions exhibit complex rate laws that do not correspond to power laws. Some examples are provided in Table 9.1. In general, rate laws are determined using experimental data. For example, we might expect that the observed rate law for the decomposition of dinitrogen pentoxide, Rxn [9.1], is second-order with respect to the concentration of N_2O_5, but the observed rate law is $(-r_{N_2O_5})=kC_{N_2O_5}$. Why does the stoichiometry not match the reaction order?

Table 9.1 Example of Reactions Not Obeying Power-Law Rate Laws

Reaction	Observed Rate Law (with k_i as constants)
Conversion of benzene to biphenyl: $2\ C_6H_6(g) \leftrightharpoons C_{12}H_{10}(g) + H_2(g)$	$R = k_1 C_{C_6H_6}^{\ 2} - k_2 C_{C_{12}H_{10}} C_{H_2}$
Decomposition of dinitrogen monoxide over platinum: $2\ N_2O(g) \xrightarrow{Pt} 2\ N_2(g) + O_2(g)$	$R = \dfrac{k_1 C_{N_2O}}{1 + k_2 C_{N_2O}}$
Catalysis of formation of carbonic acid: $CO_2(aq) + H_2O(l) \xrightarrow{\text{carbonic anhydrase}} H_2CO_3(aq)$	$R = \dfrac{k_1 C_{CO_2}}{k_2 + C_{CO_2}}$
CVD of germanium films: $GeCl_4(g) + H_2(g) \rightarrow Ge(s) + Cl_2(g) + 2\ HCl(g)$	$R = \dfrac{k_1 C_{GeCl_4} C_{H_2} C_{Cl_2}^{\ 2}}{(C_{Cl_2} + k_2 C_{GeCl_4})^3}$

To explain the observed kinetics of reactions such as Rxn [9.1] and the ones in Table 9.1, we need to examine a mechanism for the reaction describing exactly how the reaction occurs at the molecular level. Note that we cannot prove that the mechanism is correct but only that it is consistent with experimental observations. We define a reaction mechanism as a sequence of elementary steps for an overall (observed) reaction in which each elementary step is a single, irreducible microscopic process involving no detectable intermediates. Such steps proceed at the molecular level exactly as they are written. We define the molecularity for each step as the number of molecules/species participating as reactants in the step. Molecularity equals the overall reaction order for an elementary reaction, such that reaction order does match stoichiometry for an elementary reaction. Note also that the products never appear in the rate law for an elementary step. Commonly observed molecularities are unimolecular (one molecule overstretches a bond and falls apart) and bimolecular (two molecules collide), while trimolecular (three molecules collide simultaneously) are not as common. A molecularity of four or greater is rarely if ever observed (think of the likelihood that a car accident involves four or more cars hitting each other at the same time, as opposed to two cars colliding and then third and fourth cars colliding).

The decomposition of dinitrogen pentoxide has been proposed to proceed by the following mechanism, where NO and NO_3 are intermediates:

Step 1: $2 \times \{ N_2O_5 \underset{k_{-1}}{\overset{k_1}{\rightleftharpoons}} NO_2 + NO_3 \}$

Step 2: $NO_2 + NO_3 \xrightarrow{k} NO_2 + O_2 + NO$

Step 3: $NO + NO_3 \xrightarrow{k} 2\ NO_2$

Overall: $2\ N_2O_5(g) \xrightarrow{k} 4\ NO_2(g) + O_2(g)$

Note that Step 1 is unimolecular in the forward direction and bimolecular in the reverse direction and that Steps 2 and 3 are bimolecular and proposed to occur only in the forward direction. The brackets and factor of 2 around Step 1 indicate that this step occurs twice as many times as Steps 1 and 3.

To proceed further we need to invoke one of two assumptions. With the rate-limiting step (RLS) assumption we assume one elementary step occurs at a significantly slower rate than the other steps, such that the rate at which this key step occurs governs the overall rate. In contrast, with the pseudo-steady state hypothesis (PSSH) we assume the net rate of consumption and production for each intermediate is approximately zero and the concentration of each intermediate is relatively small but stable after a short induction period. This latter assumption is appropriate when highly reactive intermediates are present and relatively long reaction times occur.

Let's apply the PSSH to the proposed mechanism for the decomposition of N_2O_5. If no step is rate limiting, the observed rate of disappearance of the reactant is the difference between the rate for Step 1 in the forward and reverse directions:

$$(-r_{N_2O_5}) = -\frac{dC_{N_2O_5}}{dt} = k_1 C_{N_2O_5} - k_{-1} C_{NO_2} C_{NO_3} \quad [9.25]$$

Apply the PSSH on the intermediate NO_3:

$$\frac{dC_{NO_3}}{dt} = k_1 C_{N_2O_5} - k_{-1} C_{NO_2} C_{NO_3} - k_2 C_{NO_2} C_{NO_3} - k_3 C_{NO} C_{NO_3} \cong 0 \quad [9.26]$$

and on the other intermediate NO:

$$\frac{dC_{NO}}{dt} = k_2 C_{NO_2} C_{NO_3} - k_3 C_{NO} C_{NO_3} \cong 0 \quad [9.27a]$$

$$C_{NO} = \frac{k_2 C_{NO_2}}{k_3} \quad [9.27b]$$

Substitute Equation [9.27b] into Equation [9.26] and rearrange to solve for the concentration of NO_3:

$$k_1 C_{N_2O_5} - k_{-1} C_{NO_2} C_{NO_3} - k_2 C_{NO_2} C_{NO_3} - k_3 \left(\frac{k_2 C_{NO_2}}{k_3}\right) C_{NO_3} = 0 \quad [9.28a]$$

$$(k_{-1} + 2k_2 C_{NO_2}) C_{NO_2} C_{NO_3} = k_1 C_{N_2O_5} \quad [9.28b]$$

$$C_{NO_3} = \frac{k_1 C_{N_2O_5}}{(k_{-1} + 2k_2) C_{NO_2}} \quad [9.28c]$$

Substitute Equation [9.28c] into Equation [9.25] to yield:

$$(-r_{N_2O_5}) = -\frac{dC_{N_2O_5}}{dt} = k_1 C_{N_2O_5} - k_{-1} C_{NO_2} \left[\frac{k_1 C_{N_2O_5}}{(k_{-1} + 2k_2) C_{NO_2}}\right] = \left(\frac{2k_1 k_2}{k_{-1} + 2k_2}\right) C_{N_2O_5} \quad [9.29]$$

We identify that the observed rate constant can be related to the rate constants for the elementary steps as $k = \frac{2k_1 k_2}{k_{-1} + 2k_2}$.

We now might ask how to determine which assumption, the RLS approximation or the PSSH, is correct? We cannot! It can, however, be shown that an equivalent expression for the rate can be obtained by assuming that Step 2 is rate limiting.

9.6 Catalysis and Catalytic Mechanisms

Understanding reaction mechanisms provides a basis for understanding catalysis. A catalyst is a substance that alters the observed rate of reaction by providing an alternative pathway for the overall reaction without experiencing a net change during the overall reaction. Note that a catalyst is neither a reactant nor a product—a catalyst is neither consumed nor produced. The net effect of a catalyst is to decrease the overall activation energy, thereby increasing the observed rate constant without increasing the temperature.

We classify catalysts into two categories, based on the phase of matter they occupy relative to the phases of the reactant(s) and product(s). Homogeneous catalysts are in the same phase as the reactant(s) and product(s). Examples of homogeneous catalysts include trace amounts of acids for some reactions and enzymes in solution. The Michaelis-Menten mechanism offers a two-step description for enzyme-mediated catalysis in which a soluble substrate (S) binds reversibly with an enzyme (E) to form an enzyme-substrate complex (E-S) in Step 1, with the product formed in a subsequent step that releases the enzyme for subsequent binding:

Step 1: $E + S \underset{k_{-1}}{\overset{k_1}{\rightleftharpoons}} E - S$

Step 2: $E - S \xrightarrow{k} E + P$

Overall: $S \xrightarrow{k} P$

In this model the enzyme exhibits specificity and selectivity for particular substrates (Figure 9.7a) and allows the reaction to occur more quickly because the overall activation energy has been lowered (Figure 9.8).

We can identify the rate law for enzyme-mediated catalysis by applying the pseudo-steady state hypothesis (PSSH) to the intermediate E-S:

$$\frac{dC_{E\text{-}S}}{dt} = k_1 C_E C_S - k_{-1} C_{E\text{-}S} - k_2 C_{E\text{-}S} \cong 0 \qquad [9.30a]$$

$$C_{E\text{-}S} = \frac{k_1 C_E C_S}{k_{-1} + k_2} \qquad [9.30b]$$

The observed rate of reaction can be determined as the rate of Step 2:

$$(-r_S) = r_P = k_2 C_{E\text{-}S} = \frac{k_1 k_2 C_E C_S}{k_{-1} + k_2} \qquad [9.31]$$

But we cannot easily determine the concentration of the free enzyme, C_E. Rather, we can measure and readily set the total concentration of enzyme, $C_{E,total}$, which is related to the concentrations of free enzyme and the enzyme-substrate complex through a conservation law:

$$C_{E,total} = C_E + C_{E\text{-}S} \qquad [9.32]$$

Substituting Equation [9.30b] and rearranging to solve for the concentration of free enzyme yields:

$$C_{E,total} = C_E + \frac{k_1 C_E C_S}{k_{-1} + k_2} = C_E\left[1 + \frac{k_1 C_S}{k_{-1} + k_2}\right] = C_E\left[\frac{k_{-1} + k_2 + k_1 C_S}{k_{-1} + k_2}\right] \qquad [9.33a]$$

$$C_E = \frac{(k_{-1} + k_2) C_{E,total}}{k_{-1} + k_2 + k_1 C_S} \qquad [9.33b]$$

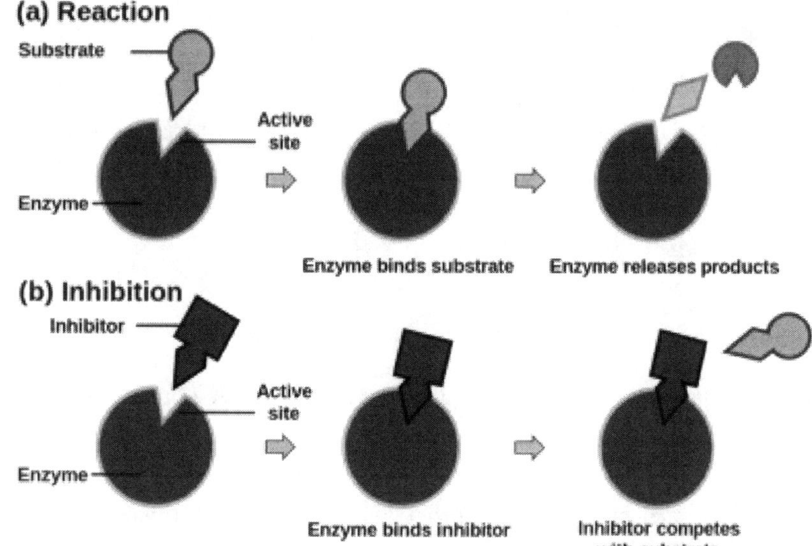

Figure 9.7. Conceptual schematic of enzyme-mediated catalysis. (a) Basic mechanism; (b) competitive inhibition.

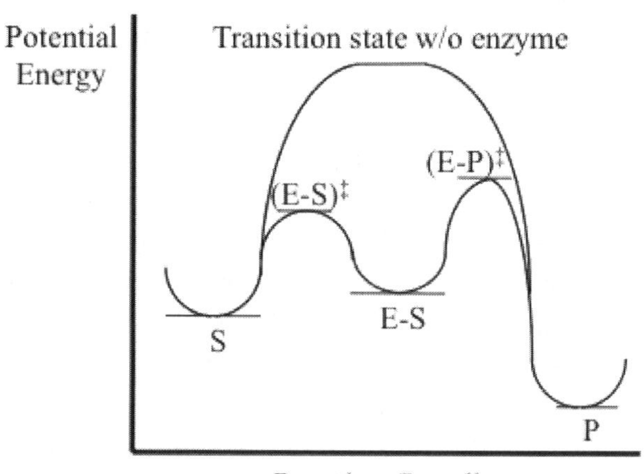

Figure 9.8. Reaction profile for enzyme-mediated catalysis, as described by the Michaelis-Menten mechanism.

Substitute Equation [9.33b] into Equation [9.31] to find:

$$(-r_S) = \frac{k_1 k_2 \left(\frac{(k_{-1}+k_2)C_{E,total}}{k_{-1}+k_2+k_1 C_S}\right) C_S}{k_{-1}+k_2} = \frac{k_1 k_2 C_{E,total} C_S}{k_{-1}+k_2+k_1 C_S} = \frac{k_2 C_{E,total} C_S}{\left(\frac{k_{-1}+k_2}{k_1}\right)+C_S} = \frac{v_{max} C_S}{K_m + C_S} \quad [9.34]$$

where $v_{max} \equiv k_2 C_{E,total}$ and $K_m \equiv \frac{k_{-1}+k_2}{k_1}$ is the Michaelis-Menten constant. We observe that the rate law does not obey a power-law form (Figure 9.9): the rate appears to be first order with respect to the substrate at relatively low concentrations of substrate (*i.e.*, for $C_S << K_m$) and zeroth-order with respect to the substrate at relatively high concentrations of substrate (*i.e.*, for $C_S >> K_m$), where the maximum rate of reaction is v_{max}.

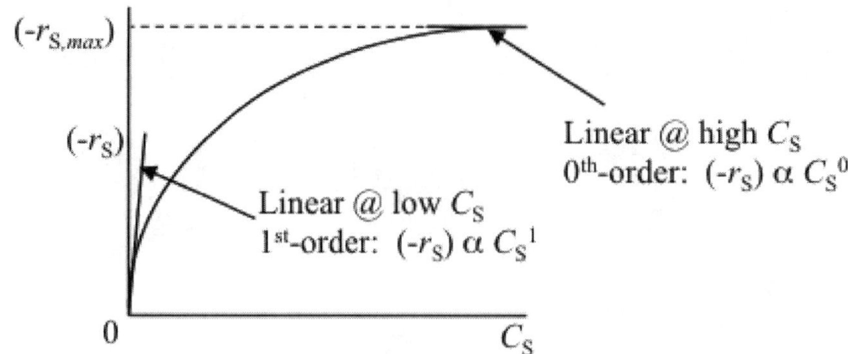

Figure 9.9. Plot of rate law for enzyme-mediated catalysis based on Michaelis-Menten mechanism.

The activity of enzymes are frequently regulated by reversible binding with inhibitors. For example, salicylic acid, a non-steroidal anti-inflammatory drug (NSAID) and the active ingredient in aspirin, inhibits the synthesis of prostaglandin, a molecule involved in sensitivity to pain. Figure 9.7b depicts one type of inhibition, called competitive inhibition.

Heterogeneous catalysts are in a different phase than the reactant(s) and/or product(s). Examples of heterogeneous catalysts include corrosion and supported metal catalysts onto which the reactant and product adsorb reversibly. For example, consider the decomposition of nitrogen dioxide mediated by a supported Pt catalyst in a catalytic converter:

$$2\ NO_2(g) \xrightarrow{Pt} N_2(g) + 2\ O_2(g) \qquad \text{Rxn [9.2]}$$

The platinum catalyst is supported on the surface of a much-less-expensive bulk metal. This reaction occurs very slowly through a gas-phase transition state in the absence of the platinum catalyst; in the presence of the catalyst a transition state adsorbed on the platinum has a decreased energy relative to that of the gas-phase transition state and mediates a faster reaction.

A generic reaction mechanism for a heterogeneous catalyst is depicted in Figure 9.10. Typically, the conversion of adsorbed reactant(s) to product(s) is the rate-limiting step that occurs much more slowly than the reversible adsorption of reactant(s) and product(s). We can frequently describe the processes of reversible adsorption/desorption using a Langmuir isotherm in which species adsorb only at a finite number of independent sites on the surface of the catalyst. Depending on the molecularity of the actual surface reaction and whether the reactant(s) and product(s) compete for the same surface sites, complicated kinetic behavior that leads to complex rate laws that do not obey power-law kinetics can result. For example, analysis of the Langmuir-Hinshelwood mechanism for a rate-limiting bimolecular surface reaction between a pair of adsorbates that compete for the same sites surface predicts that the rate actually decreases as the concentration of a reactant increases for relatively high concentrations of this reactant. Experiments in which the concentration of reactant(s) is varied provide insight into distinguishing possible mechanisms.

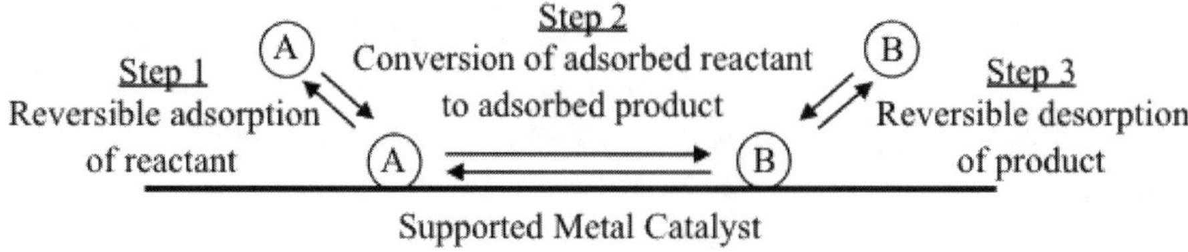

Figure 9.10. Generic mechanism for heterogeneous catalysis. Step 2 is typically the rate-limiting step.

10: Chemical Equilibrium

Learning Objectives

By the end of this chapter you should be able to:

1) Apply the law of mass action to relate equilibrium constants and equilibrium conditions.
2) Determine equilibrium constants from equilibrium concentrations or partial pressures.
3) Identify the direction in which a chemical reaction will proceed spontaneously for a given set of reactant and product concentrations or partial pressures.
4) Calculate the equilibrium concentrations or partial pressures of all species in a reaction given a set of starting conditions.
5) Use LeChâtelier's principle to predict the effect of changing the volume, changing the temperature, or removing or adding a reactant or product on a reaction at equilibrium.
6) Apply the Brønsted-Lowry model to identify conjugate acid/base pairs for an acid/base reaction.
7) Use the pH scale to express the acidity of a solution and determine the *pH* and *pOH* of solutions of strong and weak acids and bases.
8) Identify the consequences of non-ideal behavior on reaction equilibrium.
9) Determine the solubility of sparingly-soluble salts in aqueous solutions.

10.1 The Law of Mass Action Describes Chemical Equilibrium

In our examination of kinetics and reaction mechanisms in Chapter 9 we recognized that a reaction may be reversible and both convert reactant(s) to product(s) in the forward direction as well as product(s) to reactant(s) in the reverse direction. Reversibility can explain in part why actual yields are less than theoretical yields based on stoichiometric calculations that assume the reaction goes to completion and is irreversible (see Section 7.4). Chemical equilibrium, as introduced in Section 8.1, occurs for a reversible reaction when the concentration of each species participating in the reaction does not change with time because there are no net driving forces for the reaction: the rate of conversion of reactant(s) to product(s) equals the rate of conversion of product(s) to reactant(s). This type of behavior can be visualized in an animation (available at http://www.youtube.com/watch?v=LMIbJ-B92Ho&feature=related) of the reversible decomposition of calcium carbonate to form calcium oxide and carbon dioxide:

$$CaCO_3(s) \leftrightarrows CaO(s) + CO_2(g) \qquad \text{Rxn [10.1]}$$

Alternatively, we can visualize chemical equilibrium as a see-saw between reactants and products (Figure 9.1). For example, nitrogen dioxide, a brown-colored gas associated with smog (see Example 9.2), will dimerize to form dinitrogen tetroxide, a clear-colored gas, in the following elementary reaction with rate constants k_f and k_r in the forward and reverse directions, respectively:

$$2\ NO_2(g) \underset{k_r}{\overset{k_f}{\rightleftarrows}} N_2O_4(g) \qquad \text{Rxn [10.2]}$$

If we fill a container with pure $NO_2(g)$, the rate of conversion of NO_2 to N_2O_4 is infinitely greater than the rate of conversion of N_2O_4 to NO_2, and the color of the gas in the container fades. But as the concentration of NO_2 decreases and N_2O_4 increases over time, the rate of conversion of NO_2 to N_2O_4 decreases as the rate of conversion of N_2O_4 to NO_2 increases until the forward and reverse rates are identical. We observe that the color of the gas in the container never becomes clear because some NO_2 always remains.

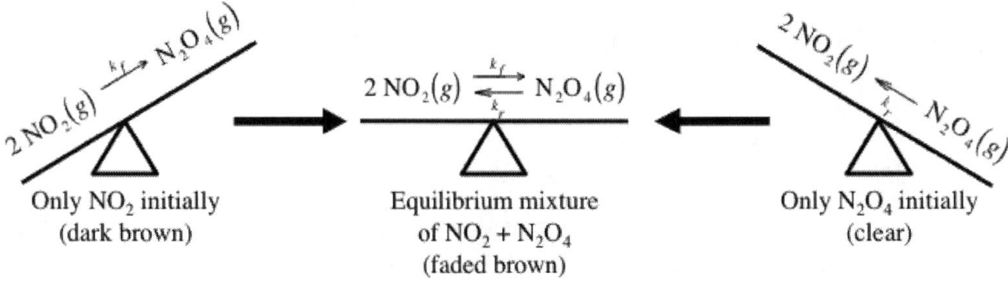

Figure 10.1. Conceptual model for dynamic equilibrium for Rxn [10.2].

When chemical equilibrium is reached, the concentration of each gas will have reached its equilibrium value, $C_{NO_2,eq}$ and $C_{N_2O_4,eq}$, respectively. Chemical equilibrium, however, is a dynamic—not a static—situation: the reaction has not stopped, but rather reactants are still converted to products and products converted to reactants, only at matching rates:

$$k_f C_{NO_2,\ eq}^{\ 2} = k_r C_{N_2O_4,\ eq} \qquad [10.1]$$

We rearrange this expression and define an <u>equilibrium constant in terms of concentrations</u>, K_c:

$$K_c = \frac{k_f}{k_r} = \frac{C_{N_2O_4,\ eq}}{C_{NO_2,\ eq}^{\ 2}} \qquad [10.2]$$

We can apply our formulation for Equation [10.2] for the specific reaction Rxn [10.2] to the generalized reaction structure used in Sections 8.7 and 7.1, where we now allow the possibility that generic Rxn [7.3] is reversible:

$$\upsilon_A\ A + \upsilon_B\ B + \ldots \leftrightarrows \upsilon_C\ C + \upsilon_D\ D + \ldots \qquad \text{Rxn [10.3]}$$

We generalize Equation [10.2] to provide a generic definition for the equilibrium constant in terms of concentrations in terms of the <u>law of mass action</u>:

$$K_c = C_{A,eq}^{\ \upsilon A} C_{B,eq}^{\ \upsilon B} \mathrm{K}\ C_{C,eq}^{\ \upsilon C} C_{D,eq}^{\ \upsilon D} \mathrm{K} \qquad [10.3]$$

where the equilibrium constant in terms of concentrations is a unitless quantity equal in value to the product of the equilibrium concentration of each dissolved or gaseous product, expressed in units of molarity (M) and raised to its corresponding stoichiometric coefficient, divided by the product of the equilibrium concentration of each dissolved or gaseous reactant, also expressed in units of molarity (M) and raised to its corresponding stoichiometric coefficient. Species that are solids or liquids do not appear in the law of mass action (Equation [10.3]) because their concentrations reflect only their densities and show no change upon reaction.

Note that for a reaction involving gases it may be more convenient to specify the equilibrium partial pressure for each gas. For example, when Rxn [10.2] reaches equilibrium, the partial pressures of NO_2 and N_2O_4 will be $P_{NO_2,eq}$ and $P_{N_2O_4,eq}$, respectively. For a reaction such as Rxn [10.2] that involves no dissolved species, we can define an alternative version for the equilibrium constant, an equilibrium constant in terms of partial pressures, K_p:

$$K_p = \frac{P_{N_2O_4,eq}}{P_{NO_2,eq}^{\;2}} \qquad [10.4]$$

where the equilibrium partial pressures re expressed in units of bar. The formulation for Equation [10.4] for the specific reaction Rxn [10.2] also can be generalized for our generic reversible Rxn [10.3]—provided only that there are no soluble reactants or products!—as an alternative statement for the law of mass action expressed in terms of equilibrium partial pressures:

$$K_p = P_{A,eq}^{\;\nu_A} P_{B,eq}^{\;\nu_B} \ldots P_{C,eq}^{\;\nu_C} P_{D,eq}^{\;\nu_D} \ldots \qquad [10.5]$$

where the equilibrium constant in terms of partial pressures is a unitless quantity equal in value to the product of the equilibrium partial pressure of gaseous product, expressed in units of bar and raised to its corresponding stoichiometric coefficient, divided by the product of the equilibrium partial pressure of each gaseous reactant, also expressed in units of bar and raised to its corresponding stoichiometric coefficient. A we observed for the law of mass action expressed in terms of the equilibrium concentrations, species that are solids or liquids do not appear in Equation [10.5] because their partial pressures are undefined.

If we assume that each gas behaves as an ideal gas, with:

$$P_{i,eq} = \frac{n_{i,eq}RT}{V} = C_{i,eq}RT \qquad [10.6]$$

we can relate the two formulations for the equilibrium constant. For example, for Rxn [10.2] we find upon substitution of Equation [10.6] for NO_2 and N_2O_4 into Equation [10.4]:

$$K_p = \frac{P_{N_2O_4,eq}}{P_{NO_2,eq}^{\;2}} = \frac{(C_{N_2O_4,eq}RT)}{(C_{NO_2,eq}RT)^2} = \left(\frac{C_{N_2O_4,eq}}{C_{NO_2,eq}^{\;2}}\right)(RT)^{-1} = \frac{K_c}{RT} \qquad [10.7]$$

where temperatures are in K and we apply the gas constant expressed as $R = 8.314\times10^{-2}$ bar-L/mol-K because partial pressures are expressed in units of bar and concentrations in units of mol/L. Equation [10.6] can be formulated for our generic reaction [10.3] as:

$$K_p = (C_{A,eq}RT)^{\nu_A}(C_{B,eq}RT)^{\nu_B}(C_{C,eq}RT)^{\nu_C}(C_{D,eq}RT)^{\nu_D}$$

$$K_p = (C_{A,eq}^{\;\nu_A} C_{B,eq}^{\;\nu_B} C_{C,eq}^{\;\nu_C} C_{D,eq}^{\;\nu_D})(RT)^{\nu_A+\nu_B+\nu_C+\nu_D} = K_c(RT)^{\Delta n_{gas}} \qquad [10.8]$$

where we define:

$$\Delta n_{gas} \equiv \sum_{\text{Gas species only}} \nu_i \qquad [10.9]$$

where the summation is only of the stoichiometric coefficients for vapor-phase reactants and products.

Example 10.1: Applying the Law of Mass Action in Terms of Concentrations and Partial Pressures

For each of the following set of reversible reactions apply the law of mass action and write expressions for equilibrium constants in terms of concentrations and in terms of partial pressures (if possible) and identify the relationship between two forms for the equilibrium constant.

$$SO_2(g) + \tfrac{1}{2} O_2(g) \leftrightarrows SO_3(g) \qquad \text{Rxn [10.E1]}$$

$$SO_3(g) + H_2O(l) \leftrightarrows H_2SO_4(aq) \qquad \text{Rxn [10.E2]}$$

$$PbF_2(s) \leftrightarrows Pb^{2+}(aq) + 2\,F^{-}(aq) \qquad \text{Rxn [10.E3]}$$

Rxn [10.E1] is important in atmospheric pollution, Rxn [10.E2] is important in acidification of water bodies, and lead fluoride in Rxn [10.E3] is an anti-reflective coating used in IR transmitting optical fibers.

Solutions

For Rxn [10.E1] applying Equations [10.3] and [10.5], respectively, we find $K_c = \frac{C_{SO_3,eq}}{C_{SO_2,eq} C_{O_2,eq}^{1/2}}$ and $K_p = \frac{P_{SO\ ,eq}}{P_{SO\ ,eq} P_{O\ ,eq}^{1/2}}$, such that applying Equation [10.8] we find x. For Rxn [10.E2], applying Equation [10.3] we find $K_c = \frac{C_{H_2SO_4,eq}}{C_{SO_3,eq}}$; for this reaction involving a solute the law of mass action expressed in terms of partial pressures in not applicable. For Rxn [10.E3], applying Equation [10.3] we find $K_c = C_{Pb^+,eq} C_{F^-,eq}^{2}$; for this reaction also involving a solute the law of mass action expressed in terms of partial pressures in not applicable.

10.2 Evaluating Equilibrium Constants Based on Data for Equilibrium Conditions

We can determine the value for an equilibrium constant based on data for equilibrium conditions (*i.e.*, equilibrium concentrations and/or partial pressures). Such problems can be divided into two general classes. For Class 1 problems we are provided a complete set of information regarding equilibrium conditions. In contrast, for Class 2 problems we are provided an incomplete set of information regarding equilibrium conditions. Class 2 problems, which tend to be more difficult and involved than Class 1 problems, are frequently addressed most easily by constructing an initial, change, equilibrium table (abbreviated ICE table).

Example 10.2: Determining Equilibrium Constants for a Class 1 Problem

Rxn [10.2] reaches equilibrium at 25. °C with concentrations of 1.96 M and 17.8 mM for NO_2 and N_2O_4, respectively.

a) What is the value for the equilibrium constant in terms of concentrations, K_c?

b) What is the value for the equilibrium constant in terms of concentrations, K_p?

c) What is the value for K_p for Rxn [10.E4]?

$$NO_2(g) \leftrightarrows \tfrac{1}{2} N_2O_4(g) \qquad \text{Rxn [10.E4]}$$

Solutions

a) Determine the value for K_c by applying the law of mass action expressed in terms of concentrations, Equation [10.3], formulated as Equation [10.2]:

$$K_c = \frac{C_{N_2O_4,eq}}{C_{NO_2,eq}^{\;2}} = \frac{\left(17.8\text{ mM} \times \frac{1\text{ M}}{10^3\text{ mM}}\right)}{(1.96\text{ M})^2} = 4.61 \times 10^{-3}$$

We remind ourselves that values for K_c are unitless, even though this property is calculated using values for concentration expressed in M.

b) Determine the value for K_p by applying Equation [10.8] formulated as Equation [10.7]:

$$K_p = \frac{(4.61 \times 10^{-3})}{(8.314 \times 10^{-2}\text{ bar} - \text{L/mol} - \text{K})(298.\text{ K})} = 1.86 \times 10^{-4}$$

We remind ourselves that values for K_p are also unitless, even though this property reflects equilibrium partial pressures expressed in bar.

c) Apply the law of mass action expressed in terms of partial pressures to Rxn [10.E4]:

$$K_{p,\text{Rxn[E10.4]}} = \frac{P_{N_2O_4,eq}^{\;1/2}}{P_{NO_2,eq}} = \sqrt{\frac{P_{N_2O_4,eq}}{P_{NO_2,eq}^{\;2}}} = \sqrt{K_{p,\text{Rxn[10.2]}}} = \sqrt{(1.86 \times 10^{-4})} = 1.36 \times 10^{-2}$$

Note that Rxn [10.E4] is simply half of Rxn [10.2]. We generalize the behavior observed here: when the stoichiometric coefficients for a reaction are multiplied by a factor χ, the equilibrium constant is raised to the power χ. Applying to Rxn [10.3] we find:

$$(\chi\upsilon_A)\text{ A} + (\chi\upsilon_B)\text{ B} + \ldots \leftrightarrows (\chi\upsilon_C)\text{ C} + (\chi\upsilon_D)\text{ D} + \ldots \qquad \text{Rxn [10.4]}$$

$$K_{c,\text{Rxn}[10.4]} = C_{A_{eq}}^{\chi\upsilon A} C_{B_{eq}}^{\chi\upsilon B} C_{C_{eq}}^{\chi\upsilon C} C_{D_{eq}}^{\chi\upsilon D} = \left(C_{A_{eq}}^{\upsilon A} C_{B_{eq}}^{\upsilon B} C_{C_{eq}}^{\upsilon C} C_{D_{eq}}^{\upsilon D}\right)^{\chi} = K_{c,Rxn[10.3]}^{\chi} \quad [10.10a]$$

$$K_{p,\text{Rxn}[10.4]} = P_{A_{eq}}^{\chi\upsilon A} P_{B_{eq}}^{\chi\upsilon B} P_{C_{eq}}^{\chi\upsilon C} P_{D_{eq}}^{\chi\upsilon D} = \left(P_{A_{eq}}^{\upsilon A} P_{B_{eq}}^{\upsilon B} P_{C_{eq}}^{\upsilon C} P_{D_{eq}}^{\upsilon D}\right)^{\chi} = K_{p,Rxn[10.3]}^{\chi} \quad [10.10b]$$

Example 10.3: Determining Equilibrium Constants for a Class 2 Problem

An empty container is filled with pure ONCl(*g*) at 1.00 bar and Rxn [10.E5a] allowed to take place:

$$2\ \text{ONCl}(g) \leftrightarrows 2\ \text{NO}(g) + \text{Cl}_2(g) \qquad \text{Rxn [10.E5a]}$$

The partial pressure of ONCl at equilibrium is 0.90 bar at 123. °C.

a) What is the value for K_p?

b) What is the value for K_c for Rxn [10.E5b]?

$$2\ \text{NO}(g) + \text{Cl}_2(g) \leftrightarrows 2\ \text{ONCl}(g) \qquad \text{Rxn [10.E5b]}$$

Solutions

a) We apply a general strategy of constructing an ICE table, determining the full set of equilibrium conditions (for this example partial pressures), and then evaluating the equilibrium constant. Start by constructing an ICE table in terms of partial pressures (in bar), where the top line is the reaction being considered, the second line is the initial partial pressure for each species, the third line is the change in the partial pressure for each species, and the fourth line is the equilibrium partial pressure for each species. The quantities labeled with a question mark are to be determined.

	2 ONCl (bar)	⇄ 2 NO (bar)	+ Cl_2 (bar)
Initial	?	?	?
Change	?	?	?
Equilibrium	?	?	?

The empty container is filled only with pure ONCl, such that initial partial pressures can be identified.

	2 ONCl (bar)	⇄ 2 NO (bar)	+ Cl_2 (bar)
Initial	1.00	0	0
Change	?	?	?
Equilibrium	?	?	?

We are explicitly told the equilibrium partial pressure for ONCl.

	2 ONCl (bar)	⇌ 2 NO (bar)	+ Cl_2 (bar)
Initial	1.00	0	0
Change	?	?	?
Equilibrium	0.90	?	?

We can complete entries for any species by working up or down a column using the relationship Initial + Change = Equilibrium.

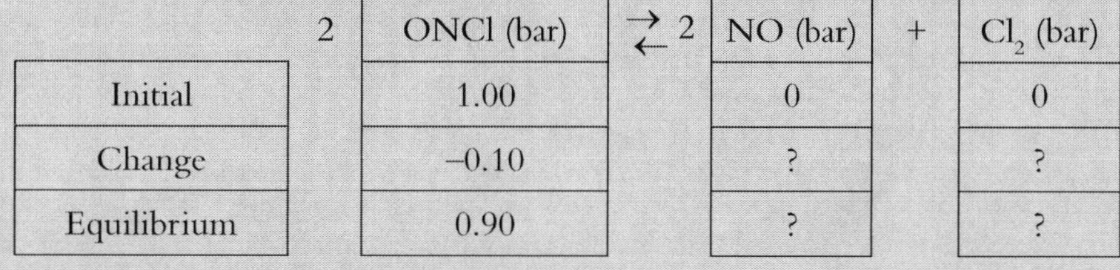

	2 ONCl (bar)	⇌ 2 NO (bar)	+ Cl_2 (bar)
Initial	1.00	0	0
Change	–0.10	?	?
Equilibrium	0.90	?	?

We can relate the quantities in the Change row based on stoichiometry—here, 2 mol ONCl = 2 mol NO = 1 mol Cl_2, such that:

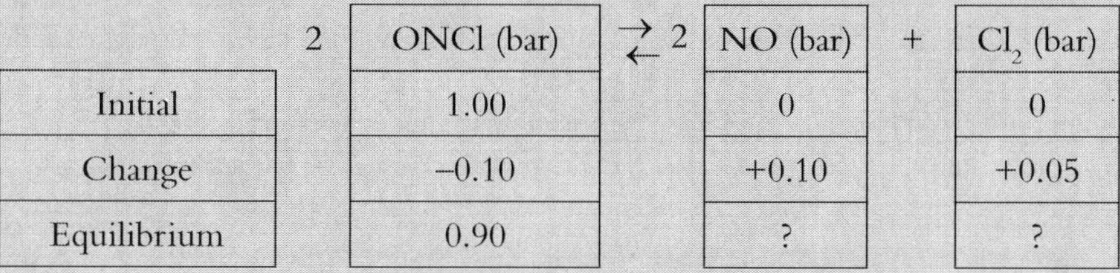

	2 ONCl (bar)	⇌ 2 NO (bar)	+ Cl_2 (bar)
Initial	1.00	0	0
Change	–0.10	+0.10	+0.05
Equilibrium	0.90	?	?

Complete the ICE table by working down the columns for NO and Cl_2.

	2 ONCl (bar)	⇌ 2 NO (bar)	+ Cl_2 (bar)
Initial	1.00	0	0
Change	–0.10	+0.10	+0.05
Equilibrium	0.90	0.10	0.05

We then determine the equilibrium constant in terms of partial pressures as:

$$K_p = \frac{P_{NO,eq}^{\;2} P_{Cl_2,eq}}{P_{ONCl,eq}^{\;2}} = \frac{(0.10 \text{ bar})^2(0.05 \text{ bar})}{(0.90 \text{ bar})^2} = 6 \times 10^{-4}$$

b) First determine the value for K_c for Rxn [10.E5a] based its value for K_p and Equations [10.8] and [10.9]:

$$\Delta n_{gas} \equiv \upsilon_{ONCl} + \upsilon_{NO} + \upsilon_{Cl_2} = (-2) + (2) + (1) = 1$$

$$K_c = K_p(RT)^{-\Delta n_{gas}} = (6 \times 10^{-4})[(8.314 \times 10^{-2} \text{ bar} - \text{L/mol} - \text{K})(396.\text{ K})]^{(1)} = 2 \times 10^{-5}$$

Now apply the law of mass action expressed in terms of concentrations to Rxn [10.E5b]:

$$K_{c,\mathrm{Rxn[E10.5b]}} = \frac{C_{\mathrm{ONCl},eq}^{\;2}}{C_{\mathrm{NO},eq}^{\;2}\,C_{\mathrm{Cl_2},eq}} = \left(\frac{C_{\mathrm{NO},eq}^{\;2}\,C_{\mathrm{Cl_2},eq}}{C_{\mathrm{ONCl},eq}^{\;2}}\right)^{-1} = K_{c,\mathrm{Rxn[E10.5a]}}^{-1} = (2\times10^{-5})^{-1} = 5\times10^{4}$$

Note that Rxn [10.E5b] is simply the inverse of Rxn [10.E5a]. We generalize the behavior observed here: when a reaction is reversed, the equilibrium constant for the reverse reaction is the reciprocal of the equilibrium constant for the original reaction. Applying to Rxn [10.3] we find:

$$\upsilon_C\,\mathrm{C} + \upsilon_D\,\mathrm{D} + \ldots \leftrightarrows \upsilon_A\,\mathrm{A} + \upsilon_B\,\mathrm{B} + \ldots \qquad \text{Rxn [10.5]}$$

$$K_{c,\mathrm{Rxn\,[10.5]}} = C_{C_{eq}}^{\;\upsilon C}\,C_{D_{eq}}^{\;\upsilon D}\,C_{A_{eq}}^{\;\upsilon A}\,C_{B_{eq}}^{\;\upsilon B} = \left(C_{A_{eq}}^{\;-\upsilon A}\,C_{B_{eq}}^{\;-\upsilon B}\,C_{C_{eq}}^{\;-\upsilon C}\,C_{D_{eq}}^{\;-\upsilon D}\right)^{-1} = \frac{1}{K_{c,\mathrm{Rxn\,[10.3]}}} \qquad [10.10a]$$

$$K_{p,\mathrm{Rxn\,[10.5]}} = P_{C_{eq}}^{\;\upsilon C}\,P_{D_{eq}}^{\;\upsilon D}\,P_{A_{eq}}^{\;\upsilon A}\,P_{B_{eq}}^{\;\upsilon B} = \left(P_{A_{eq}}^{\;-\upsilon A}\,P_{B_{eq}}^{\;-\upsilon B}\,P_{C_{eq}}^{\;-\upsilon C}\,P_{D_{eq}}^{\;-\upsilon D}\right)^{-1} = \frac{1}{K_{p,\mathrm{Rxn\,[10.3]}}} \qquad [10.10b]$$

10.3 Determining the Direction a Reaction Proceeds to Reach Equilibrium

We commonly encounter problems in which, given initial conditions (*i.e.*, concentrations and/or partial pressures) and a value for an equilibrium constant, we seek to identify the direction a reaction proceeds—from the reactant(s) toward the product(s), from the product(s) toward the reactant(s), or neither—as well as the equilibrium conditions—the set of equilibrium concentrations and/or partial pressures. We start by examining what the magnitude of the equilibrium constant tells us about equilibrium. Any equilibrium constant is the ratio of products to reactant and a function of temperature only. For example, adding a catalyst has no effect on equilibrium but rather only decreases the time required to reach equilibrium because the catalyst increases the rate in both the forward and reverse directions!

We can now attach a physical significance to the magnitude of an equilibrium constant by comparing its value to unity. Equilibrium constants can never have negative values. If an equilibrium constant is significantly greater than one (Figure 10.2a), products are greatly favored over reactants at equilibrium, and we say that the reaction essentially goes to completion. Such an assessment is appropriate for an equilibrium constant greater than 10^{+35}. An example of such a reaction is the complete combustion of ethanol. In contrast, if an equilibrium constant is significantly less than one (Figure 10.2b), reactants are greatly favored over products at equilibrium, and we say that the reaction essentially does not occur. Such an assessment is appropriate for an equilibrium constant less than 10^{-35}. An example of such a reaction is the formation of metallic sodium and water from sodium hydroxide and hydrogen gas; however, if an equilibrium constant is approximately one (Figure 10.2c), neither products nor reactants are greatly favored at equilibrium, and at equilibrium we expect a mixture of leftover reactants and products. Such an assessment is appropriate for an equilibrium constant between 10^{-35} and 10^{+35}. For such a reaction equilibrium calculations are warranted.

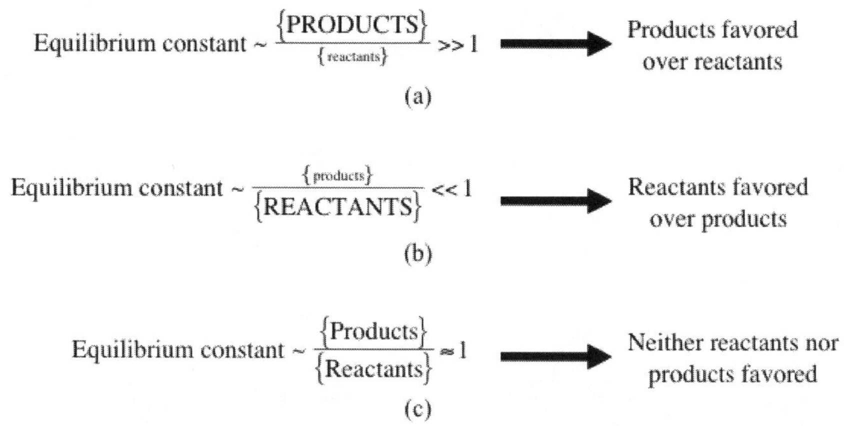

$$\text{Equilibrium constant} \sim \frac{\{\text{PRODUCTS}\}}{\{\text{reactants}\}} >> 1 \longrightarrow$$ Products favored over reactants

(a)

$$\text{Equilibrium constant} \sim \frac{\{\text{products}\}}{\{\text{REACTANTS}\}} << 1 \longrightarrow$$ Reactants favored over products

(b)

$$\text{Equilibrium constant} \sim \frac{\{\text{Products}\}}{\{\text{Reactants}\}} \approx 1 \longrightarrow$$ Neither reactants nor products favored

(c)

Figure 10.2. Comparison between amounts of products vs. reactants at equilibrium. (a) Large value for equilibrium constant; (b) small value for equilibrium constant; (c) equilibrium constant of ~1.

To predict the direction of net change given a set of initial conditions, we define a unitless reaction quotient based on the law of mass action applied to the initial conditions. In particular, for the generic Rxn [10.3] we define a reaction quotient in terms of concentrations as:

$$Q_c = C_A^{DA} C_B^{DB} \mathrm{K}\, C_C^{DC} C_D^{DD} \mathrm{K} \qquad [10.11]$$

where the concentrations are the initial concentrations of solutes and gases expressed in M (with solids and liquids omitted), and a reaction quotient in terms of partial pressures as:

$$Q_p = P_A^{DA} P_B^{DB} \mathrm{K}\, P_C^{DC} P_D^{DD} \mathrm{K} \qquad [10.12]$$

where the partial pressures are the initial partial pressures of the gases expressed in M (with solids and liquids omitted). We recognize that the reaction quotient in terms of partial pressures is only applicable for reactions not involving dissolved species.

We then determine the direction a reaction proceeds from its initial conditions towards equilibrium by comparing the values for the reaction quotient and the equilibrium constant (Figure 10.3). If the reactant quotient is less than the equilibrium constant, the reaction is not at equilibrium because excess reactants are present, and the reaction proceeds towards the products. In contrast, if the reactant quotient equals the equilibrium constant, the reaction is at equilibrium, and no net reaction takes place; however, if the reactant quotient is greater than the equilibrium constant, the reaction is not at equilibrium because excess products are present, and the reaction proceeds towards reactants.

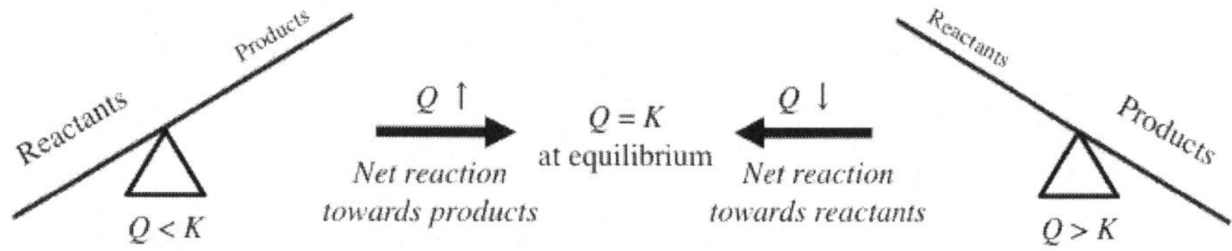

Figure 10.3. The direction a reaction proceeds to reach equilibrium can be identified by comparing the reaction quotient with the equilibrium constant. Here *Q* and *K* are generic representations for the reaction quotient and the equilibrium constant, respectively.

10.4 Determining Equilibrium Conditions Based on Initial Conditions and an Equilibrium Constant

The following pair of examples illustrate features of solving problems requiring us to determine equilibrium concentrations and partial pressures, including the use of ICE tables formulated in terms of a positively-valued "dummy" variable.

Example 10.4: The Equilibrium Dissociation of Dinitrogen Tetroxide

An empty container is filled with pure $N_2O_4(g)$ at 2.17 bar and Rxn [10.E6] allowed to take place at 661 K, with $K_p = 0.315$:

$N_2O_4(g) \leftrightarrows 2\ NO_2(g)$ Rxn [10.E6]

a) In what direction does the reaction proceed?

b) What are the equilibrium partial pressures (in bar) of N_2O_4 and NO_2?

Solutions

a) Compare the reaction quotient in terms of partial pressures, calculated by applying Equation [10.12], with the equilibrium constant in terms of partial pressures:

$$Q_p = \frac{P_{NO_2,o}{}^2}{P_{N_2O_4,o}} = \frac{(0\text{ atm})^2}{(2.17\text{ atm})} = 0 < K_p$$

Because $Q_p < K_p$, we are not at equilibrium, and the reaction will proceed towards the products until $Q_p = K_p$. This conclusion is reasonable given the initial absence of products!

b) Construct an ICE table in terms of partial pressures (in bar) to find the equilibrium partial pressures.

	N_2O_4 (bar)	$\rightleftarrows$ 2	NO_2 (bar)
Initial	2.17		0
Change	?		?
Equilibrium	?		?

Introduce a positively-valued "dummy" variable, x, as the extent of reaction in terms of partial pressures, $x \equiv -\Delta P_{N_2O_4} > 0$ (where x has units of bar), and recognize that the 1:2 stoichiometry for N_2O_4:NO_2 means that the partial pressure of the product changes by $+2x$.

With each equilibrium partial pressure as the sum of the corresponding initial partial pressure and change in partial pressure, complete the ICE table in terms of *x*.

	N_2O_4 (bar)	$\rightleftarrows$ 2	NO_2 (bar)
Initial	2.17		0
Change	–x		+2x
Equilibrium	2.17–x		2x

Set bounds on *x* as 0 bar < *x* < 2.17 bar by requiring the partial pressures of both species to be positive at equilibrium. Apply the law of mass action expressed in terms of partial pressures with the known value for the equilibrium constant in terms of partial pressures and rearrange to obtain a quadratic equation in terms of *x*:

$$K_p = \frac{P_{NO_2,eq}^{\ 2}}{P_{N_2O_4,eq}} = \frac{(2x)^2}{(2.17 - x)} = \frac{4x^2}{2.17 - x}$$
$$K_p(2.17 - x) = 2.17K_p - K_p x = 4x^2$$
$$4x^2 + K_p x - 2.17K_p = 0$$

Solving for *x* (*e.g.*, by considering its roots as a quadratic equation or using a solver tool on a computational machine, as described in Appendix A) yields *x* = 0.374 bar or *x* = -0.453 bar. We recognize that only the solution *x* = 0.374 bar is physically realizable because the other root lies outside the allowable bounds for *x*. The equilibrium partial pressures are then:

$$P_{N_2O_4,eq} = 2.17 \text{ bar} - (0.374 \text{ bar}) = 1.80 \text{ bar}$$
$$P_{NO_2,eq} = 2x = 2(0.374 \text{ bar}) = 0.748 \text{ bar}$$

A good practice is to check our answer by showing by substitution that the predicted equilibrium partial pressures satisfy the law of mass action for this reaction.

Example 10.5: The Decomposition of Phosphorus Pentachloride

For Rxn [10.E7] $K_c = 1.00$ at an unspecified temperature. If the initial concentrations of phosphorus pentachloride, phosphorus trichloride and molecular chlorine are 0.25 M, 0.20 M, and 2.25 M, respectively:

a) In what direction does the reaction proceed (if any)?

b) What is the equilibrium concentration (in M) for each species?

$$PCl_5(g) \leftrightarrows PCl_3(g) + Cl_2(g) \qquad \text{Rxn [10.E7]}$$

Solutions

a) Compare the reaction quotient in terms of concentrations with the equilibrium constant in terms of concentrations:

$$Q_c = \frac{C_{PCl_3,o} C_{Cl_2,o}}{C_{PCl_5,o}} = \frac{(0.20\text{ M})(2.25\text{ M})}{(0.25\text{ M})} = 1.8 > K_c$$

Because $Q_c > K_c$, we are not at equilibrium, and the reaction will proceed towards the reactant until $Q_c = K_c$. Note that this behavior is less obvious than the identification of the direction in which Rxn [10.E6] proceeded to reach equilibrium in Example 10.4.

b) Construct an ICE table in terms of concentrations (in M) to find the equilibrium partial pressures.

	PCl_5 (M)	$\rightleftarrows$	PCl_3 (M)	+	Cl_2 (M)
Initial	0.25		0.20		2.25
Change	?		?		?
Equilibrium	?		?		?

Introduce a positively-valued dummy variable, x, as the negative of the extent of reaction, $x \equiv +\Delta[PCl_5] > 0$ (with x in M), and recognize that 1:1:1 stoichiometry for PCl_5:PCl_3:Cl_2 means the concentrations of both products change by $-x$. With each equilibrium concentration as the sum of the corresponding initial concentration and change in concentration, complete the ICE table in terms of x.

	PCl_5 (M)	$\rightleftarrows$	PCl_3 (M)	+	Cl_2 (M)
Initial	0.25		0.20		2.25
Change	$+x$		$-x$		$-x$
Equilibrium	$0.25+x$		$0.20-x$		$2.25-x$

Set bounds on x as given $0\text{ M} < x < 0.20\text{ M}$ by requiring the concentrations of all species to be positive at equilibrium. Apply the law of mass action with the known value for the equilibrium constant in terms of concentrations and rearrange to obtain a quadratic equation in terms of x:

$$K_c = \frac{C_{PCl_3,eq} C_{Cl_2,eq}}{C_{PCl_5,eq}} = \frac{(0.20-x)(2.25-x)}{(0.25+x)} = \frac{0.45 - 2.45x + x^2}{0.25+x} = 1$$

$$0.45 - 2.45x + x^2 = 0.25 + x$$

$$x^2 - 3.45x + 0.20 = 0$$

Solving for x (*e.g.*, by considering its roots as a quadratic equation or using a solver tool on a computational machine) yields $x = 3.39$ M or $x = 0.06$. We recognize that only the solution $x = 0.06$ M is physically realizable because the other root lies outside the allowable bounds for x. The equilibrium concentrations then are:

$$C_{\mathrm{PCl_5},eq} = 0.25\ \mathrm{M} + (0.06\ \mathrm{M}) = 0.31\ \mathrm{M}$$
$$C_{\mathrm{PCl_3},eq} = 0.20\ \mathrm{M} - (0.06\ \mathrm{M}) = 0.14\ \mathrm{M}$$
$$C_{\mathrm{Cl_2},eq} = 2.25\ \mathrm{M} - (0.06\ \mathrm{M}) = 2.19\ \mathrm{M}$$

Again, a good practice is to check our answer by showing (*e.g.*, by substitution) that the predicted equilibrium partial pressures satisfy the law of mass action for this reaction.

10.5 LeChâtelier's Principle

What happens when a system in which a reaction has reached equilibrium is perturbed? LeChâtelier's principle predicts that a reaction at equilibrium will respond to a change in volume, temperature, or addition or removal of a reactant or product by reaching a new equilibrium that partially offsets the change. For example, consider the reaction that occurs when carbon dioxide dissolves in water to form carbonic acid:

$$CO_2(g) + H_2O(l) \leftrightarrows H_2CO_3(aq)\,\Delta h^\circ_{rxn} = -10.39\ \mathrm{kJ/mol} \qquad \text{Rxn [10.6]}$$

In Example 6.7 we considered this type of process from the perspective of Henry's law. We now seek to explain what happens when a bottle of carbonated water is opened, an unopened bottle of carbonated water is heated to 50 °C, or the concentration of $CO_2(g)$ is increased by introducing a chunk of $CO_2(s)$.

Opening a bottle of carbonated water results in an increase in volume for the system. According to LeChâtelier's principle the system will respond such that the increase in volume is partially offset, which can be accomplished by a partial reaction in the direction that fills the increased volume, specifically in the direction that increases the amount of gas. Thus, a partial reaction will occur towards the reactants to generate more carbon dioxide. We observe this response as the formation of additional gas bubbles.

We can generalize this behavior for any reaction at equilibrium in a system experiencing a change in volume. We predict that if the volume increases, a partial reaction will occur in the direction for which the number of moles of gas increases to partially fill the increased volume. We also predict that if the volume decreases instead, a partial reaction will occur in the direction for which the number of moles of gas decreases to partially accommodate the decreased volume.

Heating a bottle of carbonated water results in an increase in temperature for the system. According to LeChâtelier's principle the system will respond such that the increase in temperature is partially offset, which can be accomplished by a partial reaction in the direction that absorbs heat, specifically in the endothermic direction. Thus, a partial reaction will occur towards the reactants because the reaction as written (reactants going towards product) is exothermic. We observe this response as the formation of additional gas bubbles.

We can generalize this behavior for any reaction at equilibrium in a system experiencing a change in temperature. We predict that if the temperature increases, a partial reaction will occur in the endothermic direction to absorb some heat and partially offset the increase in temperature. We also predict that if the temperature decreases instead, a partial reaction will occur in the exothermic direction to release some heat and partially offset the increase in temperature.

Increasing the concentration of carbon dioxide results in an increase in the amount of reactants. According to LeChâtelier's principle the system will respond such that the increase in amount of reactants is partially offset, which can be accomplished by a partial reaction in the direction that consumes reactants, specifically towards the product. We observe this response as the formation of additional dissolved $H_2CO_3(aq)$, which we can detect by opening the bottle and observing increased formation of gas bubbles.

We can also generalize the effect of adding or removing a reactant or product on a reaction at equilibrium. We predict that if the number of moles of reactants increases and/or the number of moles of products decreases, a partial reaction will occur towards the products to consume some of the additional reactants/replace some of the removed products. We also predict that if, instead, the number of moles of reactants decreases and/or the number of moles of products increases, a partial reaction will occur towards the reactants to replace some of the removed reactants/consume some of the additional products. Note that this response is expected only if the reactant or product altered is a gas or a solute (*i.e.*, appears as a term in the law of mass action and expression for the equilibrium constant). In particular, the addition or removal of a solid, liquid, or inert substance (*e.g.*, Ar(g)) does not affect equilibrium.

Note also that adding a catalyst has no effect on equilibrium. For example, opening a bottle of carbonated water results in such a large increase in volume that we expect most of the dissolved carbonic acid to be converted to carbon dioxide gas and water at the new equilibrium. We can accelerate the rate at which the new equilibrium is reached by adding a tablet of Mentos® as a catalyst for nucleating bubbles.

10.6 The Brønsted-Lowry Model for Acids and Bases

Reactions between acids and bases can be viewed as special cases of chemical equilibrium. Based on Brønsted-Lowry theory we identify an acid as a proton donor and a base as a proton acceptor. For example, ethanoic (acetic) acid is an acid that can donate the proton from its carboxyl group to form the ethanoate (acetate) ion (Figure 10.4). Acids and bases in general form such conjugate pairs, consisting of a conjugate acid and a conjugate base, in which the conjugate acid and conjugate base differ only by one proton.

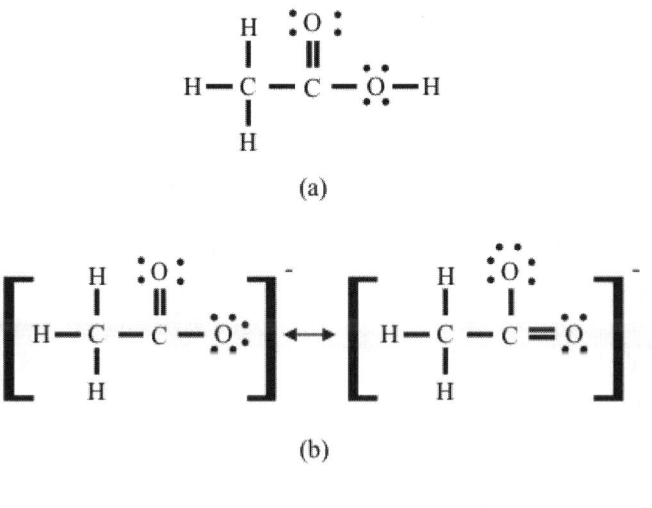

(a)

(b)

Figure 10.4. Conjugate pair of (a) the acid ethanoic (acetic) acid and (b) the base ethanoate (acetate) ion

We observe that acid/base reactions consist of a reaction between an acid and a base and involve two sets of conjugate pairs. For example, consider the reaction of acetic acid with water to form the acetate ion and the hydronium ion (H_3O^+):

$$CH_3COOH(aq) + H_2O(l) \leftrightharpoons CH_3COO^-(aq) + H_3O^+(aq) \qquad \text{Rxn [10.7]}$$

The two conjugate pairs involved in this reaction are the conjugate pair CH_3COOH and CH_3COO^- and the conjugate pair H_3O and H_2O. Notice that each side of the reaction involves an acid and a base that are not members of the same conjugate pair.

Application of the law of mass action expressed in terms of concentrations to Rxn [10.7] yields:

$$K_a = \frac{C_{CH_3COO^-,eq}C_{H_3O^+,eq}}{C_{CH_3COOH,eq}} \qquad [10.13]$$

We define this equilibrium constant as the acid dissociation (ionization) constant, K_a, which we recognize is just a special name for the equilibrium constant in terms of concentrations for the dissociation of an acid in water. We generalize this formulation for the generic reaction for dissociation of an acid HA in water:

$$HA(aq) + H_2O(l) \leftrightharpoons A^-(aq) + H_3O^+(aq) \qquad \text{Rxn [10.8]}$$

$$K_a = \frac{C_{A^-,eq}C_{H_3O^+,eq}}{C_{HA,eq}} \qquad [10.14]$$

Ammonia is a weak base that associates with water to form the ammonium ion and the hydroxide ion:

$$NH_3(aq) + H_2O(l) \leftrightharpoons NH_4^+(aq) + OH^-(aq) \qquad \text{Rxn [10.9]}$$

The two conjugate pairs involved in this reaction are the conjugate pair NH_4^+ and NH_3 and the conjugate pair H_2O and OH^-. Notice that once again each side of the reaction involves an acid and a base that are not members of the same conjugate pair. Applying the law of mass action expressed in terms of concentrations to Rxn [10.9] yields:

$$K_b = \frac{C_{NH_4^+,eq}C_{OH^-,eq}}{C_{NH_3,eq}} \qquad [10.15]$$

We define this equilibrium constant as the base association (ionization) constant, K_b, which we recognize is just a special name for the equilibrium constant in terms of concentrations for the association of a base in water. We generalize this formulation for the generic reaction for the association of a base B in water:

$$B(aq) + H_2O(l) \leftrightharpoons BH^+(aq) + OH^-(aq) \qquad \text{Rxn [10.10]}$$

$$K_b = \frac{C_{BH^+,eq}C_{OH^-,eq}}{C_{B,eq}} \qquad [10.16]$$

Data for both acid and base ionization constants are available in the *CRC Handbook of Chemistry and Physics* (available at http://www.hbcpnetbase.com/).

We observe that water can be either an acid or a base; such substances are termed amphiprotic or amphoteric. For example, the bicarbonate ion (HCO_3^-) is also amphiprotic, as we observe in Rxns [10.11a] and [10.11b]:

$$H_2CO_3(aq) + H_2O(l) \leftrightharpoons HCO_3^-(aq) + H_3O^+(aq) \qquad \text{Rxn [10.11a]}$$

$$HCO_3^-(aq) + H_2O(l) \leftrightharpoons CO_3^{2-}(aq) + H_3O^+(aq) \qquad \text{Rxn [10.11b]}$$

where in Rxn [10.11a] carbonic acid and the bicarbonate ion form a conjugate pair and in Rxn [10.11b] the bicarbonate ion and the carbonate ion form a conjugate pair. Note that the bicarbonate ion acts as a base in Rxn [10.11a] and as an acid in Rxn [10.11b].

Example 10.6: Identifying Conjugate Acid/Base Pairs

For each of the following reactions identify the two sets of conjugate acid/base pairs, including which in each set is an acid and which is a base.

$HNO_3(aq) + H_2O(l) \leftrightarrows H_3O^+(aq) + NO_3{-}(aq)$ Rxn [10.E8a]

$HSO_4^-(aq) + NO_3^-(aq) \leftrightarrows SO_4^{2-}(aq) + HNO_3(aq)$ Rxn [10.E8b]

$F^-(aq) + H_2O(l) \leftrightarrows HF(aq) + OH{-}(aq)$ Rxn [10.E8c]

$H_2PO_4^-(aq) + ClO_4^-(aq) \leftrightarrows HPO_4^{2-}(aq) + HClO_4(aq)$ Rxn [10.E8d]

Solutions

For Rxn [10.E8a] the acid HNO_3 and the base NO_3^- form one conjugate pair, and the acid H_3O^+ and the base H_2O form a second conjugate pair.

For Rxn [10.E8b] the acid HSO_4^- and the base SO_4^{2-} form one conjugate pair, and the acid HNO_3 and the base NO_3^- form a second conjugate pair.

For Rxn [10.E8c] the acid HF and the base F^- form one conjugate pair, and the acid H_2O and the base OH^- form a second conjugate pair.

For Rxn [10.E8d] the acid $H_2PO_4^-$ and the base HPO_4^{2-} form one conjugate pair, and the acid $HClO_4$ and the base ClO_4^- form a second conjugate pair.

10.7 Evaluating the Acidity and Alkalinity of Solutions of Acids and Bases

Perhaps the simplest acid/base reaction is the self-ionization (auto-ionization) of water, in which one molecule of water acts as an acid and a second molecule of water acts as a base:

$$H_2O(l) + H_2O(l) \leftrightarrows H_3O^+(aq) + OH^-(aq) \qquad \text{Rxn [10.12]}$$

We express the law of mass action and the equilibrium constant in terms of concentrations for Rxn [10.12] as:

$$K_w = C_{H_3O^+,eq} C_{OH^-,eq} \qquad [10.17]$$

where K_w is defined as the auto-ionization constant of water. We find that $K_w = 1.00\times10^{-14}$ at 25.00 °C and 1.2×10^{-15} at 0.00 °C, indicating that most water molecules are not ionized (*i.e.*, $C_{H_3O^+,eq}$ and $C_{OH^-,eq}$ are relatively small. Also note that because ionization increases as temperature increases, application of LeChâtelier's principle tells us that the auto-ionization of water is endothermic.

We define scales for <u>acidity</u> and <u>alkalinity</u> based on pH and pOH, respectively:

$$pH \equiv -\log_{10} C_{H_3O^+} \tag{10.18a}$$

$$pOH \equiv -\log_{10} C_{OH^-} \tag{10.18b}$$

A $pH < 7$ is <u>acidic</u>, a $pH > 7$ <u>basic</u>, and a pH of ~7 <u>neutral</u>, regardless of temperature. Figure 10.5 depicts the pH of commonly-encountered solutions. Note that the definitions of pH and pOH apply for both equilibrium and non-equilibrium conditions; however, because the auto-ionization of water rapidly reaches equilibrium upon perturbations at temperatures for which water is a liquid, we can typically reasonably assume that measurements of acidity and alkalinity with pH and pOH, respectively, represent equilibrium conditions. We can confirm that pure water, prepared as freshly distilled, deionized water, is neutral at room temperature, with $C_{H_3O^+,eq} = C_{OH^-,eq}$ based on Rxn [10.12]:

$$K_w = C_{H_3O^+,eq}^{\;2}$$
$$C_{H_3O^+,eq} = \sqrt{K_w} = \sqrt{(1.00 \times 10^{-14})} = 1.00 \times 10^{-7}\ \text{M}$$
$$pH = -\log_{10} C_{H_3O^+} = -\log_{10}(1.00 \times 10^{-7}\ \text{M}) = 7.00$$

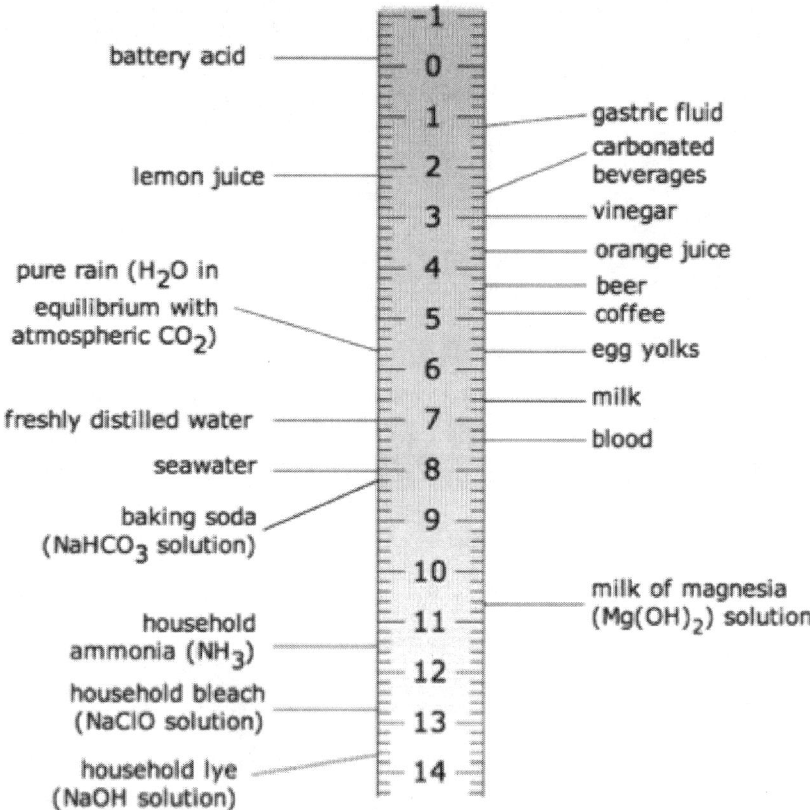

Figure 10.5. Examples of commonly-encountered solutions with different pH.

Example 10.7: Relating *pH* and *pOH*

If the *pH* of an aqueous solution is 3.14 at 25 °C, what are $C_{H_3O^+}$ (in M), C_{OH^-} (in M), and *pOH*?

Solutions

Determine the concentration of hydronium ions by rearranging the definition of *pH*, Equation [10.18a]:

$$C_{H_3O^+} = 10^{-pH} = 10^{-(3.14)} = 7.24 \times 10^{-4}\ M$$

Now determine the concentration of hydroxide ions by rearranging Equation [10.17] describing the auto-ionization of water, assuming our concentrations are equilibrium values:

$$C_{OH^-,eq} = \frac{K_w}{C_{H_3O^+,eq}} = \frac{(1.00 \times 10^{-14})}{(7.24 \times 10^{-4}\ M)} = 1.38 \times 10^{-11}\ M$$

Lastly, determine the *pOH* by its definition, Equation [10.18b]:

$$pOH = -\log_{10} C_{OH^-,eq} = -\log_{10}(1.38 \times 10^{-11}\ M) = 10.86$$

What can we conclude regarding the sum of the *pH* and *pOH*? We can relate this sum to the auto-ionization constant for water:

$$pH + pOH = -\log_{10} C_{H_3O^+,eq} - \log_{10} C_{OH^-,eq} = -\log_{10}\left(C_{H_3O^+,eq} C_{OH^-,eq}\right) = -\log_{10} K_w \qquad [10.19]$$

Consequently, the sum of the *pH* and *pOH* is 14 only at 25 °C because $-\log_{10} K_w = 14$ only at this temperature.

We are frequently interested in comparing the relative strengths of different acids and bases. Such a comparison can be made based on values for the acid and base ionization constants, K_a and K_b, respectively, using a *pH*-like scale:

$$pK_a \equiv -\log_{10} K_a \qquad [10.20a]$$

$$pK_b \equiv -\log_{10} K_b \qquad [10.20b]$$

We categorize the relative strengths of acids and bases as either strong or weak based on the extent of dissociation of an acid and association of a base. For a strong acid, equilibrium strongly favors the dissociation of the acid into its conjugate base and the hydronium ion by Rxn [10.8]. We define a strong acid as an acid for which $K_a > 1$ (*i.e.*, $pK_a < 0$) and assume that a strong acid completely dissociates into individual ions, such that no equilibrium calculations (nor an ICE table) are required. Examples of strong acids are hydrochloric acid, hydrobromic acid, nitric acid (HNO_3), and the ionization of sulfuric acid (H_2SO_4) to the bisulfate ion (HSO_4^-). Note that the subsequent ionization of HSO_4^- to SO_4^{2-} does not correspond to a strong acid.

Example 10.8: Determining the Acidity of a Solution of a Strong Acid

What is the pH of an aqueous solution formed when 7.30 g of HCl is dissolved in pure water to form a 2.0 L solution?

Solutions

Because hydrochloric acid is a strong acid, we describe its dissociation as irreversible:

$$HCl(aq) + H_2O(l) \rightarrow Cl^-(aq) + H_3O^+(aq) \qquad \text{Rxn [10.E9]}$$

With 1:1:1:1 stoichiometry we find:

$$C_{H_3O^+,f} = C_{HCl,i} = \frac{7.30 \text{ g HCl} \times \frac{1 \text{ mol HCl}}{36.5 \text{ g HCl}}}{2.0 \text{ L}} = 0.10 \text{ M}$$

$$pH = -\log_{10} C_{H_3O^+,f} = -\log_{10}(0.10 \text{ M}) = 1.0$$

Note that we did not need to specify the temperature!

For a strong base, equilibrium similarly strongly favors the association of the base to form its conjugate acid and the hydroxide ion. We define a strong base as a base for which $K_b > 1$ (*i.e.*, $pK_b < 0$) and assume that a strong base completely associates into individual ions, such that no equilibrium calculations (nor an ICE table) are required. Examples of strong acids are salts of Group 1A and Group 2A cations with hydroxide.

Example 10.9: Determining the Acidity of a Solution of a Strong Base

What is the pH of an aqueous solution formed when 7.30 g of NaOH is dissolved in pure water to form a 2.0 L solution at 25 °C?

Solutions

Because NaOH is a strong base, we describe its ionization as irreversible:

$$NaOH(aq) \rightarrow Na^+(aq) + OH^-(aq) \qquad \text{Rxn [10.E10]}$$

where water participates in Rxn [10.E10] solely as a solvent. The increase in the concentration of hydroxide ions, however, results in a decrease in the concentration of hydronium ions relative to pure water. With 1:1:1 stoichiometry we find:

$$C_{OH^-,f} = C_{NaOH,i} = \frac{7.30 \text{ g NaOH} \times \frac{1 \text{ mol NaOH}}{40.0 \text{ g NaOH}}}{2.0 \text{ L}} = 9.1 \times 10^{-2} \text{ M} = C_{OH^-,eq}$$

$$C_{H_3O^+,eq} = \frac{K_w}{C_{OH^-,eq}} = \frac{(1.0 \times 10^{-14})}{(9.1 \times 10^{-2} \text{ M})} = 1.1 \times 10^{-13} \text{ M}$$

$$pH = -\log_{10} C_{H_3O^+,eq} = -\log_{10}(1.1 \times 10^{-13} \text{ M}) = 13.$$

Note that for this analysis we did need to specify the temperature!

We define a <u>weak acid</u> as an acid for which $K_a < 1$ (*i.e.*, $pK_a > 0$), such that all acids that are *not* strong acids are weak acids. We assume that a weak acid partially dissociates into individual ions, such that equilibrium calculations (and an ICE table) are required. We find that as the acid ionization constant decreases (*i.e.*, pK_a increases), the acid becomes weaker and less dissociation is observed.

Example 10.10: Determining the pH for a Solution of a Weak Acid

What is the *pH* of an aqueous solution of 0.10 M CH_3COOH ($K_a = 1.8 \times 10^{-5}$) at 25 °C?

Solutions

Our overall strategy is to set up an ICE table for the equilibrium dissociation of ethanoic acid as a weak acid (introduced as Rxn [10.71]) to determine $C_{H_3O^+,eq}$ and then determine the *pH* from its definition. Assume only ethanoic acid is initially present.

	CH_3COOH (M)	$+ H_2O \rightleftarrows$	CH_3COO^- (M)	+	H_3O^+ (M)
Initial	0.10		0.00		0.00
Change	–x		+x		+x
Equilibrium	0.10–x		x		x

The absence of any products tells us the reaction must proceed partially towards the products, with 1:1:1 stoichiometry for CH_3COOH:CH_3COO^-:H_3O^+ and an extent of reaction $x \equiv \Delta C_{H_3O^+}$. Bounds of 0 M < *x* < 0.10 M are set by requiring the concentration of all species to be positive at equilibrium. Apply the law of mass action as Equation [10.13] with the given value for the acid ionization constant, K_a, and rearrange to obtain a quadratic equation in terms of *x*:

$$K_a = \frac{C_{CH_3COO^-,eq} C_{H_3O^+,eq}}{C_{CH_3COOH,eq}} = \frac{(x)(x)}{(0.10 - x)} = \frac{x^2}{0.10 - x}$$

$$x^2 + K_a x - 0.10 K_a = 0$$

Solving for x numerically (see Appendix A) or using the solution for the quadratic equation yields $x = 1.3\times10^{-3}$ M or $x = -1.4\times10^{-3}$ M. The physically possible solution is $x \equiv C_{H_3O^+,eq} = 1.3\times10^{-3}$ M, such that the value for pH can be found from Equation [10.18a]:

$$pH = -\log_{10} C_{H_3O^+,eq} = -\log_{10}(1.3\times10^{-3}\text{ M}) = 2.9$$

Example 10.11: Determining the pK_a for a Weak Acid Based on Measurement of pH

What is the pK_a for fluoroacetic acid (CH_2FCOOH) if a 0.318 M solution of this acid in pure water has a pH of 1.56 at 25 °C?

Solutions

Start by determining the concentration of hydronium ions corresponding to the given pH:

$$pH = -\log_{10} C_{H_3O^+,eq}$$
$$C_{H_3O^+,eq} = 10^{-pH} = 10^{-1.56} = 2.75\times10^{-2}\text{ M}$$

Next, set up an ICE table to determine the equilibrium concentrations of fluoroacetic acid, its conjugate base, the fluoroacetate ion (CH_2FCOO^-), and the hydronium ion based on assuming an initial concentration of 0.318 M for fluoroacetic acid and 0.000 M for both the fluoroacetate and hydronium ions and an equilibrium concentration of 2.75×10^{-2} M for the hydronium ion.

	CH_2FCOOH (M)	$+ H_2O \rightleftarrows$	CH_2FCOO^- (M)	+	H_3O^+ (M)
Initial	0.318		0.000		0.000
Change	?		?		?
Equilibrium	?		?		2.75×10^{-2}

The change in concentration of the hydronium ion is its equilibrium concentration less its initial concentration, such that our ICE table becomes:

	CH_2FCOOH (M)	$+ H_2O \rightleftarrows$	CH_2FCOO^- (M)	+	H_3O^+ (M)
Initial	0.318		0.000		0.000
Change	?		?		2.75×10^{-2}
Equilibrium	?		?		2.75×10^{-2}

We can determine the change in concentrations of fluoroacetic acid and the fluoroacetate ion, recognizing the 1:1:1 stoichiometry for CH_2FCOOH:CH_2COO^-:H_3O^+, and then determine the equilibrium concentrations of fluoroacetic acid and the fluoroacetate ion based on the sum for their respective initial and change in concentrations.

	CH_2FCOOH (M)	$+ H_2O \rightleftarrows$	CH_2FCOO^- (M)	+	H_3O^+ (M)
Initial	0.318		0.000		0.000
Change	–2.75×10^{-2}		2.75×10^{-2}		2.75×10^{-2}
Equilibrium	0.290		2.75×10^{-2}		2.75×10^{-2}

The acid ionization constant, K_a, and pK_a can then be determined by applying Equations [10.14] and [10.20a]:

$$K_a = \frac{C_{CH_2FCOO^-,eq}C_{H_3O^+,eq}}{C_{CH_2FCOOH,eq}} = \frac{(2.75\times 10^{-2}\text{ M})(2.75\times 10^{-2}\text{ M})}{(0.290\text{ M})} = 2.61\times 10^{-3}$$

$$pK_a = -\log_{10}K_a = -\log_{10}(2.61\times 10^{-3}) = 2.58$$

We define a <u>weak base</u> as a base for which $K_b < 1$ (*i.e.*, $pK_b > 0$), such that all bases that are not strong bases are weak bases. We assume that a base acid partially associates into individual ions, such that equilibrium calculations (and an ICE table) are required. We find that as the base ionization constant decreases (*i.e.*, pK_b increases), the base becomes weaker and less association (*i.e.*, ionization) is observed.

Example 10.12: Determining the pH for a Solution of a Weak Acid

What is the *pH* of an aqueous solution of 0.38 M ethylamine ($C_2H_5NH_2$, $pK_b = 3.4$) at 25 °C?

Solutions

This example can be compared with Example 10.10. Our overall strategy is to set up an ICE table for the equilibrium association of the weak base to determine $C_{OH^-,eq}$, evaluate the auto-ionization of water to determine $C_{H_3O^+,eq}$, and then determine the *pH* from its definition. We assume only ethylamine is initially present.

	$C_2H_5NH_2$ (M)	$+ H_2O \rightleftarrows$	$C_2H_5NH_3^+$ (M)	+	OH^- (M)
Initial	0.38		0.00		0.00
Change	–x		+x		+x
Equilibrium	0.38–x		x		x

The absence of any products tells us the reaction must proceed partially towards the products, with 1:1:1 stoichiometry for $C_2H_5NH_2$:$C_2H_5NH_3^+$:OH^- and an extent of reaction $x \equiv \Delta C_{OH^-}$. The bounds 0 M < x < 0.38 M are set by requiring the concentration of all species to be positive at equilibrium. Apply the law of mass action for the association of this weak base with water, Equation [10.16], and rearrange to obtain a quadratic equation in terms of x.

$$K_b = \frac{C_{C_2H_5NH_3^+,eq}C_{OH^-,eq}}{C_{C_2H_5NH_2,eq}} = \frac{(x)(x)}{(0.38-x)} = \frac{x^2}{0.38-x}$$

$$x^2 + K_b x - 0.38K_b = 0$$

The base ionization constant can be determined by rearranging Equation [10.20b]:

$$K_b = 10^{-pK_b} = 10^{-(3.4)} = 4.0\times 10^{-4}$$

Solving for x numerically or using the solution for the quadratic equation yields $x = 1.2\times10^{-2}$ M or $x = -1.2\times10^{-2}$, with the physically possible solution of $x = C_{OH^-,eq} = 1.2\times10^{-2}$ M. The concentration of hydronium ions is found by evaluating the auto-ionization of water:

$$K_w = C_{H_3O^+,eq}C_{OH^-,eq} = 1.00\times 10^{-14}$$

$$C_{H_3O^+,eq} = \frac{1.00\times10^{-14}}{C_{OH^-,eq}} = \frac{1.00\times10^{-14}}{(1.2\times10^{-2}\ \text{M})} = 8.3\times10^{-13}\ \text{M}$$

and the pH using its definition:

$$pH = -\log_{10}C_{H_3O^+,eq} = -\log_{10}(8.3\times10^{-13}\ \text{M}) = 12.1$$

10.8 Non-ideality in Chemical Equilibrium and Quantitative Descriptions for the Solubility of Salts

From the perspective of thermodynamics, equilibrium constants in terms of concentration and partial pressure assume that each species behaves ideally. Although we have previously considered the concept for ideality for gases, what does it mean for a solute to behave "ideally? An ideal solution is one in which each of the species in the solution, including all solutes and the solvent, experience similar strengths for IM forces. For such a solution there are no changes in enthalpy or volume upon mixing. This description is appropriate for solutions of hydrocarbons. In contrast, a non-ideal solution is one in which different species experience different strengths for IM forces, such that each species interacts energetically with itself and others differently. For such a solution mixing may be associated with the absorption or release of heat, and the volume of the solution may differ from the sum of the volumes of the individual component. This description is appropriate for many important systems, including mixtures of water and ethanol and dissolved aqueous solutions.

To incorporate non-ideality into our descriptions for chemical equilibrium, we introduce a third formulation for the equilibrium constant and a corresponding reaction quotient by expressing the law of mass action for our generic Rxn [10.3] in terms of a unitless equilibrium constant in terms of activities:

$$K = a_{A_{eq}}{}^{DA} a_{B_{eq}}{}^{DB} \mathrm{K}\, a_{C_{eq}}{}^{DC} a_{D_{eq}}{}^{DD} \mathrm{K} \qquad [10.21]$$

describing the equilibrium state and in terms of a unitless reaction quotient in terms of activities:

$$Q = a_A{}^{DA} a_B{}^{DB} \mathrm{K}\, a_C{}^{DC} a_D{}^{DD} \mathrm{K} \qquad [10.22]$$

describing the current (*e.g.*, initial) state. We define the activity for a species based on the phase for the species. The activity for a liquid or solid i is $a_i \equiv 1$, for a gas i $a_i \equiv \frac{P_i}{P^o}$, and for a non-ionizing solute i $a_i \equiv \frac{\gamma_i C_i}{C^o}$, where $C^o \equiv 1$ M is standard molarity and γ_i is the activity coefficient for the solute. A modified convention is used for ions: the activity for an ion i is $a_i \equiv \frac{\gamma_\pm C_i}{C^o}$, where $\gamma_\pm$ is the mean activity coefficient of the ion pair corresponding to the ion i. The mean activity coefficient is introduced because it is impossible to evaluate γ_{cation} and γ_{anion} separately given the impossibility of creating a solution in which there is an excess of cations over anions (or *vice versa*).

Solutions behave ideally in the limit of infinite dilution, with the activity coefficient for the solute approaching one. Activity coefficients greater than one are observed when solute and solvent experience net repulsive IM forces and prefer not to mix. In contrast, activity coefficients less than one occur when solute and solvent experience net attractive IM forces and prefer to mix. An activity coefficient equal to one indicates that attractive and repulsive IM forces between solute and solvent approximately balance. Extensive efforts have been expended to measure and model activity coefficients for both ionizing and non-ionizing solute.

We can apply the law of mass action expressed in terms of activities to understand the solubility of strong electrolytes in water. You may recall that we discussed the solubility of salts from the perspective of precipitation reactions in Chapter 7, applying a set of solubility rules to predict whether a salt dissolves in water. But this qualitative description over-simplifies the situation and is inappropriate quantitatively: all ionic compounds can dissolve to some (albeit sometimes very small) extent in water. Consequently, compounds that we identify as "insoluble" based on Table 7.1 really are "sparingly" soluble and dissolve to a very small extent.

To quantitatively describe the dissolution of a sparingly-soluble ionic solid in water, represented by the generic reaction:

$$A_mB_n(s) \leftrightharpoons mA^{j+}(aq) + nB^{k-}(aq) \qquad \text{Rxn } [10.13]$$

we apply the formulation of the law of mass action expressed in terms of activities and define the unitless equilibrium constant in terms of activities as a unitless solubility product, K_{sp}:

$$K_{sp} = a_{A^{j+},eq}{}^{m} a_{B^{k-},eq}{}^{n} \qquad [10.22]$$

and the corresponding reaction quotient in terms of activities:

$$Q_{sp} = a_{A^{j+}}{}^{m} a_{B^{k-}}{}^{n} \qquad [10.23]$$

where concentrations are expressed in units of molality. By applying the framework introduced in Sections 10.2–10.5, we now have a basis to evaluate the solubility of "soluble" salts (*e.g.*, NaCl) as well as "insoluble" salts (*e.g.*, $CaCO_3$). For example, calcium carbonate is not completely insoluble (it just has a very small solubility product), and neither is sodium chloride infinitely soluble (it just has a very large solubility product). Precipitates form if the initial concentrations of ions are such that $Q_{sp} > K_{sp}$. We identify saturated solutions, such as encountered with sodium chloride in Examples 6.2 and 6.3, as solutions with equilibrium concentration of dissolved solutes. Data for solubility products at room temperature are accessible in the *CRC Handbook of Chemistry and Physics* (available at http://www.hbcpnetbase.com/).

As an example of our approach, consider the dissolution of silver chloride in water. Silver chloride is a sparingly-soluble salt, with $K_{sp} = 1.80\times10^{-10}$ at 25 °C, that dissolves in water as:

$$AgCl(s) \leftrightarrows Ag^{+}(aq) + Cl^{-}(aq) \qquad \text{Rxn [10.E11]}$$

We commonly denote the product pair as "AgCl(*aq*)," although all we can really say is that there is one silver cation for each chloride anion (*i.e.*, we do not find individual ions paired in a solution of a strong electrolyte). The solubility product can be related to the concentration of each ion and the mean activity coefficient based on Equation [10.22] and the definition for the activity of an ion:

$$K_{sp} = a_{Ag^{+},eq}a_{Cl^{-},eq} = \left(\frac{\gamma_{\pm}C_{Ag^{+},eq}}{C^{\circ}}\right)\left(\frac{\gamma_{\pm}C_{Cl^{-},eq}}{C^{\circ}}\right) = \frac{\gamma_{\pm}^{2}C_{Ag^{+},eq}C_{Cl^{-},eq}}{C^{\circ 2}}$$

When a solid sample of silver chloride is added to pure water, there is no source of silver and chloride ions other than what dissolves from the solid, such that the 1:1 stoichiometry of the salt means $C_{Ag^{+},eq} = C_{Cl^{-},eq} = s$, where s is defined as the solubility of the salt (*i.e.*, the number of moles of the ionic compound dissolved per L of solution). Because the salt is sparingly soluble, we expect its solubility to be small, such that we can reasonably assume that the solution behaves ideally and $\gamma_{\pm} \cong 1$. We then find the solubility of silver chloride in pure water as:

$$K_{sp} = \frac{\gamma_{\pm}^{2}s^{2}}{C^{\circ 2}}$$

$$s = \frac{C^{\circ}\sqrt{K_{sp}}}{\gamma_{\pm}} = \frac{(1\ \text{M})\sqrt{(1.80\times10^{-10})}}{(1)} = 1.34\times10^{-5}\ \text{M}$$

Two factors complicate our evaluation if the sparingly-soluble salt is not dissolved in pure water. If the ionic solid is added to an aqueous solution that contains appreciable amounts of another salt (*e.g.*, a highly soluble salt such as sodium chloride), the solution will not behave ideally and it is inappropriate to assume a mean activity coefficient of one. Under such circumstances the solution is "globally" electrically neutral, but locally one is more likely to find an ion cloud of counter-ions rather than co-ions surrounding a given ion (Figure 10.6). By representing the concentration of ions as an ionic strength, quantitative models of this situation based on electrostatics predict that the mean activity coefficient decreases for small to moderate increases in ionic strength, increasing the solubility of the sparingly-soluble salt, but increases as for relatively large increases in ionic strength, decreasing the solubility of the sparingly-soluble salt. These effects are known as "salting in" and "salting out" and can be profoundly useful for separating charged proteins in biotech processes.

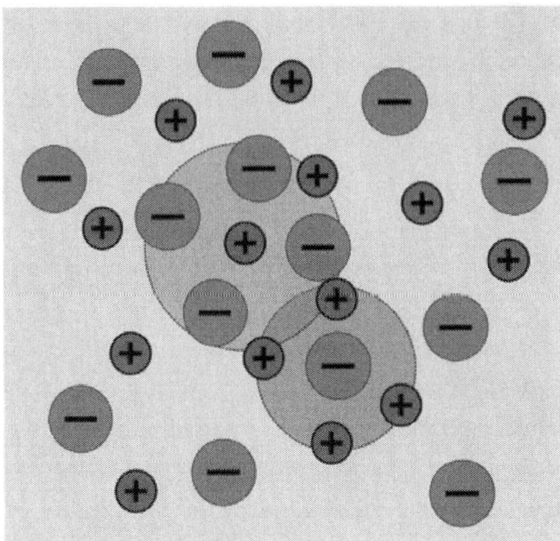

Figure 10.6. Idealized representation of the ion cloud associated with a solution of a 1:1 strong electrolyte.

A second factor that complicates our evaluation of solubility arises if an ionic solid is added in a solution that already contains one of the ions. Under these circumstances, beyond effects associated with the mean activity coefficient, the common ion effect results: the presence of additional ions for one species lowers the number of ions for both species that dissolve, and the solubility of the salt is less than it would be for a solution that did not already contain one of the ions.

Example 10.13: Dissolution of Calcium Phosphate in Water

What is the solubility in pure water at 25 °C of calcium phosphate, $Ca_3(PO_4)_2$, a sparingly soluble salt ($K_{sp} = 2.08\times10^{-33}$) that is the principle mineral component of bone in the form of hydroxyapatite? Assume ideal behavior.

Solutions

Calcium phosphate dissolves in water according to the reaction:

$$Ca_3(PO_4)_2(s) \leftrightarrows 3\,Ca^{2+}(aq) + 2\,PO_4^{3-}(aq) \qquad \text{Rxn [10.E11]}$$

with a solubility product related to the concentration of each ion and the mean activity coefficient based on Equation [10.22] and the definition for the activity of an ion:

$$K_{sp} = a^3_{Ca^{2+},eq}\,a^2_{PO_4^{3-},eq} = \left(\frac{\gamma_\pm C_{Ca^{2+},eq}}{C^\circ}\right)^3\left(\frac{\gamma_\pm C_{PO_4^{3-},eq}}{C^\circ}\right)^2 = \frac{\gamma_\pm^5 C_{Ca^{2+},eq}^3 C_{PO_4^{3-},eq}^2}{C^{\circ 5}}$$

Calcium phosphate dissolves in water as a strong electrolyte with a 3:2 stoichiometry, such that the concentrations of calcium and phosphate ions are related to the solubility as $C_{Ca^{2+},eq} = 3s$ and $C_{PO_4^{3-},eq} = 2s$. Assuming that the solution behaves ideally means $\gamma_\pm \cong 1$. We then find the solubility of silver chloride in pure water as:

$$K_{sp} = \left[(1)\frac{[3s]}{C^{\circ}}\right]^3\left[(1)\frac{[2b_{Ca_3(PO_4)_2}]}{b^{\circ}}\right]^2 = \frac{3^3 \times 2^2 s^5}{C^{\circ 5}}$$
$$s = C^{\circ}\sqrt[5]{\frac{K_{sp}}{3^3 \times 2^2}} = (1\text{ M})\left[\frac{(2.08 \times 10^{-33})}{3^3 \times 2^2}\right]^{1/5} = 1.14 \times 10^{-7}\text{ M}$$

Example 10.14: Mixing Streams of Dissolved Salts (Example 6.5 Revisited)

Reconsider the mixing of 1.0000×10^3 L of 3.0000×10^{-2} M Na_2CO_3(*aq*) with 2.0000×10^3 L of 6.0000×10^{-3} M $Ca(NO_3)_2$(*aq*) in a wastewater treatment plant at room temperature previously examined in Example 6.5. The solubility product of calcium carbonate is 2.8×10^{-9} at 25 °C. Assume all solutions behave ideally.

a) Does a precipitate of $CaCO_3$(*s*) form?

b) What are the equilibrium concentrations (in M) of Ca^{2+}(*aq*) and CO_3^{2-}(*aq*)?

Solutions

a) Consider the solubility equilibrium for $CaCO_3$:

$CaCO_3(s) \leftrightharpoons Ca^{2+}(aq) + CO_3^{2-}(aq)$ Rxn [10.E12]

Determine the concentrations of calcium and carbonate ions immediately after mixing the solutions of 3.0000×10^{-2} M Na_2CO_3(*aq*) and 6.0000×10^{-3} M Ca $(NO_3)_2$(*aq*):

$$C_{Ca^{2+},o} = \frac{(2.0000 \times 10^3\text{ L})(6.0000 \times 10^{-3}\text{ M})}{(1.0000 \times 10^3\text{ L}) + (2.0000 \times 10^3\text{ L})} = 4.0000 \times 10^{-3}\text{ M}$$
$$C_{CO_3^{2-},o} = \frac{(1.0000 \times 10^3\text{ L})(3.0000 \times 10^{-2}\text{ M})}{(1.0000 \times 10^3\text{ L}) + (2.0000 \times 10^3\text{ L})} = 1.0000 \times 10^{-2}\text{ M}$$

The reaction quotient now can be determined with $\gamma_{\pm} \cong 1$ as:

$$Q_{sp} = a_{Ca^{2+},o}a_{CO_3^{2-},o} = \left(\frac{\gamma_{\pm}C_{Ca^{2+},o}}{C^{\circ}}\right)\left(\frac{\gamma_{\pm}C_{CO_3^{2-},o}}{C^{\circ}}\right) = \frac{C_{Ca^{2+},o}C_{CO_3^{2-},o}}{C^{\circ 2}}$$
$$Q_{sp} = (4.0000 \times 10^{-3}\text{ M})(1.0000 \times 10^{-2}\text{ M}) = 4.0000 \times 10^{-5}$$

Because $Q_{sp} > K_{sp}$, we are not at equilibrium, and the reaction will proceed towards the reactant (*i.e.*, a precipitate forms) until $Q_{sp} = K_{sp}$.

b) Construct an ICE table in terms of concentrations (in M) for the ions. We ignore the solid because its activity is one, such that it does not appear in the law of mass action.

$CaCO_3 \rightleftarrows$	Ca^{2+} (M)	+	CO_3^{2-} (M)
Initial	4.0000×10^{-3}		1.0000×10^{-2}
Change	?		?
Equilibrium	?		?

Introduce a dummy variable, x, as the negative of the extent of reaction as $x \equiv -\Delta C_{Ca^{2+}} = -\Delta C_{CO_3^{2-}}$. With each equilibrium concentration as the sum of the corresponding initial concentration and change in concentration, complete the ICE table in terms of x.

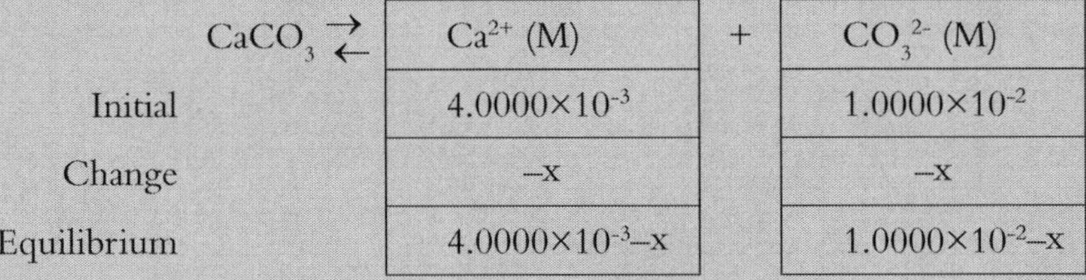

$CaCO_3 \rightleftarrows$	Ca^{2+} (M)	+	CO_3^{2-} (M)
Initial	4.0000×10^{-3}		1.0000×10^{-2}
Change	$-x$		$-x$
Equilibrium	$4.0000\times10^{-3}-x$		$1.0000\times10^{-2}-x$

The bounds 0 M < x < 4.0000×10^{-3} M are set by requiring the concentration for both species to be positive at equilibrium. Apply the law of mass action expressed in terms of activities with the known value for the solubility product, assume ideality (*i.e.*, $\gamma_\pm \cong 1$), and rearrange to obtain a quadratic equation in terms of x.

$$K_{sp} = a_{Ca^{2+},eq}\,a_{CO_3^{2-},eq} = \left(\frac{\gamma_\pm C_{Ca^{2+},eq}}{C^\circ}\right)\left(\frac{\gamma_\pm C_{CO_3^{2-},eq}}{C^\circ}\right) = \frac{C_{Ca^{2+},eq}C_{CO_3^{2-},eq}}{C^{\circ 2}}$$

$$K_{sp} = \frac{(4.0000\times10^{-3}-x)(1.0000\times10^{-2}-x)}{C^{\circ 2}}$$

$$K_{sp}C^{\circ 2} = x^2 - 1.4000\times10^{-2}x + 4.0000\times10^{-5}$$

$$x^2 - 1.4000\times10^{-2}x + 4.0000\times10^{-5} - K_{sp}C^{\circ 2} = 0$$

Solving for x (*e.g.*, by considering its roots as a quadratic equation or using a solver feature on a computational machine) yields $x = 1.0000\times10^{-2}$ M or $x = 3.9995\times10^{-3}$ M, with only the solution $x = 3.9995\times10^{-3}$ M being physically possible (the other root is outside the allowable bounds for x). The equilibrium concentrations are then:

$$C_{Ca^{2+},eq} = 4.0000\times10^{-3}\text{ M} - (3.9995\times10^{-3}\text{ M}) = 5.\times10^{-7}\text{ M}$$

$$C_{CO_3^{2-},eq} = 1.0000\times10^{-2} - (3.9995\times10^{-3}\text{ M}) = 6.000\times10^{-3}\text{ M}$$

We remind ourselves that a good practice is to check our answer by showing (*e.g.*, by substitution) that the predicted equilibrium concentrations satisfy the law of mass action for this reaction.

11: Spontaneity and the Directionality of Reactions

Learning Objectives

By the end of this chapter you should be able to:

1) Identify that processes occur in directions in which microscopic disorder increases, with entropy providing a macroscopic metric reflecting the amount of microscopic disorder.
2) Predict the qualitative change in entropy for systems undergoing different chemical or physical processes and quantitatively calculate standard-state changes in entropy for chemical processes based on standard molar entropies.
3) Evaluate the performance of heat engines based on the second law of thermodynamics.
4) Apply the criterion for the spontaneity of a process based on changes in Gibbs free energy to identify temperature ranges in which a particular process is spontaneous.
5) Determine changes in Gibbs free energy at standard state for chemical processes from standard molar Gibbs free energies of formation.
6) Identify whether a process is spontaneous and its equilibrium properties by applying relationships between changes in Gibbs free energy at standard and non-standard states, reaction quotients, and equilibrium constants.
7) Characterize how the temperature dependence of an equilibrium is constant.

11.1 Directionality, Microscopic Disorder, and Entropy

We know from our practical experience that certain processes occur only in certain directions. For example, when partitions separating a gas from a vacuum are removed, we observe that the gas expands into the vacuum (Figure 11.1). Although it is possible that all of the particles in the gas return to their initial, compressed state, the likelihood of finding the gas in this initial state decreases strongly as the number of gas particles increases beyond a small number, and for any macroscopic-sized sample this likelihood is vanishingly small. Note that the final state exhibits greater positional disorder than the initial state. The directionality in this process can be termed mechanical directionality.

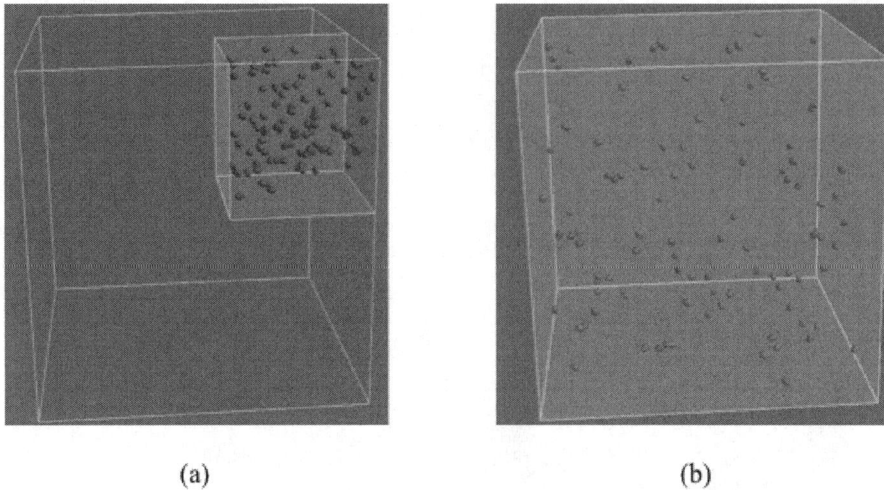

(a) (b)

Figure 11.1. Expansion of a gas into a vacuum is associated with mechanical directionality. (a) Initial state in which $N_2(g)$ is separated from a vacuum by partitions; (b) final state in which $N_2(g)$ has expanded into larger compartment. Images created using the molecular modeling software *Odyssey*.

What do we expect to happen when an impermeable partition separating a pair of pure gases is removed? The process of mixing that results (Figure 11.2) can be conceived as a mutual, simultaneous expansion of both gases into a common volume. Note that the final state exhibits greater chemical disorder than the initial state. The likelihood of the gases "unmixing" is vanishingly small provided that each gas consists of more than a small number of particles. The process of mixing can be considered an example of chemical directionality.

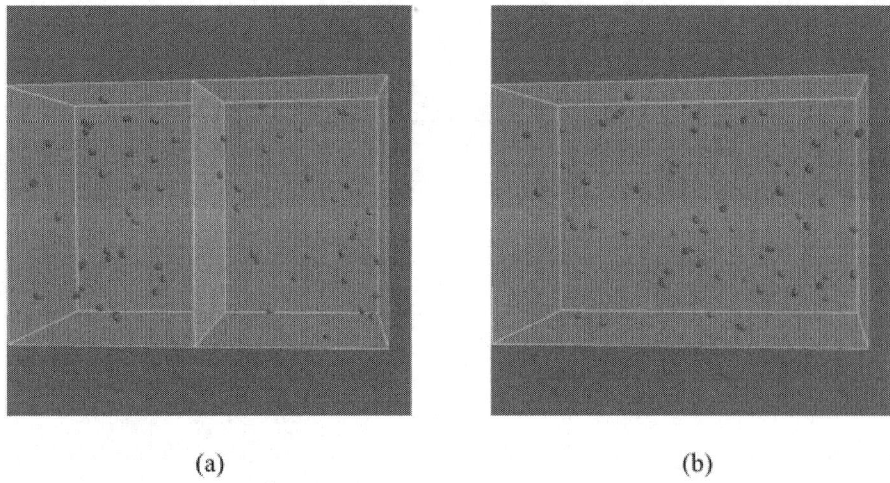

(a) (b)

Figure 11.2. Mixing of different pure gases is associated with chemical directionality. (a) Initial state in which $CO_2(g)$ in left compartment is separated from $N_2(g)$ in right compartment by an impermeable barrier; (b) final state in which gases have mixed upon removal of barrier. Images created using the molecular modeling software *Odyssey*.

What happens when a hot object is placed in thermal contact with a cold object, as we considered in Chapter 8? We expect that the energy of the hot object decreases while the energy of the cold object increases. If neither object changes phase, we further expect the temperature of the hot object to decrease while the temperature of the cold object increases. This process occurs by particles (*i.e.*, molecules and atoms) having relatively high translational, vibrational, and/or rotational kinetic energies in the hot object colliding with particles having relatively low kinetic energies in the cold object, with energy being transferred from particles with relatively high kinetic energy to particles with relatively low kinetic energy. We would never expect the temperature of the hot object to increase and/or the temperature of the cold object to decrease, for this process would correspond to collisions between pairs of particles with high and low respective kinetic energies resulting in loss of energy by less energetic particles and gain of energy by more energetic particles! This process is associated with thermal directionality: heat always flows from regions of high temperature to regions of low temperature, never in reverse. We also conclude that heat flow results in a redistribution of energies at the molecular level.

Statistical descriptions of matter offer a basis for understanding mechanical, chemical, and thermal directionality by providing schemes for quantifying disorder at the molecular level. With such an approach we identify microstates characterizing the positions and energies of individual atoms and molecules. Each microstate represents one possible snapshot or arrangement of microscopic properties for the system. Statistical thermodynamics seeks to relate averages for ensembles (statistical collections) of microstates to macroscopic observables for a system based on properties of individual microstates and their relative likelihoods of being observed. The ergodic hypothesis states that a time-average of the properties of a system, such as what might measure experimentally for pressure or temperature, is equivalent to an instantaneous average of these properties over an ensemble of available microstates.

As an illustration consider the application of a lattice model, in which particles only occupy discrete spatial positions, to describe the mixing of a pair of different ideal gases between two compartments (Figure 11.3). To simplify our visualization and analysis assume this pair of gases can be described as four black particles and four white particles. The particles are initially separated by a partition into compartments with identical volumes and at identical temperatures. Upon removal of the partition the particles can exchange positions, and we count the number of ways in which different configurations of particles can result. As particles for an ideal gas exert negligible forces on each other, the probability of finding a particular type of particle at any specific position is independent of the occupancy of neighboring positions.

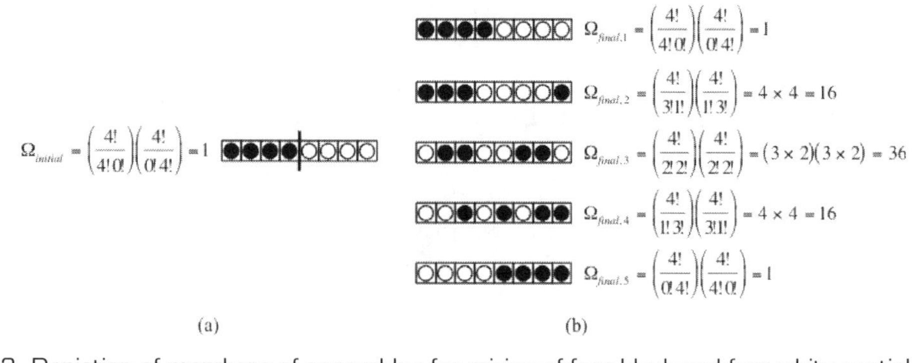

$$\Omega_{initial} = \left(\frac{4!}{4!0!}\right)\left(\frac{4!}{0!4!}\right) = 1$$

$$\Omega_{final,1} = \left(\frac{4!}{4!0!}\right)\left(\frac{4!}{0!4!}\right) = 1$$

$$\Omega_{final,2} = \left(\frac{4!}{3!1!}\right)\left(\frac{4!}{1!3!}\right) = 4 \times 4 = 16$$

$$\Omega_{final,3} = \left(\frac{4!}{2!2!}\right)\left(\frac{4!}{2!2!}\right) = (3 \times 2)(3 \times 2) = 36$$

$$\Omega_{final,4} = \left(\frac{4!}{1!3!}\right)\left(\frac{4!}{3!1!}\right) = 4 \times 4 = 16$$

$$\Omega_{final,5} = \left(\frac{4!}{0!4!}\right)\left(\frac{4!}{4!0!}\right) = 1$$

(a) (b)

Figure 11.3. Depiction of members of ensembles for mixing of four black and four white particles. (a) Initial state; (b) final state.

What distribution of particles is most likely upon the removal of the partition? We use multiplicity (denoted with the symbol Ω) to represent the number of possible different spatial microstates for the collection of articles. We intuitively expect that more disordered configurations are more likely than more ordered configurations. Combinatorics tells us the number of configurations with N distinguishable particles arranged into M categories with n_i particles in category i, such that $N = \sum_{i=1}^{M} n_i$, is:

$$\Omega = \frac{N!}{n_1! \times n_2! \times \ldots \times n_M!} = \frac{N!}{\prod_{i=1}^{M} n_i!} \qquad [11.1]$$

where $n_i! = n_i \times (n_i - 1) \times (n_i - 2) \ldots \times 2 \times 1$, with $0! = 1$. The multiplicity for a configuration j for our system is the product of the multiplicities for the right and left compartments:

$$\Omega_j = \Omega_{j,right} \Omega_{j,left} \qquad [11.2]$$

For an individual compartment k with $N = 4$ positions and $M = 2$ categories there are $n_{b,j,k}$ black and $n_{w,j,k}$ white particles, respectively, such that:

$$\Omega_{j,k} = \frac{N!}{n_{b,j,k}!\, n_{w,j,k}!} \qquad [11.3]$$

The initial state is an ensemble with one configuration, one microstate, and $\Omega_{initial} = 1$. The final state is an ensemble with $\frac{8!}{4!4!} = \frac{8 \times 7 \times 6 \times 5 \times 4 \times 3 \times 2 \times 1}{(4 \times 3 \times 2 \times 1) \times (4 \times 3 \times 2 \times 1)} = 2 \times 7 \times 5 = 70$ microstates consisting of five configurations, with a sample microstate for each configuration depicted in Figure 11.3. The most probable configuration is the one with the maximum multiplicity, the configuration in which the particles are equally mixed between the pair of compartments, such that $\Omega_{mp} = 36$. This conclusion is consistent with our expectation that ideal gases mix, and we observe in general that processes proceed in directions towards equilibria corresponding to the most probable configurations.

Boltzmann proposed that the maximum multiplicity at the molecular level, corresponding to the multiplicity of the most probable configuration, can be related to the macroscopic property molar entropy (s), a relationship known as the Boltzmann equation:

$$s = R \ln \Omega_{mp} \qquad [11.4]$$

We find from this equation that molar entropy has the same units as the gas constant (*i.e.*, J/mol-K), such that the total entropy of a sample of matter has units of J/K.

11.2 Evaluating Changes in Entropy for Physical and Chemical Processes

Processes that we observe proceed in directions towards greater molecular-level disorder and increased multiplicity, such that entropy increases in processes that actually occur. This interpretation of entropy is sometimes conceived macroscopically as a reflection of "time's arrow": systems evolve over time such that entropy increases. Processes in which the entropy increases are exoentropic; processes in which entropy decreases are endoentropic. From a practical perspective, evaluating changes in entropy typically offers a more feasible approach for specifying the direction a process—such a reaction—occurs than counting the number of microscopic configurations and identifying the configuration with the greatest multiplicity.

The molar entropy of a substance depends on its composition. The molar entropy of a pure substance increases with increased structural complexity, corresponding to a greater multiplicity in the atomic arrangements, for a given molar mass. For example, pyrolidine (C_4H_9N, with a cyclic structure formed by a five-member ring of carbon and nitrogen atoms) (Figure 11.4a), has a greater molar entropy than molecular chlorine (Figure 11.4b), although both substances have similar molar masses of ~71 g/mol. Structural complexity is a form of chemical disorder. The molar entropy for a mixture is greater than the sum of the weighted averages (based on mole fractions) of the molar entropies of the corresponding pure components because disorder increases upon mixing. Mixing is a second form of chemical disorder.

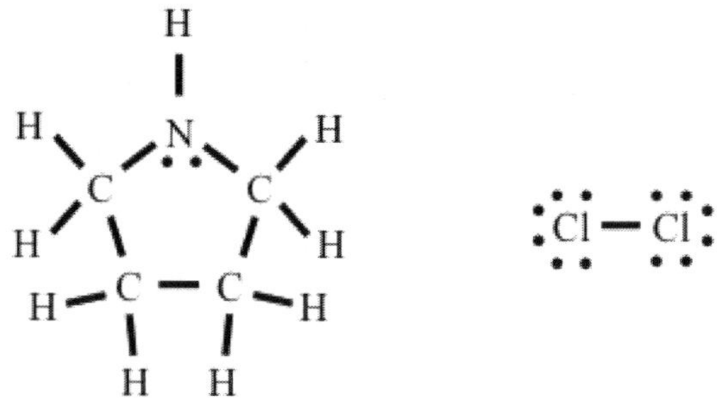

(a) (b)

Figure 11.4. Structures of two compounds with similar molar masses of ~71 g/mol but significantly different molar entropies. Pyrolidine (a) has a more complex structure and larger molar entropy than molecular chlorine (b).

The molar entropy of a substance also depends on its state. Molar entropy increases as temperature increases because particles can occupy a broader distribution of energy levels at higher temperature (*i.e.*, greater thermal disorder results). Molar entropy increases as pressure decreases because particles exhibit greater positional disorder at lower temperatures. This effect, however, is significant only for gases. Molar entropy also depends on the state of matter: at the triple point the molar entropy of a gas is greater than the molar entropy of the liquid, which in turn is greater than the molar entropy of the solid, as illustrated in Figure 11.5. The difference between any two pairs of phases also applies at any common temperature and pressure at which the two phases coexist. The third law of thermodynamics provides a lower limit on the absolute molar entropy: for a pure, perfect crystal (*i.e.*, matter in the limit of perfect ordering and no microscopic motion) $\lim_{T \to 0\ \mathrm{K}} s = 0$ J/mol – K.

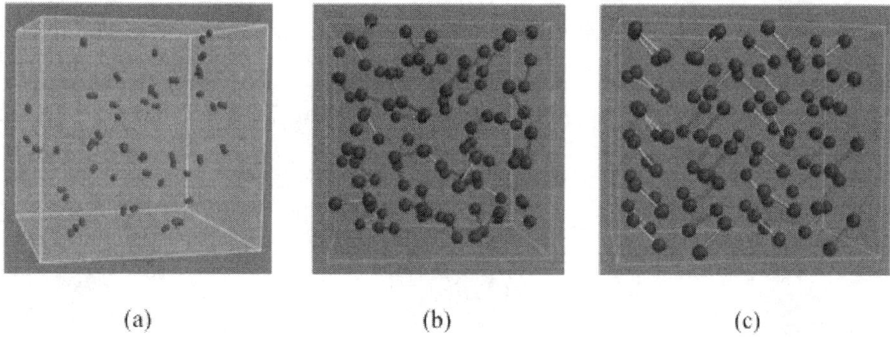

Figure 11.5. Microscopic organization of different phases of I_2 at the triple point. (a) Gas; (b) liquid; (c) solid.

We can reach a set of qualitative conclusions regarding changes in entropy for different physical and chemical processes. Qualitative evaluation of whether molar entropy increases or decreases is more straightforward for physical processes. For example, for mixing $\Delta s_{mixing} > 0$, for endothermic phase transitions (*e.g.*, fusion, vaporization, and sublimation) $\Delta s_{trans} > 0$, but for exothermic phase transitions (*e.g.*, freezing, condensation, and deposition) $\Delta s_{trans} < 0$. Yet qualitatively evaluating changes in entropy associated with chemical reactions can be difficult. Assuming similar levels of molecular complexity in reactants and products, increases in the number of moles of gas are associated with $\Delta s_{rxn} > 0$ and decreases in the number of moles of gas with $\Delta s_{rxn} < 0$. Further qualitative evaluation is problematic when a reaction, as it typically does, involves changes in the level of molecular complexity.

Example 11.1: Qualitative Evaluation of Changes in Molar Entropy for Physical Processes

For each of the following processes determine whether Δs is positive or negative.

a) Condensation of tetrachloromethane vapor at its normal boiling point.

b) Mixing of liquid ethanol and liquid water.

c) Compressing neon gas isothermally from 0.5 atm to 1.5 atm.

d) Heating solid copper from 275 K to 295 K.

e) Grinding a crystal of sodium chloride into a fine powder.

f) Stretching a rubber band made of an entangled network of cross-linked polyisoprene chains.

Solutions

a) Because the molar entropy of the gas is greater than the molar entropy of the corresponding liquid at the boiling point, $\Delta s < 0$.

b) Because mixing always involves an increase in entropy, $\Delta s > 0$.

c) Because molar entropy decreases with increasing pressure at a fixed temperature and phase, $\Delta s < 0$.

d) Because molar entropy increases with temperature (and the molar entropy of a solid does not significantly depend on pressure), $\Delta s > 0$.

e) Because grinding a crystal is associated with disrupting the crystalline lattice and an increase in molecular-level disorder, $\Delta s > 0$.

f) Because stretching a rubber band causes the chains to become more aligned and ordered, $\Delta s < 0$.

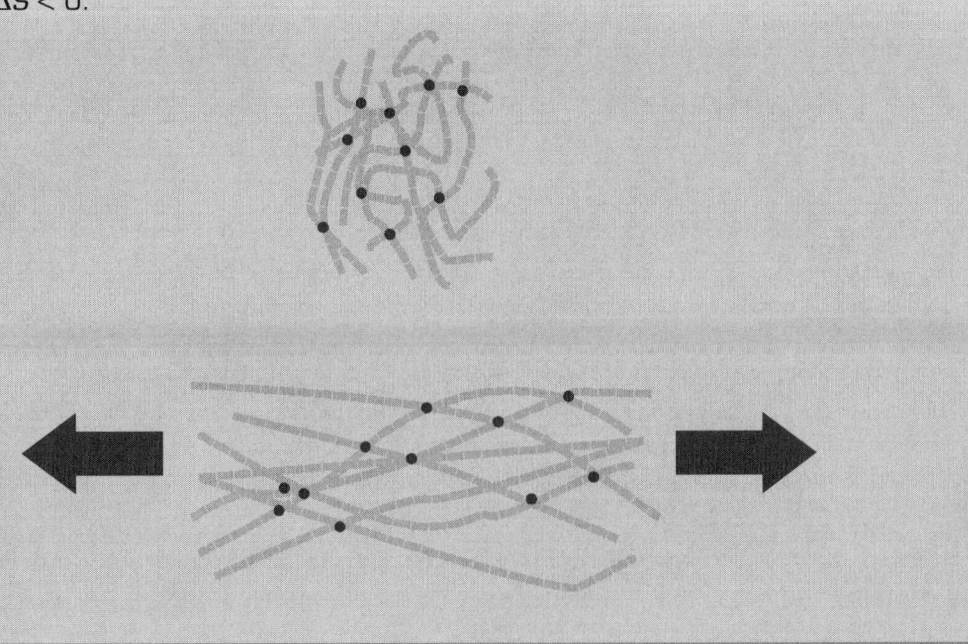

We can quantitatively evaluate the change in molar entropy for a reaction based on data for standard molar entropies, s_i° (expressed in J/mol-K), for reactant(s) and product(s) at standard state (*i.e.*, $P = P^o =$ 1 bar, $C = C^o = 1$ M, and each reactant and product is present in its pure and reference form at 1 bar and the temperature of interest) using the equivalent of Hess's law applied to entropy (because entropy, like enthalpy, is a state property):

$$\Delta s_{rxn}^\circ = \sum \upsilon_i s_i^\circ \qquad [11.5]$$

where the summation is over both reactants and products and analogous to Equation [8.13] for determining molar changes in enthalpy of reaction based on standard molar enthalpies of formation. Data for standard molar entropies are tabulated at 298.15 K and are accessible in the *CRC Handbook of Chemistry and Physics* (available at http://www.hbcpnetbase.com/) and through the NIST Chemistry Webbook (http://webbook.nist.gov). Note that all substances have positively-valued standard molar entropies at room temperature as a consequence of the third law of thermodynamics and the effect of temperature on molar entropy. Similar to what we discussed for the effect of temperature and pressure on reaction enthalpy in Section 8.7, we typically assume that reaction entropy, Δs_{rxn}°, is independent of temperature and pressure.

Example 11.2: Evaluating the Reaction Entropy for Oxidation of Metallic Aluminum

Consider the reaction that might occur on the surface of a freshly-machined aluminum part:

$$4\ Al(s) + 3\ O_2(g) \rightarrow 2\ Al_2O_3(s) \qquad \text{Rxn [11.E1]}$$

What is Δs_{rxn}° (in J/mol-K) if at 298.15 K $s_{Al(s)}^\circ = 28.33$ J/mol − K, $s_{O_2(g)}^\circ = 205.1$ J/mol − K, and $s_{Al_2O_3(s)}^\circ = 50.92$ J/mol − K?

Solutions

Apply Equation [11.5] with the given stoichiometry and values for standard molar entropies:

$$\Delta s_{rxn}^\circ = \upsilon_{Al} s_{Al(s)}^\circ + \upsilon_{O_2} s_{O_2(g)}^\circ + \upsilon_{Al_2O_3} s_{Al_2O_3(s)}^\circ$$

$$= (-4)(28.33\ \text{J/mol} - \text{K}) + (-3)(205.1\ \text{J/mol} - \text{K}) + (2)(50.92\ \text{J/mol} - \text{K})$$

$$\Delta s_{rxn}^\circ = -626.8\ \text{J/mol} - \text{K}$$

We see that the qualitative effect of decreasing the number of moles of gas upon reaction on the change in entropy upon reaction is more significant than the effect of increasing structural complexity.

11.3 The Second Law of Thermodynamics and Heat Engines

We have concluded thus far that processes occur in directions that result in increased molecular-level disorder, and that evaluating changes in entropy provides a macroscopic basis for representing the consequences of changes in molecular-level disorder. We seek to apply these conclusions in a practical fashion to distinguish spontaneous processes—ones that occur eventually when the system is left undisturbed—from non-spontaneous processes—ones that will never occur unless there is a continuous transfer of energy in the form of heat and/or work with the surroundings. For example, we can compress a gas and create a vacuum if we do work on a gas by increasing the pressure in the surroundings, but the gas will expand once the pressure is dropped back to a vacuum. Note that identifying that a process is spontaneous tells us absolutely nothing about how fast the process occurs.

The second law of thermodynamics provides a quantitative basis for specifying whether a process is spontaneous:

$$\Delta S_{uni} = \Delta S_{sys} + \Delta S_{surr} \begin{cases} > 0 \text{ for a spontaneous process} \\ = 0 \text{ at equilibrium} \\ < 0 \text{ for a non-spontaneous process} \end{cases} \quad [11.6]$$

where for any process $\Delta S_{surr} = \frac{q_{surr}}{T_{surr}}$. An alternative statement of the second law, the Kelvin-Planck formulation, is that any engine using heat to generate work must include both a heat source and a heat sink: "It is impossible for a system to undergo a cyclic process whose sole effects are the flow of heat into the system and the performance of an equivalent amount of work by the system." This statement of the second law tells us that perpetual motion machines are impossible. To consider why compare the heat engines, cycles in which work is performed as a result of the flow of heat (*e.g.*, by the release of heat by combustion of a fuel as a heat source), depicted in Figure 11.6. We evaluate the capability of heat engines based on their thermodynamic efficiency, defined as the magnitude of the amount of work they perform per amount of heat transferred from the heat source:

$$\eta \equiv \frac{|w|}{q_{hot}} \quad [11.7]$$

where any engine of possible utility produces work (*i.e.*, $w < 0$). Applying the first law of thermodynamics to the implausible engine with heat transfer $q_{hot} > 0$ from the heat source to the engine yields:

$$\Delta U_{cycle} = q_{hot} + w = 0$$
$$w = -q_{hot} < 0$$

such that its efficiency is 100%. In contrast, applying the first law to the plausible engine with $q_{hot} > 0$ and heat transfer $q_{cold} < 0$ from the engine to the heat sink:

$$\Delta U_{cycle} = q_{hot} + q_{cold} + w = 0$$
$$w = -(q_{hot} + q_{cold}) < 0$$

implying that $|q_{hot}| > |q_{cold}|$ (*i.e.*, more heat is transferred from the hot reservoir to the engine than is transferred from the engine to the cold reservoir) and that the efficiency is less than 100%:

$$\eta = \frac{|-(q_{hot} + q_{cold})|}{q_{hot}} = 1 + \frac{q_{cold}}{q_{hot}} < 1$$

The upper limit on efficiency, corresponding to the heat engine producing the maximum amount of work as a Carnot cycle, occurs when the cycle is operated reversibly (as introduced in Section 8.4), is substantially less than 100%, and depends on the temperature T_{hot} and T_{cold} for the heat source and heat sink, respectively, where these temperatures are expressed in K:

$$\eta_{rev} = \frac{T_{hot} - T_{cold}}{T_{hot}} \quad [11.8]$$

In practice, a gasoline-powered engine operating as an irreversible Otto cycle has an ideal efficiency of .20–0.30. A diesel-powered engine—which uses a high compression ratio and operates without spark plugs but at higher temperatures—has a greater ideal efficiency of 0.30–0.35.

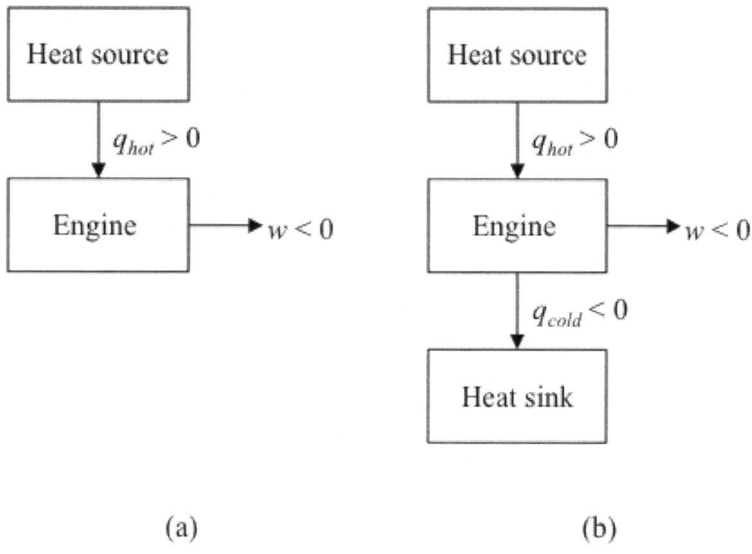

Figure 11.6. Comparison of (a) an implausible heat engine and (b) a plausible heat engine.

Example 11.3: Evaluating a Gasoline-Powered Engine as a Heat Engine

A typical street-car engine operates with an engine temperature of 85. °C and a maximum power of $|\dot{w}_{max}| = 140\text{ hp}$, where 1 hp = 746 W was defined by Joule as the power a "hard-working horse" expends lifting a 330-pound mass 100 feet in 1 minute.

a) What is the maximum efficiency for the engine, assuming the surroundings at 20. °C act as the heat sink.

b) What is the minimum rate (in g/s) at which gasoline, with a fuel value of 47.9 kJ/g, must be burned when the engine is delivering maximum power? *Hint: efficiency can be expressed in terms of power as* $\eta = \frac{|\dot{w}|}{\dot{q}_{hot}}$.

Solutions

a) Assume the engine has a maximum efficiency corresponding to a reversible Carnot engine according to Equation [11.8] with T_{hot} = 85. °C = 358. K and T_{cold} = 20. °C = 293. K:

$$\eta_{max} = \eta_{rev} = \frac{T_{hot} - T_{cold}}{T_{hot}} = \frac{(358.\text{ K}) - (293.\text{ K})}{(358.\text{ K})} = 0.18$$

b) Apply the definition for efficiency, Equation [11.7], expressed in terms of rates of work (power produced), $\dot{w}_{max}$, and rate of heat transfer from the heat source, $\dot{q}_{hot}$. The minimum rate at which heat is transferred and gasoline is burned corresponds to the maximum efficiency, with the rate of heat transfer given as the product of the specific rate of consumption of gasoline, $\dot{m}$, and the fuel value (equal to the negative of the specific heat of reaction, $\Delta\hat{h}_{rxn}$). Rearrange to solve for the minimum specific rate of consumption of gasoline:

$$\eta_{max} = \frac{|\dot{w}|}{\dot{q}_{hot,min}} = \frac{|\dot{w}|}{\dot{m}_{min}\Delta\hat{h}_{rxn}}$$

$$\dot{m}_{min} = \frac{|\dot{w}|}{\eta_{max}\Delta\hat{h}_{rxn}} = \frac{\left(140\text{ hp}\times\frac{746\text{ W}}{1\text{ hp}}\times\frac{1\text{ kJ/s}}{10^3\text{ W}}\right)}{(0.18)(47.9\text{ kJ/g})} = 12.\text{ g/s}$$

This rate of consumption roughly corresponds to a full 15-gallon tank of gasoline per hour!

11.4 The Change in Gibbs Free Energy as a Criterion for Spontaneity at Constant Temperature and Pressure

It would be more convenient to evaluate whether a process is spontaneous by evaluating only changes experienced by the system and ignoring what occurs in the surroundings. For processes occurring at constant temperature and pressure, which is a relevant description for reactions, the second law of thermodynamics can be formulated in terms of changes in the system's Gibbs free energy, defined as $G \equiv H - TS$ (where temperature is expressed in K. At constant temperature the change in Gibbs free energy can be expressed as:

$$\Delta G = \Delta(H - TS) = \Delta H - \Delta(TS)$$

$$\Delta G = \Delta H - T\Delta S \qquad [11.9]$$

where Equation [11.9] is known as the Gibbs-Helmholtz equation. For a process occurring at constant temperature and pressure, for which $q_P = \Delta H$ (Equation [8.7]), we find:

$$q_{surr} = -q_{sys} = -\Delta H_{sys}$$

$$\Delta S_{surr} = \frac{q_{surr}}{T_{surr}} = \frac{q_{surr}}{T} = -\frac{\Delta H_{sys}}{T}$$

$$\Delta H_{sys} = -T\Delta S_{surr}$$

$$\Delta G_{sys} = \Delta H_{sys} - T\Delta S_{sys} = -T\Delta S_{surr} - T\Delta S_{sys} = -T(\Delta S_{surr} + \Delta S_{sys})$$

$$\Delta G_{sys} = -T\Delta S_{uni}$$

Thus, we can formulate the second law of thermodynamics in terms of Gibbs free energy to reactions conducted at constant temperature and pressure:

$$\Delta G_{sys}\begin{cases} < 0 \text{ for a net reaction of reactants} \rightarrow \text{products} \\ = 0 \text{ for no net reaction (\textit{i.e.}, equilibrium)} \\ > 0 \text{ for a net reaction of products} \rightarrow \text{reactants} \end{cases} \qquad [11.10]$$

The Gibbs free energy increases in exergonic processes and decreases in endergonic processes.

Example 11.4: Evaluating the Change in Entropy and Disorder Upon Melting Ice

Consider the equilibrium between solid and liquid water, $H_2O(s) \leftrightarrows H_2O(l)$, at the normal melting point of ice, $T_{fus} = 0.00\ °C$.

a) If $\Delta h_{fus} = 6.01\ \text{kJ/mol}$, what is Δs_{fus} (in J/mol-K) for ice at its normal melting point? Assume both Δh_{fus} and Δs_{fus} are independent of temperature.

b) What is the value of the ratio of the most probable number of configurations for water in its liquid state to water in its ice state, $\frac{\Omega_{H_2O(l),mp}}{\Omega_{H_2O(s),mp}}$, at its normal melting point?

Solutions

a) Apply the Gibbs-Helmholtz equation, Equation [11.9], expressed on a molar basis and the definition for equilibrium based on Equation [11.10]:

$$\Delta g_{fus}(T = T_{fus}) = \Delta h_{fus} - T_{fus}\Delta s_{fus} = 0$$

$$\Delta s_{fus} = \frac{\Delta h_{fus}}{T_{fus}} = \frac{(6.01\ \text{kJ/mol})}{(273.15\ \text{K})} \times \frac{10^3\ \text{J}}{1\ \text{kJ}} = 22.0\ \text{J/mol} - \text{K}$$

b) Apply the Boltzmann Equation, Equation [11.4], to the initial and final states of ice and liquid water and rearrange to solve for the ratio of the most-probable number of configurations:

$$\Delta s_{fus} = s_{H_2O(l)} - s_{H_2O(s)} = (R\ln\Omega_{H_2O(l),mp}) - (R\ln\Omega_{H_2O(s),mp}) = R\ln\left(\frac{\Omega_{H_2O(l),mp}}{\Omega_{H_2O(s),mp}}\right)$$

$$\frac{\Omega_{H_2O(l),mp}}{\Omega_{H_2O(s),mp}} = e^{\frac{\Delta s_{fus}}{R}} = e^{\frac{(22.00\ \text{J/mol-K})}{(8.314\ \text{J/mol-K})}} = 14.10$$

We conclude there are more than fourteen times as many ways to arrange water molecules in the liquid state at the normal melting point compared with the solid state.

Based on the formulation of the second law in terms of Gibbs free energy, we can identify how temperature affects whether a process is spontaneous (assuming that changes in enthalpy and entropy for the process are independent of temperature). Consider the form for the Gibbs-Helmholtz equation, Equation [11.9], expressed on a molar basis:

$$\Delta g = \Delta h - T\Delta s \qquad [11.11]$$

We observe that the sign for the change in molar Gibbs free energy, Δg, depends on the values and signs for the change in enthalpy, the change in entropy, and the temperature (Table 11.1). Exothermic processes that also are exoentropic (*i.e.*, $\Delta h < 0$ and $\Delta s > 0$) are spontaneous, and endothermic processes that also are endoentropic (*i.e.*, $\Delta h > 0$ and $\Delta s < 0$) are non-spontaneous, regardless of temperature. Exothermic processes that are also endoentropic (*i.e.*, $\Delta h < 0$ and $\Delta s < 0$), however, are spontaneous only at relatively low temperatures. Similarly, endothermic processes that also are exoentropic (*i.e.*, $\Delta h > 0$ and

$\Delta s < 0$) are spontaneous only at relatively high temperatures. The temperature at which a process changes from being spontaneous to non-spontaneous or vice versa is termed the crossover temperature. Figure 11.7 depicts these behaviors graphically. At the crossover temperature the process is at equilibrium. Note that a non-spontaneous process can occur if it is coupled to a spontaneous process such that the overall change in Gibbs free energy for the combined processes is negative. Living organisms take advantage of this coupling to power biological processes.

Table1.1. Effect of Temperature on Spontaneity of Process Occurring at Constant Temperature and Pressure

Sign for: Δh	Sign for: Δs	Range for T	Sign for Δg	Spontaneous?	Example
–	+	All	–	Yes	$2\ Na(s) + 2\ H_2O(l) \rightarrow 2\ NaOH(aq) + H_2(g)$
+	–	All	+	No	$3\ O_2(g) \rightarrow 2\ O_3(g)$
–	–	Low	–	Yes	$H_2O(l) \rightarrow H_2O(s)$
		High	+	No	
+	+	Low	+	No	$H_2O(l) \rightarrow H_2O(g)$
		High	–	Yes	

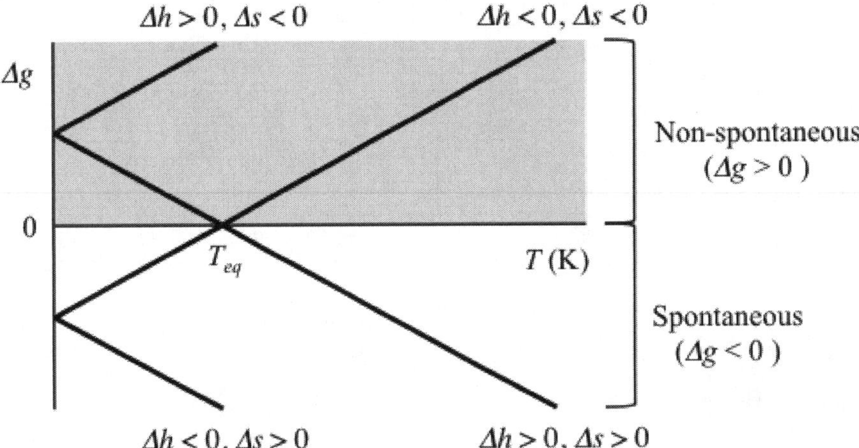

Figure 11.7. Effect of temperature on the spontaneity of a process based on Equation [11.11]. The slope of plots of Δg vs. T is Δs with a y-intercept of Δh. The crossover temperature is T_{eq}. For purpose of illustration each process has values for Δh with similar magnitudes and values for Δs with similar magnitudes.

Two strategies are available for us to evaluate the change in Gibbs free energy for reactions conducted at standard state (*i.e.*, all gases are present at standard pressures of 1 bar and all solutes are present at standard concentrations of 1 M) If we have available values for the change in enthalpy and change in entropy for the reaction at standard state (which to a first approximation we assume are independent of temperature), we can determine the change in Gibbs free energy for the reaction at standard state based on the Gibbs-Helmholtz equation:

$$\Delta g^{\circ}_{rxn} = \Delta h^{\circ}_{rxn} - T\Delta s^{\circ}_{rxn} \qquad [11.12]$$

Note that it is never appropriate to assume that changes in Gibbs free energy are independent of temperature!

Example 11.5: The Oxidation of Metallic Aluminum

Reconsider the oxidation of metallic aluminum, Rxn [11.E1], introduced in Example 11.2. The standard molar enthalpy of formation of aluminum oxide is -1676. kJ/mol.

a) Does aluminum oxidize spontaneously under standard-state conditions at 25.00 °C?

b) What is the crossover temperature for the oxidation of metallic aluminum?

Solutions

a) Apply Equation [11.12] with $\Delta h^{\circ}_{rxn} = -1676.\ \text{kJ/mol}$, $\Delta s^{\circ}_{rxn} = -626.8\ \text{J/mol}-\text{K}$ (from Example 11.2), and an absolute temperature of T = 298.15 K, converting units as necessary:

$$\Delta g^{\circ}_{rxn} = \Delta h^{\circ}_{rxn} - T\Delta s^{\circ}_{rxn} = (-1676.\ \text{kJ/mol}) - (298.15\ \text{K})\left(-626.8\ \text{J/mol}-\text{K} \times \frac{1\ \text{kJ}}{10^3\ \text{J}}\right)$$

$$\Delta g^{\circ}_{rxn} = -1489.\ \text{kJ/mol}$$

We expect metallic aluminum to spontaneously oxidize at standard state and room temperature.

b) Determine the crossover temperature, T_{eq}, as the temperature for which Equation [11.12] is equal to zero:

$$\Delta g^{\circ}_{rxn} = \Delta h^{\circ}_{rxn} - T_{eq}\Delta s^{\circ}_{rxn} = 0$$

$$T_{eq} = \frac{\Delta h^{\circ}_{rxn}}{\Delta s^{\circ}_{rxn}} = \frac{(-1676.\ \text{kJ/mol})}{\left(-626.8\ \text{J/mol}-\text{K} \times \frac{1\ \text{kJ}}{10^3\ \text{J}}\right)} = 2674.\ \text{K}$$

As the melting points of metallic aluminum and aluminum oxide are 934 K and 2327 K, respectively, we always expect solid aluminum to oxidize at standard state, regardless of the temperature.

A second strategy for evaluating the change in Gibbs free energy for reactions conducted at standard state is based on using the equivalent of Hess's law applied to the Gibbs free energy (the Gibbs free energy, like enthalpy and entropy, is a state property):

$$\Delta g^{\circ}_{rxn} = \sum \upsilon_i \Delta g^{\circ}_{f,i} \qquad [11.13]$$

where the summation, again, is over both reactants and products and analogous to Equations [8.13] and [11.5], and $\Delta g^{\circ}_{f,i}$ is the standard molar Gibbs free energy of formation of substance i (expressed in kJ/mol), defined as the change in Gibbs free energy for the reaction to form one mole of substance i at standard state from its constitutive pure elements in their reference form at standard state. Based on this definition $\Delta g^{\circ}_{f,i} = 0$ kJ/mol (like $\Delta h^{\circ}_{f,i} = 0$ kJ/mol) for a pure element in its reference form. Data for standard molar Gibbs free energies of formation are tabulated at 298.15 K and are accessible in the *CRC Handbook of Chemistry and Physics* (available at http://www.hbcpnetbase.com/) and through the NIST Chemistry Webbook (http://webbook.nist.gov).

Example 11.6: Dissolution of Magnesium Carbonate in Water

Magnesium carbonate dissolves in water according to the reaction:

$MgCO_3(s) \leftrightarrows Mg^{2+}(aq) + CO_3^{2-}(aq)$ Rxn [11.E2]

Is the dissolution of magnesium carbonate in water spontaneous under standard-state conditions at 298.15 K if at this temperature $\Delta g^{\circ}_{f,MgCO_3(s)} = -1012.2$ kJ/mol, $\Delta g^{\circ}_{f,Mg^{2+}(aq)} = -454.8$ kJ/mol, and $\Delta g^{\circ}_{f,CO_3^{2-}(aq)} = -527.8$ kJ/mol?

Solutions

Apply Equation [11.13] with the given stoichiometry and values for standard molar Gibbs free energies of formation:

$$\Delta g^{\circ}_{rxn} = \upsilon_{MgCO_3}\Delta g^{\circ}_{f,MgCO_3(s)} + \upsilon_{Mg^{2+}}\Delta g^{\circ}_{f,Mg^{2+}(aq)} + \upsilon_{CO_3^{2-}}\Delta g^{\circ}_{f,CO_3^{2-}(aq)}$$
$$= (-1)(-1012.2 \text{ kJ/mol}) + (1)(-454.8 \text{ kJ/mol}) + (1)(-527.8 \text{ kJ/mol})$$
$$\Delta g^{\circ}_{rxn} = 29.6 \text{ kJ/mol}$$

The solubilization of magnesium carbonate is non-spontaneous at standard state and 298.15 K.

11.5 Relating Thermodynamic Descriptions of Spontaneity and Chemical Equilibrium

Thermodynamics can provide a quantitative link between the concepts of spontaneity and chemical equilibrium. It can be shown that we can determine the change in Gibbs free energy for a reaction under non-standard state conditions (*e.g.*, not all gases are present at a standard pressure of 1 bar and/or not all solutes are present at standard concentrations of 1 M) based on the change in Gibbs free energy at standard state and the reaction quotient in terms of activities, Q, from Chapter 10:

$$\Delta g_{rxn} = \Delta g^{\circ}_{rxn} + RT\ln Q \qquad [11.14]$$

Given that equilibrium corresponds to $\Delta g_{rxn,eq} = 0$ and $Q = K$, where K is the equilibrium constant in terms of activities from Chapter 10:

$$\Delta g_{rxn,eq} = 0 = \Delta g^{\circ}_{rxn} + RT\ln K$$

$$\Delta g^{\circ}_{rxn} = -RT\ln K \qquad [11.15]$$

$$\ln K = -\frac{\Delta g^\circ_{rxn}}{RT} \qquad [11.16]$$

$$K = e^{-\frac{\Delta g^\circ_{rxn}}{RT}} \quad [11.17]$$

Example 11.7: Evaluating The Equilibrium Constants for the Oxidation of Metallic Aluminum and Dissolution of Magnesium Carbonate in Water

Determine the value for the equilibrium constant in terms of activities, *K*, for the following pair of reactions:

a) The dissolution of magnesium carbonate in water, Rxn [11.E2] (previously considered in Example 11.6), at 25.00 °C.

b) The oxidation of metallic aluminum, Rxn [11.E1] (previously considered in Examples 11.2 and 11.5), at 200. °C.

Solutions

a) Apply Equation [11.17] using $\Delta g^\circ_{rxn} = 29.6 \text{ kJ/mol}$ at *T* = 298.15 K from Example 11.6, expressing the changes in Gibbs free energy of reaction at standard state in units of J/mol:

$$K = \exp\left(\frac{-\Delta g^\circ_{rxn}}{RT}\right) = \exp\left[\frac{-\left(29.6 \text{ kJ/mol} \times \frac{10^3 \text{ J}}{1 \text{ kJ}}\right)}{(8.314 \text{ J/mol} - \text{K})(298.15 \text{ K})}\right] = 6.52 \times 10^{-6}$$

Equilibrium does not favor the dissolution of magnesium carbonate in water at room temperature. This salt is sparingly soluble in water at this temperature and a contributor to precipitates formed with "hard" water, as introduced in Section 7.2.

b) To evaluate the equilibrium constant at 200. °C using Equation [11.16] we first need to recall that the change in Gibbs free energy depends on temperature, such that we cannot use the value found in Example 11.5 for room temperature at a different temperature. Determine the change in molar Gibbs free energy for the reaction at standard state at *T* = 473. K using Equation [11.12] and the data $\Delta h^\circ_{rxn} = -1676.\ \text{kJ/mol}$ and $\Delta s^\circ_{rxn} = -626.8 \text{ J/mol} - \text{K}$ from Examples 11.5 and 11.2, respectively:

$$\Delta g^\circ_{rxn} = \left(-1676.\ \text{kJ/mol} \times \frac{10^3 \text{ J}}{1 \text{ kJ}}\right) - (473.\ \text{K})(-626.8 \text{ J/mol} - \text{K}) = -1.380 \times 10^6 \text{ J/mol}$$

where we express our value for the change in molar Gibbs free energy in J/mol for convenience for the next part of this problem. Now apply Equation [11.16]:

$$K = \exp\left[\frac{-(-1.380 \times 10^6 \text{ J/mol})}{(8.314 \text{ J/mol} - \text{K})(473.\ \text{K})}\right] = 2.24 \times 10^{152}$$

The enormous value for the equilibrium constant tells us that at standard-state conditions—corresponding to a partial pressure of molecular oxygen of 1 bar—at 200. °C, the product of Rxn [11.E2], aluminum oxide, is favored overwhelmingly compared with the reactants, metallic aluminum and oxygen (*i.e.*, we should expect this reaction goes to completion).

We now attach a physical significance linking values for the change in Gibbs free energy for a reaction at standard state and the equilibrium constant in terms of activities at a given temperature. If $\Delta g^{\circ}_{rxn} < 0$, we find $K > 1$ and the products are favored over the reactants at equilibrium. In particular, we specify that if $\Delta g^{\circ}_{rxn} \leq -100\ \text{kJ/mol}$, $K >> 1$ and the reaction essentially goes to completion. This condition typically describes combustion reactions. In contrast, if $\Delta g^{\circ}_{rxn} > 0$, $K < 1$ and the reactants are favored over the products at equilibrium. In particular, we specify that if $\Delta g^{\circ}_{rxn} \geq +100\ \text{kJ/mol}$, $K << 1$ and the reaction essentially does not occur. An example of such a reaction is the electrolysis of water:

$$H_2O(l) \rightarrow H_2(g) + \tfrac{1}{2}\, O_2(g) \qquad \text{Rxn [11.1]}$$

a reaction we will consider further in Chapter 12 as a source of hydrogen for fuel cells. If $\Delta g^{\circ}_{rxn} \sim 0$, however, $K \sim 1$ and the products and reactants are roughly equally favored at equilibrium. In particular, we specify that if $-100\ \text{kJ/mol} \leq \Delta g^{\circ}_{rxn} \leq +100\ \text{kJ/mol}$, equilibrium calculations are warranted.

Example 11.8: The Carbonation of Water

Reconsider the carbonation of water, previously examined in Section 10.5 from the perspective of LeChâtelier's principle:

$$CO_2(g) + H_2O(l) \leftrightarrows H_2CO_3(aq) \quad \Delta h^{\circ}_{rxn} = -10.39\ \text{kJ/mol} \qquad \text{Rxn [10.6]}$$

If $\Delta g^{\circ}_{rxn} = 8.43\ \text{kJ/mol}$ at 298.15 K, is Rxn [10.6] at equilibrium at 298.15 K if $P_{CO_2} = 2.0$ bar (typical of bottled soda) and $C_{H_2CO_3} = 1.0 \times 10^{-2}$ M?

Solutions

There are two possible strategies we can apply: compare the value for the reaction quotient in terms of activities, *Q*, vs. the value for the equilibrium constant in terms of activities, *K*, based on the criteria introduced in Chapter 10, or evaluate the change in molar Gibbs free energy for the reaction at the given temperature $\Delta G_{rxn,m}$, based on the second law formulated in terms of Gibbs free energy, Equation [11.10]. Note that both strategies require knowledge of the reaction quotient in terms of activities.

$$Q = \frac{a_{H_2CO_3(aq)}}{a_{CO_2(g)}\, a_{H_2O(l)}} = \frac{\left(\dfrac{C_{H_2CO_3}}{C^{\circ}}\right)}{\left(\dfrac{P_{CO_2}}{P^{\circ}}\right)(1)} = \frac{\left[\dfrac{(1.0 \times 10^{-2}\ \text{M})}{(1\ \text{M})}\right]}{\left[\dfrac{(2.0\ \text{bar})}{(1\ \text{bar})}\right](1)} = 5.0 \times 10^{-3}$$

Apply the strategy based on the second law:

$$\Delta g_{rxn} = \Delta g^{\circ}_{rxn} + RT\ln Q$$
$$= \left(8.43\ \text{kJ/mol} \times \frac{10^3\ \text{J}}{1\ \text{kJ}}\right) + (8.314\ \text{J/mol - K})(298.15\ \text{K})\ln(5.0\times10^{-3})$$
$$\Delta g_{rxn} = -4.7\times10^3\ \text{J/mol}$$

Because $\Delta g_{rxn} < 0$, the reaction is spontaneous (*i.e.*, not at equilibrium) and proceeds towards the product, with the reaction quotient increasing until equilibrium is reached with $\Delta g_{rxn} = 0$.

11.6 Evaluating How the Equilibrium Constant Quantitatively Depends on Temperature

Although LeChâtelier's principle can provide qualitative insight into how temperature affects chemical equilibrium, we seek a quantitative relationship between temperature and the value of the equilibrium constant. One might argue that we could use the strategy applied in Example 11.7b: evaluate the change in molar Gibbs free energy at a temperature of interest and apply the relationship between this change and the equilibrium constant, Equation [11.17]. But this strategy can be tedious for some problems. As a convenient alternative, approach this problem by first substituting the Gibbs-Helmholtz Equation for the change in molar Gibbs free energy for a reaction, Equation [11.12], into Equation [11.16]:

$$\ln K = \frac{-\Delta g^{\circ}_{rxn}}{RT} = \frac{-(\Delta h^{\circ}_{rxn} - T\Delta s^{\circ}_{rxn})}{RT} = \frac{-\Delta h^{\circ}_{rxn}}{RT} + \frac{\Delta s^{\circ}_{rxn}}{R} \qquad [11.18]$$

Typically, we encounter problems for which we need to compare values for the equilibrium constant at two different temperatures. If we identify $K_1 = K(T_1)$ and $K_2 = K(T_2)$ as the respective values for the equilibrium constant for a reaction at the pair of temperatures T_1 and T_2 and reasonably assume that the changes in molar enthalpy and entropy for the reaction at standard state are independent of temperature, subtracting Equation [11.18] evaluated at T_2 from its evaluation at T_1 yields:

$$\ln K_2 - \ln K_1 = \left(\frac{-\Delta h^{\circ}_{rxn}}{RT_2} + \frac{\Delta s^{\circ}_{rxn}}{R}\right) - \left(\frac{-\Delta h^{\circ}_{rxn}}{RT_1} + \frac{\Delta s^{\circ}_{rxn}}{R}\right) = -\frac{\Delta h^{\circ}_{rxn}}{R}\left(\frac{1}{T_2} - \frac{1}{T_1}\right)$$

$$\ln\left(\frac{K_2}{K_1}\right) = -\frac{\Delta h^{\circ}_{rxn}}{R}\left(\frac{1}{T_2} - \frac{1}{T_1}\right) \qquad [11.19]$$

Equation [11.18] is known as the van't Hoff equation for chemical equilibrium.

Note the great similarity between the van't Hoff equation for chemical equilibrium and the Arrhenius equation (Equation [9.23]) from Chapter 9. As we did with the Arrhenius equation, when using the van't Hoff equation for chemical equilibrium express the change in molar enthalpy for the reaction at standard state, Δh°_{rxn}, in units of J/mol, use the gas constant expressed as $R = 8.314$ J/mol-K, and express temperature in absolute units of K. Types of problems we can now address using Equation [11.19] include evaluating the value for the equilibrium constant at one temperature given the value of the equilibrium constant at a different temperature and the standard molar enthalpy change of reaction, determining the standard molar enthalpy change of reaction given values for the equilibrium constant at a pair of temperatures, and identifying an unknown temperature at which the equilibrium constant has a specific value given the value of the equilibrium constant at one temperature and the standard molar enthalpy change of reaction. Note that the van't Hoff equation for chemical equilibrium can be applied to any form for the equilibrium constant (*i.e.*, K_c, K_p, K_w, K_a, K_b, and K_{sp}) as long as both equilibrium constants in the expression are of the same form.

Example 11.9: The Effect of Temperature on the Equilibrium Constant for the Carbonation of Water

If the equilibrium constant in terms of activities for the carbonation of water (Example 11.8) given as Rxn [10.6] is 3.33×10^{-2} at 25. °C, what is the value for the equilibrium constant at 50. °C? Explain your result in light of the observation that carbon dioxide bubbles out of solution when the temperature is increased from room temperature.

Solutions

Apply the van't Hoff equation for chemical equilibrium, Equation [11.19], defining $T_1 = 25.$ °C = 298. K for $K_1 = 3.33 \times 10^{-2}$ and $T_2 = 50.$ °C = 323. K for the unknown value for the equilibrium constant, K_2, and rearrange symbolically to solve for K_2:

$$\frac{K_2}{K_1} = \exp\left[-\frac{\Delta h^{\circ}_{rxn}}{R}\left(\frac{1}{T_2} - \frac{1}{T_1}\right)\right]$$

$$K_2 = K_1 \exp\left[-\frac{\Delta h^{\circ}_{rxn}}{R}\left(\frac{1}{T_2} - \frac{1}{T_1}\right)\right]$$

$$K_2 = (3.33 \times 10^{-2})\exp\left[-\frac{\left(-10.39\ \text{kJ/mol} \times \frac{10^3\ \text{J}}{1\ \text{kJ}}\right)}{(8.314\ \text{J/mol}-\text{K})}\left[\frac{1}{(323.\ \text{K})} - \frac{1}{(298.\ \text{K})}\right]\right] = 2.41 \times 10^{-2}$$

Note that the equilibrium constant decreases as the temperature increases, predicting that less $CO_2(g)$ should be dissolved at equilibrium as the temperature increases, consistent with the empirical observations and their interpretation based on LeChâtelier's principle in Section 10.5.

Example 11.10: Determining the Reaction Enthalpy for the Self-Ionization of Water

As introduced in Section 10.7, the self-ionization constant for water, K_w, is 1.0×10^{-14} at 25.00 °C and 1.2×10^{-15} at 0.00 °C, with the self-ionization of water an endothermic process based on qualitative interpretation of this behavior using LeChâtelier's principle. What is Δh°_{rxn} (in kJ/mol) for the self-ionization of water based on applying the van't Hoff equation for chemical equilibrium?

Solutions

Apply Equation [11.19], defining T_1 = 25.00 °C = 298.15 K for $K_{w,1} = 1.0\times10^{-14}$ and T_2 = 0.00 °C = 273.15 K for $K_{w,2} = 1.2\times10^{-15}$, and rearrange symbolically to solve for Δh°_{rxn}:

$$\Delta h^{\circ}_{rxn} = -\frac{R\ln\left(\frac{K_{w,2}}{K_{w,1}}\right)}{\frac{1}{T_2}-\frac{1}{T_1}} = -\frac{(8.314\ \text{J/mol}-\text{K})\ln\left[\frac{(1.2\times10^{-15})}{(1.0\times10^{-14})}\right]}{\frac{1}{(273.15\ \text{K})}-\frac{1}{(298.15\ \text{K})}}$$

$$\Delta h^{\circ}_{rxn} = 5.7\times10^{4}\ \text{J/mol}\times\frac{1\ \text{kJ}}{10^{3}\ \text{J}} = 5.7\times10^{1}\ \text{kJ/mol}$$

Example 11.11: Determining the Temperature to Double the Solubility of Magnesium Carbonate

Magnesium carbonate, as qualitatively identified based on the solubility rules (Table 7.1) from Chapter 7 and quantitatively examined in Examples 11.6 and 11.7a in this chapter, is a sparingly soluble salt. If $\Delta h^{\circ}_{rxn} = -46.3\ \text{kJ/mol}$ for Rxn [11.E2], at what temperature (in °C) is the solubility of this salt twice what its solubility is at 25. °C? Assume that there are no other salts present (*e.g.*, the water is otherwise pure).

Solutions

First determine the relationship between the solubility of magnesium carbonate using the strategy introduced for silver chloride in Section 10.8. Assuming magnesium and carbonate ions behave ideally (*i.e.*, $\gamma_{\pm} \cong 1$), which should be a reasonable first approximation for a sparingly soluble salt, the solubility product can be expressed as:

$$K_{sp} = a_{Mg^{2+},eq}\,a_{CO_3^{2-},eq} = \left(\frac{\gamma_{\pm}C_{Mg^{2+},eq}}{C^{\circ}}\right)\left(\frac{\gamma_{\pm}C_{CO_3^{2-},eq}}{C^{\circ}}\right)$$

$$K_{sp} = \frac{s^2}{C^{\circ 2}} \qquad [11.E1]$$

where the solubility of the salt is $s = C_{Mg^{2+},eq} = C_{CO_3^{2-},eq}$. Defining T_1 = 25. °C = 298. K with corresponding equilibrium constant $K_{sp,1}$ and solubility s_1, we seek to identify the temperature T_2 with corresponding equilibrium constant $K_{sp,2}$ at which the solubility $s_2 = 2s_1$. Apply Equation [11.19], substitute Equation [11.E1]:

$$-\frac{\Delta h^{\circ}_{rxn}}{R}\left(\frac{1}{T_2}-\frac{1}{T_1}\right)=\ln\left(\frac{K_{sp,2}}{K_{sp,1}}\right)=\ln\left[\frac{\left(\frac{s_2^2}{C^{\circ 2}}\right)}{\left(\frac{s_1^2}{C^{\circ 2}}\right)}\right]=\ln\left[\left(\frac{s_2}{s_1}\right)^2\right]=\ln\left[\left[\frac{(2s_1)}{s_1}\right]^2\right]=\ln 4$$

$$\frac{1}{T_2}-\frac{1}{T_1}=-\frac{R\ln 4}{\Delta h^{\circ}_{rxn}}$$

$$\frac{1}{T_2}=\frac{1}{T_1}-\frac{R\ln 4}{\Delta h^{\circ}_{rxn}}$$

$$T_2=\left(\frac{1}{T_1}-\frac{R\ln 4}{\Delta h^{\circ}_{rxn}}\right)^{-1}=\left[\frac{1}{(298.\ \mathrm{K})}-\frac{(8.314\ \mathrm{J/mol-K})\ln 4}{\left(-46.3\ \mathrm{kJ/mol}\times\frac{10^3\ \mathrm{J}}{1\ \mathrm{kJ}}\right)}\right]^{-1}=277.\ \mathrm{K}=4.\ ^{\circ}\mathrm{C}$$

12: Electrochemistry

Learning Objectives

By the end of this chapter you should be able to:

1) Distinguish between different types of batteries, fuel cells, and electrolytic devices.

2) Assign oxidation states to the atoms in a substance given its chemical formula.

3) Determine the reductant and oxidant in a redox reaction and combine half reactions to obtain a balanced overall redox reaction.

4) Identify the cathode, anode, reduction and oxidation half-reactions, and direction of electron flow in galvanic and electrolytic cells.

5) Calculate the standard voltage of an electrochemical cell based on standard reduction potentials for its half-cell reactions.

6) Apply standard reduction potentials to predict the spontaneity for a redox reaction and relate cell potentials and electrolyte concentrations for electrochemical cells under non-standard conditions using the Nernst equation.

7) Use Faraday's laws to predict the quantities of substances produced or consumed at the electrodes of electrochemical cells from the total charge passing through the circuit.

12.1 Introduction to Electrochemical Systems

Electrochemical systems (also sometimes termed electrochemical cells) are devices in which chemical reactions in which electrons are transferred occur. We can classify these devices and their applications from the perspective of energy conversion, spontaneity, and reversibility, as either galvanic cells or electrolytic cells. Galvanic cells are devices that convert stored chemical energy to electrical energy spontaneously (*i.e.*, these devices involve electrochemical reactions for which the change in molar Gibbs free energy is negative). There are several types of galvanic cells encountered in practice. Batteries and fuel cells are galvanic cells in which desired electrochemical reactions occur. In particular, batteries are desired galvanic cells with fixed amounts of reactants. In primary batteries, which are not rechargeable and include alkaline batteries (Figure 12.1a), irreversible electrochemical reactions occur. In secondary batteries, such as Ni-Cd (nickel-cadmium, Figure 12.1b), NiMH (nickel metal hydride, Figure 12.1c), lead acid ("car," Figure 12.1d) batteries, and lithium ion (Figure 12.2), reversible electrochemical reactions occur. When secondary batteries discharge, the reaction taking place has a negatively-valued change in molar Gibbs free energy; however, during recharge, which occurs only as long as electrical power is applied, a reaction takes place for which the change in molar Gibbs free energy is positive.

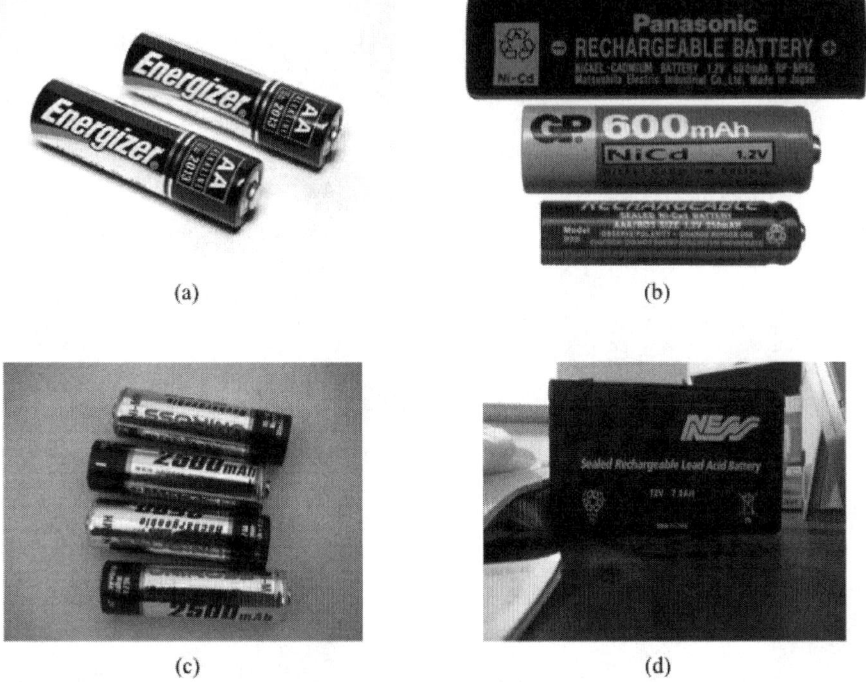

Figure 12.1. Common types of batteries. (a) Alkaline batteries; (b) Ni-Cd batteries; (c) NiMH batteries; (d) lead acid car battery. (12d: Copyright © by Bisapien. Reprinted with permission.)

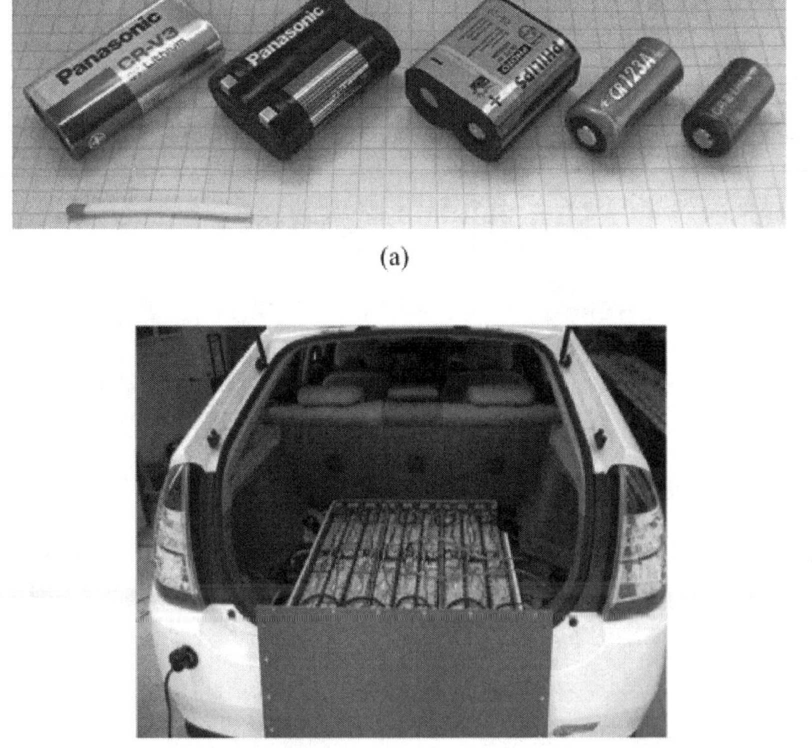

Figure 12.2. Lithium ion batteries. (a) Camera batteries; (b) battery pack in CalCars' EnergyCS/EDrive converted Toyota Prius.

Fuel cells (Figure 12.3) are desired galvanic cells with continuous flows of reactants and products. These electrochemical systems "recharge" by the replenishment of reactants and removal of waste products. Fuel cells have the advantage over batteries in that whereas batteries have limited lifetimes, fuel cells can be operated indefinitely (in theory). Further, with no moving parts, high efficiencies (due to the direct conversion of chemical to electrical energy without the generation of heat as an intermediate), low emissions, and quiet operating conditions, fuel cells have significant theoretical advantages over internal combustion systems as power supplies. The widespread adoption of hydrogen and other fuel cells, however, has been hindered by both significant technical issues and commercialization challenges that currently limit the widespread adoption of hydrogen fuel cells as power sources.

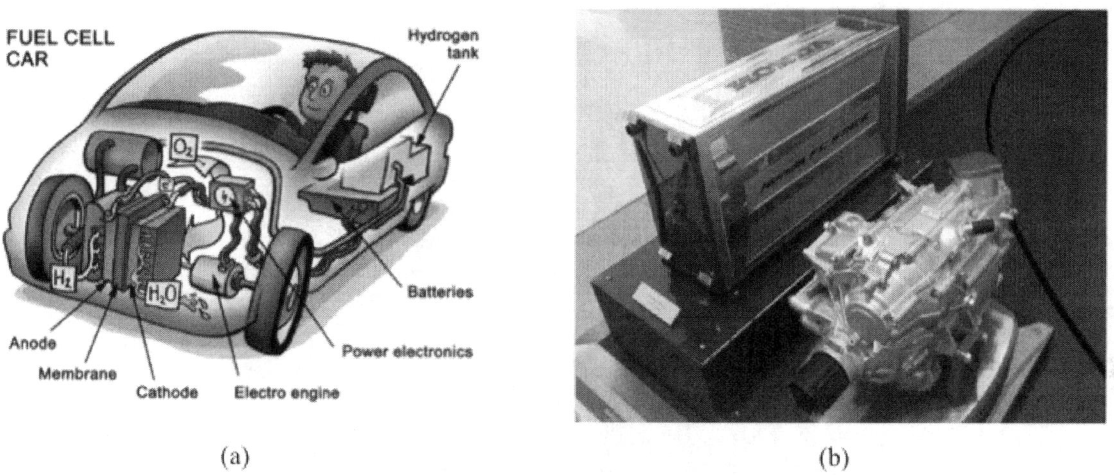

Figure 12.3. Fuel cells. (a) Concept for a vehicle powered by a fuel cell; (b) Honda FC fuel cell stack with gearbox.

Corrosion, such as the rusting of metallic iron (Figure 12.4) and tarnishing of silver, occurs in galvanic cells in which unwanted electrochemical reactions occur spontaneously at metal interfaces. In effect, when corrosion occurs microscopic batteries exist; however, no one has found a practical way to harness the flow of electrons that occurs during corrosion.

Figure 12.4. A rusted iron chain is the result of corrosion.

Electrolytic cells are devices that convert electrical energy to stored chemical energy. By definition these devices involve reactions that are non-spontaneous and have positively-valued changes in molar Gibbs free energy. Common processes involving electrolytic cells include electrolysis, the "splitting" of water to form molecular hydrogen and oxygen (Figure 12.5). Electrolytic devices provide one means to generate hydrogen for hydrogen fuel cells. Electroplating, the electrochemical deposition of metal and metal oxides, is a process involving an electrolytic cell. For example, electroplating in the form of anodization is used to deposit thin, relatively hard and wear-resistant layers on the surfaces of relatively soft metals such as aluminum. Another process that occurs in electrolytic cells is electrorefining, the electrochemical purification of metal ores. For example, although aluminum is a relatively plentiful element in the Earth's crust, until the late nineteenth century metallic aluminum was relatively rare because the electrochemical process to convert aluminum oxide to metallic aluminum was unknown.

Figure 12.5. High-pressure electrolyzer for producing hydrogen gas from water.

12.2 Representing Electron Transfer by Assigning Oxidation States

Chemical reactions involving the transfer of electrons among atoms are termed redox (reduction-oxidation) reactions. We evaluate electron transfer from the perspective of changes in oxidation state (abbreviated *O.S.*). We define the oxidation state for an element in a compound as a formal "representation" of the number of electrons an atom of that element appears to gain or lose in forming the compound.

Note that the oxidation state for an element is not necessarily the charge for the ionized atom in the compound. Further, the concept of oxidation state is merely a convenience, as we cannot measure oxidation states.

We assign an oxidation state to each element in a compound using the following set of eleven rules:

Rule 1: The oxidation state for an element not combined with other elements is 0.

Rule 2: The sum of the oxidation state over all atoms in a compound is 0; the sum of the oxidation states over all atoms in a polyatomic ion is the charge of the ion.

Rule 3: The oxidation state for a Group 1/1A element (the alkali metals, which do not include hydrogen) is +1.

Rule 4: The oxidation state for a Group 2/2A element (the alkaline metals) is +2.

Rule 5: The oxidation state for fluorine is -1.

Rule 6: The oxidation state for hydrogen is +1.

Rule 7: The oxidation state for oxygen is -2.

Rule 8: The oxidation state for a Group 17/7A element (the halogens other than fluorine, in order of chlorine, bromine, and iodine) is -1.

Rule 9: The oxidation state for a Group 16/6A element (other than oxygen, in order of sulfur, selenium, and tellurium) is -2.

Rule 10: The oxidation state for a Group 15/5A element (in order of nitrogen, phosphorus, and arsenic) is -3.

Rule 11: The oxidation state for all other elements is either the group number or the group number less eight, with more electronegative elements having more negative oxidation states.

These rules are applied sequentially, with rules higher on the list having higher priority, until the oxidation state has been assigned to each atom in the compound. For example, Rule 1 has the highest priority, Rule 2 the second highest, Rule 3 the third highest. For elemental substance we only need to apply Rule 1. For compounds consisting of two elements, we identify the element whose oxidation state can be assigned by the highest priority rule, assign its oxidation state based on this rule, and then assign the oxidation state to the other element based on Rule 2. For compounds with polyatomic ions it is frequently easiest to consider the polyatomic ion(s) first and then evaluate the oxidation states for any monatomic ions. Note that we typically do not need to apply Rule 11. For example, the oxidation state of carbon is typically assigned as the last element in the compound based on Rule 2.

Example 12.1: Assigning Oxidation States

Assign the oxidation state for each element in the following compounds:

a) O_2 b) H_2O c) H_2O_2 d) LiCl

e) $Fe(OH)_2$ f) NiO(OH) g) $PbSO_4$ h) CH_3OH

Lithium chloride is produced as lithium ion batteries discharge. Iron (II) hydroxide is the product of the first step in the corrosion of iron in basic solutions (see Example 12.3). Nickel oxyhydroxide is a reactant in Ni-Cd rechargeable batteries (see Example 12.4). Lead sulfate is one of the products in lead-acid car batteries (see Example 12.5). Methanol is a fuel for the direct methanol fuel cell (see Example 12.6).

Solutions

a) Based on Rule 1, assign $(O.S.)_O = 0$ because O_2 contains only a pair of O atoms not combined with other elements.

b) Based on the rule with the highest priority, Rule 6, assign $(O.S.)_H = +1$. Based on Rule 2, then, assign $(O.S.)_O = -2$ in order to have the sum of the oxidation states for the two hydrogen and one oxygen atoms be 0.

c) Based on Rule 6, assign $(O.S.)_H = +1$. Based then on Rule 2, assign $(O.S.)_O = -1$ in order to have the sum of the oxidation states for the two hydrogen and one oxygen atoms be 0.

d) Based on the rule with the highest priority, Rule 3, assign $(O.S.)_{Li} = +1$. Based then on Rule 2, assign $(O.S.)_{Cl} = -1$ in order to have the sum of the oxidation states for the one lithium and one chlorine atoms be 0.

e) This compound contains the polyatomic hydroxide ion, which we consider first. Based on Rule 6, assign $(O.S.)_H = +1$. Based then on Rule 2 applied to the hydroxide ion, assign $(O.S.)_O = -2$ in order to have the sum of the oxidation states for the one oxygen and one hydrogen atoms be -1. Based on Rule 2 applied to the overall compound, assign $(O.S.)_{Fe} = +2$ in order to have the sum of the oxidation states over all atoms for the compound be 0.

f) Although this compound contains the hydroxide ion, it is easiest to evaluate each element separately for the overall compound. Based on Rule 6, assign $(O.S.)_H = +1$. Based on the rule with the next highest priority, Rule 7, assign $(O.S.)_O = -2$. Based then on Rule 2, assign $(O.S.)_{Ni} = +3$ in order to have the sum of the oxidation states for the one nickel, two oxygen, and one hydrogen atoms be 0.

g) This compound contains the polyatomic sulfate ion, which we consider first. Based on Rule 7, assign $(O.S.)_O = -2$. Based then on Rule 2 applied to the sulfate ion, assign $(O.S.)_S = +6$ in order to have the sum of the oxidation states for the one sulfur and four oxygen atoms be -2. Based on Rule 2 applied to the overall compound, assign $(O.S.)_{Pb} = +2$ in order to have the sum of the oxidation states over all atoms in the compound be 0.

h) Based on Rule 6, assign $(O.S.)_H = +1$. Based on the rule with the next highest priority, Rule 7, assign $(O.S.)_O = -2$. Based then on Rule 2, assign $(O.S.)_C = -2$ in order to have the sum of the oxidation states for the one carbon, four hydrogen, and one oxygen atoms be 0.

12.3 Balancing Redox Reactions

We describe any redox reaction as a pair of <u>half-reactions</u>, one each for oxidation and reduction. In an <u>oxidation half-reaction</u>, electrons appear to be produced (often termed "lost"):

$$A \rightarrow A^{n+} + n\,e^- \qquad \text{Half-rxn [12.1]}$$

Because the species A is <u>oxidized</u> (*i.e.*, the oxidation state of A increases), A is the <u>reductant</u> (the agent causing reduction). Examples of oxidation half-reactions include:

$$Zn(s) \rightarrow Zn^{2+}(aq) + 2\ e^- \qquad \text{Half-rxn [12.2]}$$

$$Li(s) \rightarrow Li^{+}(aq) + e^- \qquad \text{Half-rxn [12.3]}$$

$$\tfrac{1}{2}\ H_2(g) + H_2O(l) \rightarrow H_3O^{+}(aq) + e^- \qquad \text{Half-rxn [12.4]}$$

In a <u>reduction half-reaction,</u> electrons appear to be consumed (often termed "gained"):

$$B^{m+} + m\ e^- \rightarrow B \qquad \text{Half-rxn [12.5]}$$

Because the species B is <u>reduced</u> (*i.e.*, the oxidation state of B decreases), B is the <u>oxidant</u> (the agent causing oxidation). Examples of reduction half-reactions include:

$$Zn^{2+}(aq) + 2\ e \rightarrow Zn(s) \qquad \text{Half-rxn [12.6]}$$

$$Al^{3+}(aq) + 3\ e^- \rightarrow Al(s) \qquad \text{Half-rxn [12.7]}$$

$$O_2(g) + 2\ H_2O(l) + 4\ e^- \rightarrow 4\ OH^-(aq) \qquad \text{Half-rxn [12.8]}$$

Some students have found the mnemonic "OILRIG"—standing for "oxidation is loss, reduction is gain"—to be a useful tool.

Note that we never observe individual half-reactions taking place by themselves: we cannot have oxidation without reduction and *vice versa* because the energy of a macroscopic amount of free electrons is enormous! The transfer (*i.e.*, flow) of electrons in the form of an electrical current can occur, however. In batteries and fuel cells we harness this current to perform electrical work; in electrolytic devices we apply a current to do electrical work on the system in the context of a non-spontaneous chemical reaction.

To balance a redox reaction and describe its accompanying electrochemical cell, we apply a four-step procedure. In the first step we identify the species being oxidized (the reductant), with its oxidation state increasing, and the species being reduced (the oxidant), with its oxidation state decreasing. For any well-formulated redox reaction there will be only one species whose oxidation state increases and only one whose oxidation state decreases. We then write one "skeleton" half-reaction each for oxidation and reduction, balancing all species except oxygen and hydrogen. In this step we do balance each half-reaction for the number of electrons transferred based on the changes in oxidation state, with electrons as products for oxidation and as reactants for reduction, but without yet balancing oxygen and hydrogen. At this point we do not have to identify the state for each species.

How we then proceed depends on whether the solution is acidic, neutral, or basic. For acidic and neutral solutions (*i.e.*, $pH \leq 7$) we balance the skeleton half-reactions for oxygen and hydrogen using $H_2O(l)$ and $H_3O^+aq)$. In contrast, for basic solutions (*i.e.*, $pH > 7$) we balance the skeleton half-reactions for oxygen and hydrogen using $H_2O(l)$ and $OH^-aq)$. Consider that we are more likely to find hydronium ions in acidic solutions and hydroxide ions in basic solutions than *vice versa*. A useful strategy to keep in mind is that water and hydronium ions differ by only one hydrogen, such that adding one species to one side of a reaction and the other species to the other side does not change the number of oxygen atoms but does incrementally alter the number of hydrogen atoms; water and hydroxide ions behave similarly. Also recognize that if we introduce water to one side of the oxidation half-reaction, it is likely that we introduce water on the other side of the reduction half-reaction.

In the final step we multiply each half-reaction such that the number of electrons produced by oxidation equals the number of electrons consumed by reduction. We define this number of electrons as the implicit stoichiometric coefficient, z, for the number of moles of electrons transferred per mole of reaction. The overall balanced redox reaction, then, is the sum of these multiplied half-reactions. Note that the value for z depends on the values for the other stoichiometric coefficients, such that if we multiply all of the stoichiometric coefficients by a factor we must also multiply z by this factor. We conclude by labeling the state for each reactant and product.

Example 12.2: The Reaction of Aluminum Foil with Dissolved Copper

Consider the reaction of dissolved Cu^{2+} ions in an acidic aqueous solution with metallic aluminum:

$$Al(s) + Cu^{2+}(aq) \rightarrow Al^{3+}(aq) + Cu(s) \text{ at } pH < 7$$

Determine the reductant, the oxidant, the half-reactions involved, the overall balanced redox reaction, and the number of electrons transferred (z).

Solutions

Identify the changes in oxidation state: the oxidation state of aluminum increases from 0 as metallic aluminum to +3 as Al^{3+}, and the oxidation state of copper decreases from +2 as Cu^{2+} to 0 for metallic copper. Thus, $Al(s)$ is the reductant and $Cu^{2+}(aq)$ is the oxidant. Write the skeleton half-reactions for oxidation and reduction, including the number of electrons lost by oxidation and gained by reduction:

$$\text{Oxidation half-rxn: } Al \rightarrow Al^{3+} + 3\,e^-$$

$$\text{Reduction half-rxn: } Cu^{2+} + 2\,e^- \rightarrow Cu$$

There is no need to introduce water and/or hydronium ions to balance for oxygen and hydrogen. Combine the half-reactions to obtain an overall balanced redox reaction by multiplying the oxidation half-reaction by 2× and the reduction half-reaction by 3× to balance the number of electrons transferred and adding the multiplied half-reactions:

$$\text{Oxidation half-rxn: } 2 \times [Al \rightarrow Al^{3+} + 3\,e^-]$$

$$+ \text{ Reduction half-rxn: } 3 \times [Cu^{2+} + 2\,e^- \rightarrow Cu]$$

$$2\,Al + 3\,Cu^{2+} + 6\,e^- \rightarrow 2\,Al^{3+} + 3\,Cu + 6\,e^-$$

Reintroduce the state of matter for each species to obtain the overall balanced redox reaction:

$$2\,Al(s) + 3\,Cu^{2+}(aq) \rightarrow 2\,Al^{3+}(aq) + 3\,Cu(s) \qquad \text{Rxn [12.E1]}$$

There are $z = 6\,e^-$ transferred (*i.e.*, there are six moles of electrons transferred for every two moles of metallic aluminum reacted with three moles of copper ions to produce two moles of aluminum ions and three moles of metallic copper).

Example 12.3: The First Step in the Corrosion of Iron

The corrosion of iron (Figure 12.4) occurs as the following overall reaction:

$$4\ Fe(s) + 3\ O_2(g) \rightarrow 2\ Fe_2O_3(s) \qquad \text{Rxn [12.E2]}$$

This reaction occurs as a series of steps, including a first step in which metallic iron reacts with molecular oxygen and water to produce iron hydroxide under basic conditions:

$$Fe(s) + O_2(g) + H_2O(l) \rightarrow Fe(OH)_2(s) \text{ at } pH > 7$$

Determine the reductant, the oxidant, the half-reactions involved, the overall balanced redox reaction, and the number of electrons transferred (z).

Solutions

Identify the changes in oxidation state: the oxidation state of iron increases from 0 as metallic iron to +2 as Fe^{2+} in $Fe(OH)_2$, and the oxidation state of oxygen decreases from 0 as molecular oxygen to -2 as OH^- in $Fe(OH)_2$. Thus, $Fe(s)$ is the reductant and $O_2(g)$ the oxidant. Note that the oxidation states of oxygen in H_2O and of hydrogen in H_2O and $Fe(OH)_2$ do not change. Write the skeleton half-reactions for oxidation and reduction, including the number of electrons lost by oxidation and gained by reduction:

Oxidation half-rxn: $Fe \rightarrow Fe(OH)_2 + 2\ e^-$

Reduction half-rxn: $O_2 + 4\ e^- \rightarrow 2\ OH^-$

Introduce water and hydroxide ions to balance for oxygen and hydrogen under basic conditions. In particular, for the reduction half-reaction balance for the excess of hydrogen on the product side by introducing water as a reactant and hydroxide ions as a product:

Oxidation half-rxn: $Fe + 2\ OH^- \rightarrow Fe(OH)_2 + 2\ e^-$

Reduction half-rxn: $O_2 + 2\ H_2O + 4\ e^- \rightarrow 4\ OH^-$

Combine the half-reactions to obtain an overall balanced redox reaction by multiplying the oxidation half-reaction by 2× to balance the number of electrons transferred and adding the multiplied half-reactions:

Oxidation half-rxn: $2 \times [Fe + 2\ OH^- \rightarrow Fe(OH)_2 + 2\ e^-]$

+ Reduction half-rxn: $O_2 + 2\ H_2O + 4\ e^- \rightarrow 4\ OH^-$

$2\ Fe + 4\ OH^- + O_2 + 2\ H_2O + 4\ e^- \rightarrow 2\ Fe(OH)_2 + 4\ OH^- + 4\ e^-$

Simplify, noting that there is no net change in the amount of hydroxide ions and that a net consumption of water occurs, as we should expect from the original unbalanced reaction. Reintroduce the state of matter for each species to obtain the overall balanced redox reaction:

$$2\ Fe(s) + O_2(g) + 2\ H_2O(l) \rightarrow 2\ Fe(OH)_2(s) \qquad \text{Rxn [12.E3]}$$

There are $z = 4\ e^-$ transferred.

Example 12.4: The Ni-Cd Rechargeable Battery

Rechargeable batteries based on nickel and cadmium (Figure 12.1b) have been marketed for a number of years; they discharge based on a reaction between metallic cadmium and nickel oxyhydroxide under basic conditions:

$$Cd(s) + NiO(OH)(s) + H_2O(l) \rightarrow Cd(OH)_2(s) + Ni(OH)_2(s) \text{ at } pH > 7$$

Determine the reductant, the oxidant, the half-reactions involved, the overall balanced redox reaction, and the number of electrons transferred (z).

Solutions

Identify the changes in oxidation state: the oxidation state of cadmium increases from 0 as metallic cadmium to +2 as Cd^{2+} in $Cd(OH)_2$, and the oxidation state of nickel decreases from +3 as Ni^{3+} in NiO(OH) to +2 as Ni^{2+} in $Ni(OH)_2$. Thus, Cd(s) is the reductant and NiO(OH)(s) the oxidant. Note that the oxidation states of oxygen and hydrogen in NiO(OH), H_2O, $Cd(OH)_2$, and $Ni(OH)_2$ do not change. Write the skeleton half-reactions for oxidation and reduction, including the number of electrons lost by oxidation and gained by reduction:

$$\text{Oxidation half-rxn: } Cd(s) \rightarrow Cd(OH)_2(s) + 2\,e^-$$

$$\text{Reduction half-rxn: } NiO(OH)(s) + e^- \rightarrow Ni(OH)_2(s)$$

Introduce water and hydroxide ions to balance for oxygen and hydrogen under basic conditions. In particular, for the reduction half-reaction balance for the excess of hydrogen on the product side by introducing water as a reactant and hydroxide ions as a product.

$$\text{Oxidation half-rxn: } Cd + 2\,OH^- \rightarrow Cd(OH)_2 + 2\,e^-$$

$$\text{Reduction half-rxn: } NiO(OH) + H_2O + e^- \rightarrow Ni(OH)_2 + OH^-$$

Combine the half-reactions to obtain an overall balanced redox reaction by multiplying the reduction half-reaction by 2× to balance the number of electrons transferred and adding the multiplied half-reactions:

$$\text{Oxidation half-rxn: } \quad Cd(s) + 2\,OH^-(aq) \rightarrow Cd(OH)_2(s) + 2\,e^-$$

$$\text{+ Reduction half-rxn: } \quad 2 \times [NiO(OH)(s) + H_2O(l) + e^- \rightarrow Ni(OH)_2(s) + OH^-(aq)]$$

$$Cd + 2\,NiO(OH) + 2\,H_2O + 2\,OH^- + 2\,e^- \rightarrow Cd(OH)_2 + 2\,Ni(OH)_2 + 2\,OH^- + 2\,e^-$$

Simplify, noting that there is no net change in the amount of hydroxide ions and that a net consumption of water occurs, as we should expect from the original unbalanced reaction. Reintroduce the state of matter for each species to obtain the overall balanced redox reaction:

$$Cd(s) + 2\,NiO(OH)(s) + 2\,H_2O(l) \rightarrow Cd(OH)_2(s) + 2\,Ni(OH)_2(s) \qquad \text{Rxn [12.E4]}$$

There are $z = 2\,e^-$ transferred.

Example 12.5: The Lead Acid Battery

The lead acid battery (Figure 12.1d) has historically been used to power the start-up of vehicles and occurs as a reaction between metallic lead, lead (IV) oxide, and sulfuric acid under acidic conditions:

$$Pb(s) + PbO_2(s) + H_2SO_4(aq) \rightarrow PbSO_4(s) + H_2O(l) \text{ at } pH < 7$$

Determine the reductant, the oxidant, the half-reactions involved, the overall balanced redox reaction, and the number of electrons transferred (z).

Solutions

Identify the changes in oxidation state: the oxidation state of lead increases from 0 as metallic lead to +2 as Pb^{2+} in $PbSO_4$, while the oxidation state of lead also decreases from +4 as Pb^{4+} in PbO_2 to +2 +as Pb^{2+} in $PbSO_4$. Thus, $Pb(s)$ is the reductant and $PbO_2(s)$ the oxidant. Note that the oxidation states of oxygen in PbO_2 , H_2SO_4, and H_2O and of hydrogen in H_2SO_4 and H_2O do not change. Write the skeleton half-reactions for oxidation and reduction, including the number of electrons lost by oxidation and gained by reduction:

$$\text{Oxidation half-rxn: } Pb + H_2SO_4 \rightarrow PbSO_4 + 2\ e^-$$

$$\text{Reduction half-rxn: } PbO_2 + H_2SO_4 + 2\ e^- \rightarrow PbSO_4$$

Introduce water and hydronium ions to balance for oxygen and hydrogen under acidic conditions. In particular, for the oxidation half-reaction balance the excess of hydrogen on the reactant side by introducing water as a reactant and hydronium ions as a product, and for the reduction half-reaction balance for the excess of hydrogen and oxygen on the reactant side by introducing hydronium ions as a reactant and water as a product.

$$\text{Oxidation half-rxn: } Pb + H_2SO_4 + 2\ H_2O \rightarrow PbSO_4 + 2\ H_3O^+ + 2\ e^-$$

$$\text{Reduction half-rxn: } PbO_2 + H_2SO_4 + 2\ H_3O^+ + 2\ e^- \rightarrow PbSO_4 + 4\ H_2O$$

Combine the half-reactions to obtain an overall balanced redox reaction by adding them:

$$\text{Oxidation half-rxn: } \quad Pb + H_2SO_4 + 2\ H_2O \rightarrow PbSO_4 + 2\ H_3O^+ + 2\ e^-$$

$$+ \text{ Reduction half-rxn: } \quad PbO_2 + H_2SO_4 + 2\ H_3O^+ + 2\ e^- \rightarrow PbSO_4 + 4\ H_2O$$

$$Pb + PbO_2 + 2\ H_2SO_4 + 2\ H_2O + 2\ H_3O^+ + 2\ e^- \rightarrow PbSO_4 + 4\ H_2O + 2\ H_3O^+ + 2\ e^-$$

Simplify, noting that there is no net change in the amount of hydronium ions and that a net production of water occurs, as we should expect from the original unbalanced reaction. Reintroduce the state of matter for each species to obtain the overall balanced redox reaction:

$$Pb(s) + PbO_2(s) + 2\ H_2SO_4(aq) \rightarrow 2\ PbSO_4(s) + 2\ H_2O(l) \qquad \text{Rxn [12.E5]}$$

There are $z = 2\ e^-$ transferred.

Example 12.6: The Direct Methanol Fuel Cell (DMFC, Figure 12.6)

Fuel cells based on methanol are potentially superior to fuel cells based on hydrogen because $CH_3OH(l)$ is a denser fuel and involves a greater density of electrons available for transfer compared with $H_2(g)$. Such devices, however, are limited at present due to significant technical and logistical issues. Consider the reaction taking place in a DMFC operated under acidic conditions:

$$CH_3OH(l) + O_2(g) \rightarrow CO_2(g) + H_2O(l) \text{ at } pH < 7$$

Determine the reductant, the oxidant, the half-reactions involved, the overall balanced redox reaction, and the number of electrons transferred (z).

Solutions

Identify the changes in oxidation state: the oxidation state of carbon increases from -2 in CH_3OH to +4 in CO_2, and the oxidation state of oxygen decreases from 0 as molecular oxygen to -2 in H_2O and CO_2. Thus, $CH_3OH(l)$ is the reductant and $O_2(g)$ the oxidant. Note that the oxidation states of oxygen in CH_3OH and of hydrogen in CH_3OOH and H_2O do not change. Write the skeleton half-reactions for oxidation and reduction, including the number of electrons lost by oxidation and gained by reduction:

$$\text{Oxidation half-reaction: } CH_3OH \rightarrow CO_2 + 6\ e^-$$

$$\text{Reduction half-reaction: } O_2 + 4\ e^- \rightarrow 2\ H_2O$$

Introduce water and hydronium ions to balance for oxygen and hydrogen under acidic condition. In particular, for the oxidation half-reaction balance for the excess of hydrogen on the reactant side and oxygen on the product side by introducing water as a reactant and hydronium ions as a product, and for the reduction half-reaction balance for the excess of oxygen on the reactant side and hydrogen on the product side by introducing hydronium ions as a reactant and water as a product.

$$\text{Oxidation half-reaction: } CH_3OH + 7\ H_2O \rightarrow CO_2 + 6\ H_3O^+ + 6\ e^-$$

$$\text{Reduction half-reaction: } O_2 + 4\ H_3O^+ + 4\ e^- \rightarrow 6\ H_2O$$

Combine the half-reactions to obtain an overall balanced redox reaction by multiplying the oxidation half-reaction by 2× and the reduction half-reaction by 3× to balance the number of electrons transferred, and adding the multiplied half-reactions:

$$\text{Oxidation half-rxn: } 2 \times [CH_3OH + 7\ H_2O \rightarrow CO_2 + 6\ H_3O^+ + 6\ e^-]$$

$$+ \text{ Reduction half-rxn: } 3 \times [O_2 + 4\ H_3O^+ + 4\ e^- \rightarrow 6\ H_2O]$$

$$2\ CH_3OH + 14\ H_2O + 3\ O_2 + 12\ H_3O^+ + 12\ e^- \rightarrow 2\ CO_2 + 12\ H_3O^+ + 18\ H_2O + 12\ e^-$$

Simplify, noting that there is no net change in the amount of hydronium ions and that a net production of water occurs, as we should expect from the original unbalanced reaction. Reintroduce the state of matter for each species to obtain the overall balanced redox reaction:

$$2\ CH_3OH(l) + 3\ O_2(g) \rightarrow 2\ CO_2(g) + 4\ H_2O(l) \qquad \text{Rxn [12.E6]}$$

There are $z = 12\ e^-$ transferred.

12.4 Structure of Electrochemical Devices

Having an understanding of the balance between oxidation and reduction that occurs in redox reactions can provide the basis for explaining the physical structure common to all electrochemical devices. Any electrochemical device has three principal components: an anodic compartment, a cathodic compartment, and a junction (Figure 12.7). The anodic compartment (also known as the anode) is where the oxidation half-reaction occurs at the interface between a conducting material—also termed the anode—and the electrolyte, a conducting solution (*e.g.*, an aqueous solution of dissolved ions. By convention the anodic compartment always is depicted on the left. The cathodic compartment (also known as the cathode) is where the reduction half-reaction occurs at the interface between a conducting material—also termed the cathode—and the electrolyte.

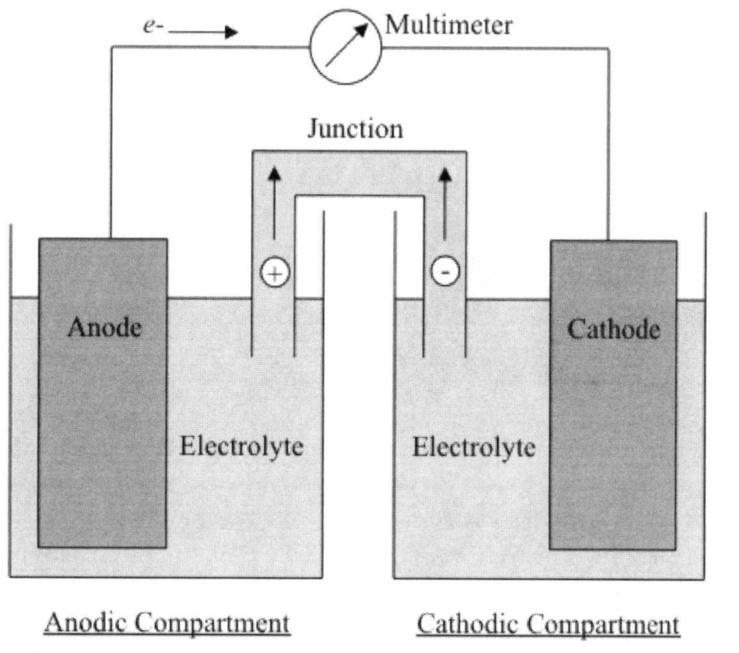

Figure 12.7. Generic structure for an electrochemical cell. Junction is depicted as a salt bridge.

The anode and cathode are connected by a conducting wire through which electrons "lost" (*i.e.*, produced) by the oxidation half-reaction at the anode flow to the cathode, where they are "gained" (*i.e.*, consumed) by the reduction half-reaction. When we operate the electrochemical device as a battery or fuel cell, we place an external device (*e.g.*, a cell phone) that we wish to power in this electrical path. We can also place a multimeter—a device that measures the cell's electric potential (*i.e.*, voltage, E_{cell}) and current (I) associated with the flow of electrons between the anode and cathode. We measure the voltage in units of volts (V), where 1 V = 1 J/C, and the current in amperes (A, amps), where 1 A = 1C/s. The total charge transferred over a period of time (t) is the product of the current and this time.

The circuit for the flow of charge is completed through a junction allowing the flow of ions between the anodic and cathodic compartments. Anions carrying negative charge flow through the junction from the cathodic compartment to the anodic compartment; cations flow through the junction in the opposite direction. In practice there are a variety of ways to form a junction, including a salt bridge (depicted in Figure 12.7), a porous separator (depicted in Figure 12.8), or a gel tube of electrolyte. Each of these features allow the flow of ions but prevent physical contact between the conducting anode and cathode, which would result in a short circuit.

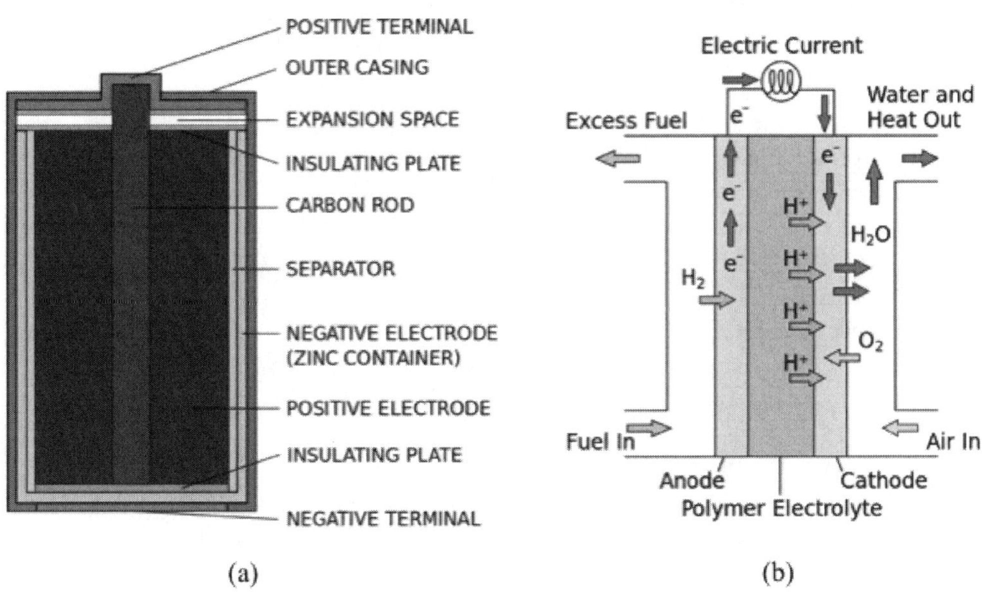

Figure 12.8. Structure of (a) an alkaline battery and (b) a hydrogen fuel cell.

12.5 Characterizing Electrochemical Devices Based on Standard Reduction Potentials

We now are in a position to address key questions regarding the operating properties of electrochemical devices, including:

- How do we identify if a given redox reaction is appropriate for a galvanic cell or an electrolytic device?
- What determines the voltage associated with an electrochemical system?
- How do we evaluate how much electrical power can be delivered by a galvanic cell or is consumed by an electrolytic device?

To determine the cell potential (*i.e.*, voltage) associated with an electrochemical device, we start by identifying an expression for the electrochemical cell potential at standard state based on characterizing the associated pair of individual half-reactions for oxidation and reduction. In particular, we define the standard reduction potential ($E°$, in V), as the electric potential associated with a reduction half-reaction occurring at standard state (*i.e.*, all the partial pressures are $P_i = P° = 1$ bar, all the concentrations are $C_i = C° = 1$ M, and each species is present in its pure form and reference form at 1 bar and the temperature of interest). Values for the standard reduction potential for a wide variety of reduction half-reactions are tabulated at 298.15 K in Appendix B. Consider that because individual half-reactions cannot be run separately, we need to define a reference half-reaction which has a standard reduction potential of 0 V—*i.e.*, we need to set a zero. The standard convention is to choose standard hydrogen electrode (abbreviated SHE) as our basis:

$$H_3O^+ + e^- \rightarrow H_2O + \tfrac{1}{2}\, H_2 \quad E° \equiv 0\,\text{V} \qquad \text{Half-rxn [12.9]}$$

The reduction of metallic silver to silver chloride electrode is also used as the basis of some reference electrodes, including as the internal reference electrode for some *pH* meters (Figure 12.9). The standard reduction potential for a specific half-reaction can then be made using an appropriate reference electrode (Figure 12.10).

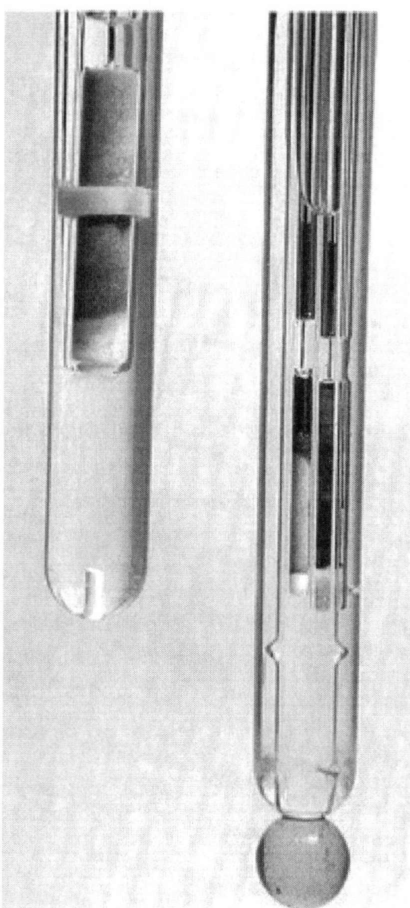

Figure 12.9. Silver-silver chloride reference electrode (left) and its implementation in a *pH* glass electrode (right).

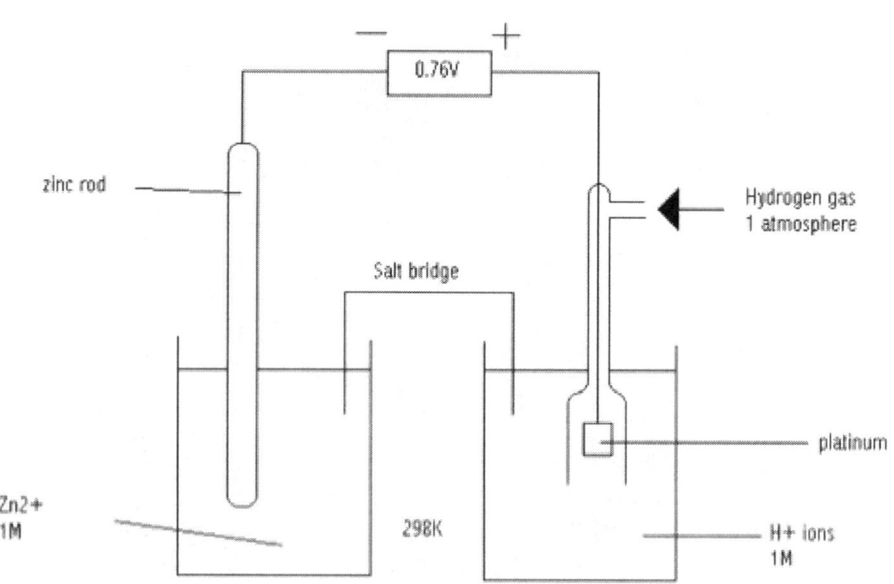

Figure 12.10. Measurement of standard reduction potential for zinc.

The standard reduction potential is defined explicitly and solely for a reduction half-reaction. We could conceive that the "standard oxidation potential," for an oxidation half-reaction is the negative of the standard reduction potential for the corresponding reduction half-reaction that occurs in the opposite direction:

$$E^{\circ}_{oxidation} = -E^{\circ} \quad [12.1]$$

In practice, there is no need to define both a standard reduction potential and a "standard oxidation potential," and "standard oxidation potentials" are not used.

The standard reduction potential is an intensive property (*e.g.*, like density) and not based on the extent or absolute stoichiometry of the half-reaction. For example, both of the following half-reactions for the reduction of ionized aluminum to metallic aluminum have the same standard reduction potential:

$Al^{3+} + 3\,e^{-} \rightarrow Al$	$E° = -1.676\,V$	Half-rxn [12.7]
$2\,Al^{3+} + 6\,e^{-} \rightarrow 2\,Al$	$E° = -1.676\,V$	Half-rxn [12.10]

The standard reduction potential provides a metric for the tendency for reduction to occur: the more positive the value for standard reduction potential, the greater the tendency for reduction. In contrast, the more negative the value for the standard reduction potential, the greater the tendency for oxidation. For example, compare the following set of reduction half-reactions based on data in Appendix B:

$Zn^{2+} + 2\,e^{-} \rightarrow Zn$	$E° = -0.7621\,V$	Half-rxn [12.6]
$Fe^{2+} + 2\,e^{-} \rightarrow Fe$	$E° = -0.440\,V$	Half-rxn [12.11]
$O_2 + 4\,H_3O^{+} + 4\,e^{-} \rightarrow 6\,H_2O$	$E° = +1.2288\,V$	Half-rxn [12.12]

When metallic zinc and metallic iron are placed in contact with an acidic solution and air, we expect that it is more likely that zinc will be oxidized than iron, and that the actual reduction half-reaction will be the reduction of molecular oxygen to water. This behavior can be applied to prevent an iron support from rusting by placing a sacrificial piece or coating of zinc in close proximity (Figure 12.11): the zinc forms an anode, the iron a cathode at which the reduction of oxygen takes place.

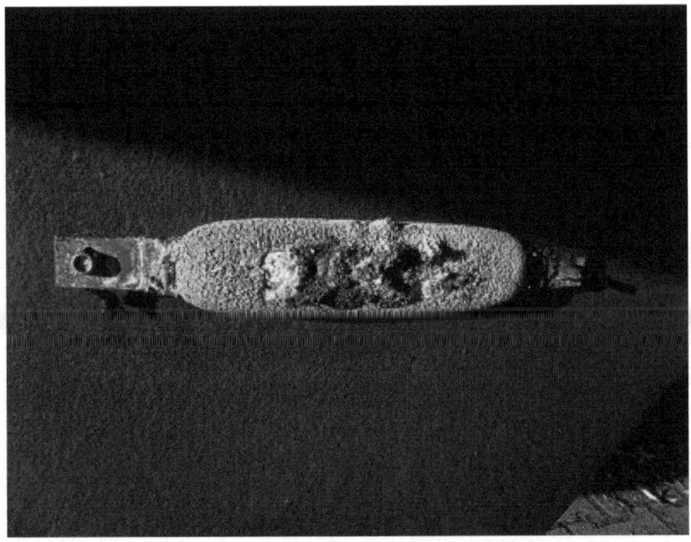

Figure 12.11. Sacrificial anode on hull of a ship. (Credit: Zwergelstern)

We determine the standard cell potential, E°_{cell}, the electric potential (*i.e.*, voltage) at standard state for a redox reaction composed of an oxidation half-reaction occurring at an anode (with standard reduction potential E°_{anode} for the corresponding reduction half-reaction) and a paired reduction half-reaction occurring at a cathode (with standard reduction potential $E^{\circ}_{cathode}$), as:

$$E^{\circ}_{cell} = E^{\circ}_{cathode} - E^{\circ}_{anode} \qquad [12.2]$$

The standard cell potential is also termed the standard electromotive force because it represents the driving force for movement of electrons under conditions of standard state.

Example 12.7: The Reaction of Aluminum Foil with Dissolved Copper

Determine the standard reduction potentials (in V) for the anode and cathode and the overall standard cell potential (in V) for the oxidation of metallic aluminum and reduction of aqueous Cu^{2+}, as introduced in Example 12.2, at room temperature.

Solutions

In Example 12.2 we balanced the overall redox reaction for the oxidation of metallic aluminum and reduction of aqueous Cu^{2+} in terms of the following pair of half-reactions, with associated standard reduction potentials from Appendix B:

Oxidation half-rxn: $Al \rightarrow Al^{3+} + 3\ e$ $\qquad E^{\circ}_{anode} = -1.676\ V$

Reduction half-rxn: $Cu^{2+} + 2\ e \rightarrow Cu$ $\qquad E^{\circ}_{cathode} = +0.3394\ V$

Note that the standard reduction potential for the oxidation half-reaction appears in Appendix B as the corresponding reduction half-reaction, $Al^{3+} + 3\ e \rightarrow Al$. The standard cell potential for this electrochemical cell is:

$$E^{\circ}_{cell} = E^{\circ}_{cathode} - E^{\circ}_{anode} = (+0.3394\ V) - (-1.676\ V) = +2.015\ V$$

Example 12.8: The First Step in the Corrosion of Iron

Determine the standard reduction potentials (in V) for the anode and cathode and the overall standard cell potential (in V) for the first step in the corrosion of metallic iron under basic conditions, as introduced in Example 12.3, at room temperature.

Solutions

In Example 12.3 we balanced the overall redox reaction for the oxidation of metallic iron and reduction of molecular oxygen under basic conditions in terms of the following pair of half-reactions, with associated standard reduction potentials from Appendix B:

Oxidation half-rxn: $Fe \rightarrow Fe^{2+} + 2\,e^-$ $E^\circ_{anode} = -0.440\ V$

Reduction half-rxn: $O_2 + 2\,H_2O + 4\,e^- \rightarrow 4\,OH^-$ $E^\circ_{cathode} = +0.401\ V$

Note that the standard reduction potential for the oxidation half-reaction appears in Appendix B as the corresponding reduction half-reaction, $Fe^{2+} + 2\,e^- \rightarrow Fe$. The standard cell potential for this electrochemical cell is:

$$E^\circ_{cell} = E^\circ_{cathode} - E^\circ_{anode} = (+0.401\ V) - (-0.440\ V) = +0.841\ V$$

Example 12.9: The Ni-Cd Rechargeable Battery

Determine the standard reduction potentials (in V) for the anode and cathode and the overall standard cell potential (in V) for the Ni-Cd rechargeable battery, introduced in Example 12.4, at room temperature.

Solutions

In Example 12.4 we balanced the overall redox reaction for the oxidation of metallic cadmium and reduction of nickel oxyhydroxide under basic conditions in terms of the following pair of half-reactions, with associated standard reduction potentials from Appendix B:

Oxidation half-rxn: $Cd + 2\,OH^- \rightarrow Cd(OH)_2 + 2\,e^-$ $E^\circ_{anode} = -0.81\ V$

Reduction half-rxn: $NiO(OH) + H_2O + e^- \rightarrow Ni(OH)_2 + OH^-$ $E^\circ_{cathode} = +0.49\ V$

Note that the standard reduction potential for the oxidation half-reaction appears in Appendix B as the corresponding reduction half-reaction, $Cd(OH)_2 + 2\,e^- \rightarrow Cd + 2\,OH^-$. The standard cell potential for this electrochemical cell is:

$$E^\circ_{cell} = E^\circ_{cathode} - E^\circ_{anode} = (+0.49\ V) - (-0.81) = +1.30\ V$$

Example 12.10: The Lead Acid Battery

Determine the standard reduction potentials (in V) for the anode and cathode and the overall standard cell potential (in V) for the lead acid battery, introduced in Example 12.5, at room temperature.

Solutions

In Example 12.5 we balanced the redox reaction for the oxidation of metallic lead and reduction of lead (IV) oxide under acidic conditions in terms of the following pair of half-reactions, with associated standard reduction potentials from Appendix B:

Oxidation half-rxn: $Pb + H_2SO_4 + 2\,H_2O \rightarrow PbSO_4 + 2\,H_3O^+ + 2\,e$ $E^{\circ}_{anode} = -0.28\text{ V}$

Reduction half-rxn: $PbO_2 + H_2SO_4 + 2\,H_3O^+ + 2\,e \rightarrow PbSO_4 + 4\,H_2O$ $E^{\circ}_{cathode} = +1.74\text{ V}$

Note that the standard reduction potential for the oxidation half-reaction appears in Appendix B as the corresponding reduction half-reaction, $PbSO_4 + 2\,H_3O^+ + 2\,e \rightarrow Pb + H_2SO_4 + 2\,H_2O$. The standard cell potential for this electrochemical cell is:

$$E^{\circ}_{cell} = E^{\circ}_{cathode} - E^{\circ}_{anode} = (+1.74\text{ V}) - (-0.28\text{ V}) = +2.02\text{ V}$$

Example 12.11: The Direct Methanol Fuel Cell (DMFC)

Determine the standard reduction potentials (in V) for the anode and cathode and the overall standard cell potential (in V) for the direct methanol fuel cell, introduced in Example 12.6, at room temperature.

Solutions

In Example 12.6 we balanced the redox reaction for the oxidation of methanol and reduction of molecular oxygen under acidic conditions in terms of the following pair of half-reactions, with associated standard reduction potentials from Appendix B:

Oxidation half-rxn: $CH_3OH + 7\,H_2O \rightarrow CO_2 + 6\,H_3O^+ + 6\,e$ $E^{\circ}_{anode} = +0.02\text{ V}$

Reduction half-rxn: $O_2 + 4\,H_3O^+ + 4\,e \rightarrow 6\,H_2O$ $E^{\circ}_{cathode} = +1.2288\text{ V}$

Note that the standard reduction potential for the oxidation half-reaction appears in Appendix B as the corresponding reduction half-reaction, $CO_2 + 6\,H_3O^+ + 6\,e \rightarrow CH_3OH + 7\,H_2O$. The standard cell potential for this electrochemical cell is:

$$E^{\circ}_{cell} = E^{\circ}_{cathode} - E^{\circ}_{anode} = (+1.2288\text{ V}) - (+0.02\text{ V}) = +1.21\text{ V}$$

12.6 Thermodynamics of Electrochemical Devices

The cell potential and the spontaneity of a redox reaction are related. Positive cell potentials are associated with galvanic cells (*i.e.*, batteries, fuel cells, and corrosion processes) and negatively-valued changes in Gibbs free energy. In contrast, negative cell potentials are associated with electrolytic cells and positively-valued changes in Gibbs free energy. Zero-valued cell potentials are associated with electrochemical systems at equilibrium (*i.e.*, $\Delta g_{rxn} = 0$).

Note that when a battery "dies," its voltage (*i.e.*, cell potential) is zero, corresponding to the redox reaction powering the battery having reached equilibrium!

This qualitative behavior can be expressed quantitatively:

$$\Delta g_{rxn} = -zFE_{cell} \qquad [12.3]$$

where z is the number of electrons transferred per mole of the reaction and F = 96,485 C/mol is defined as the Faraday constant. The Faraday constant is simply the charge per electron expressed in units of charge per mole of electrons:

$$1.602 \times 10^{-19}\ \text{C}/e^- \times \frac{6.022 \times 10^{23}\ e^-}{1\ \text{mol}\ e^-} = 96,485\ \text{C/mol}\ e^-$$

At standard state Equation [12.3] is expressed as:

$$\Delta g^\circ_{rxn} = -zFE^\circ_{cell} \qquad [12.4]$$

such that substitution into Equation [11.17] from Chapter 11 yields:

$$K = \exp\left(\frac{zFE^\circ_{cell}}{RT}\right) \qquad [12.5]$$

When applying Equation [12.5] we remind ourselves that 1 V = 1 J/C.

We can now relate the standard cell potential to the directionality of its redox reaction. If $E^\circ_{cell} > 0$, we find $K > 1$ and the products are favored over the reactants at equilibrium. In particular, we find that if $E^\circ_{cell} \geq 1\ \text{V}$, $K >> 1$ and the redox reaction essentially goes to completion. This condition is typical of most batteries and fuel cells as well as the process for corrosion. In contrast, if $E^\circ_{cell} < 0$, we find $K < 1$ and the reactants are favored over the products at equilibrium. In particular, we find that if $E^\circ_{cell} \leq -1\ \text{V}$, $K << 1$ and the redox reaction essentially does not occur. This condition is typical for many electrolytic devices.

Example 12.12: The Reaction of Aluminum Foil with Dissolved Copper

Determine the equilibrium constant at room temperature for Rxn [12.E1], the oxidation of metallic aluminum and reduction of aqueous Cu^{2+}, as examined in Examples 12.2 and 12.7.

Solutions

In Example 12.7 we found a standard cell potential of $E^\circ_{cell} = +2.015\ \text{V}$ at room temperature. In Example 12.2 we found $z = 6\ e$ for Rxn [12.E1], such that this reaction has an equilibrium constant evaluated as:

$$K = \exp\left[\frac{(6)(96,485\ \text{C/mol})(+2.015\ \text{J/C})}{(8.314\ \text{J/mol}-\text{K})(298\ \text{K})}\right] > 10^{204}$$

The reaction overwhelmingly favors the products.

Example 12.13: The Ni-Cd Rechargeable Battery

Determine the equilibrium constant at room temperature for Rxn [12.E4], the oxidation of metallic cadmium and reduction of nickel oxyhydroxide in a Ni-Cd, as examined in Examples 12.4 and 12.9.

Solutions

In Example 12.9 we found a standard cell potential of $E^{\circ}_{cell} = +1.30$ V at room temperature. In Example 12.4 we found $z = 2$ e for Rxn [12.E4], such that this reaction has an equilibrium constant evaluated as:

$$K = \exp\left[\frac{(2)(96,485 \text{ C/mol})(+1.30 \text{ J/C})}{(8.314 \text{ J/mol}-\text{K})(298 \text{ K})}\right] > 9 \times 10^{43}$$

The reaction strongly favors the products.

Example 12.14: The Direct Methanol Fuel Cell (DMFC)

Determine the equilibrium constant at room temperature for Rxn [12.E6], the oxidation of methanol and reduction of molecular oxygen in direct methanol fuel cell, as examined in Examples 12.6 and 12.11.

Solutions

In Example 12.11 we found a standard cell potential of $E^{\circ}_{cell} = +1.25$ V at room temperature. In Example 12.6 we found $z = 12$ e for Rxn [12.E6], such that this reaction has an equilibrium constant evaluated as:

$$K = \exp\left[\frac{(12)(96,485 \text{ C/mol})(+1.25 \text{ J/C})}{(8.314 \text{ J/mol}-\text{K})(298 \text{ K})}\right] > 4 \times 10^{253}$$

The reaction overwhelmingly favors the products.

For most practical applications we are not at standard state. For example, we do not typically operate devices with all gases at partial pressures of 1 bar and all solutes at concentrations of 1 . We modify our development to address this situation by substituting Equations [12.3] and [12.4] into Equation [11.14] from Chapter 11:

$$-zFE_{cell} = -zFE^{\circ}_{cell} + RT\ln Q \qquad [12.6]$$

Dividing both sides of Equation [12.6] yields the Nernst equation:

$$E_{cell} = E^{\circ}_{cell} - \left(\frac{RT}{zF}\right)\ln Q \qquad [12.7]$$

Example 12.15: Determining the Concentration of Dissolved Aluminum When Aluminum Foil Contacts Dissolved Copper

What is the concentration of Al^{3+} (in M) when Rxn [12.E1] takes place if a cell potential of 1.98 V is observed at room temperature with a concentration of 1.00×10^{-2} M Cu^{2+}?

Solutions

Express the reaction quotient in terms of activities for Rxn [12.E1] as:

$$Q = \frac{a_{Al^{3+}(aq)}^{2}\, a_{Cu(s)}^{3}}{a_{Al(s)}^{2}\, a_{Cu^{2+}(aq)}^{3}} = \frac{\left(\frac{[Al^{3+}]}{C^\circ}\right)^2 (1)^3}{(1)^2\left(\frac{[Cu^{2+}]}{C^\circ}\right)^3} = \frac{C^\circ[Al^{3+}]^2}{[Cu^{2+}]^3}$$

Apply the Nernst equation, Equation [12.7], with $E_{cell} = +1.98\text{ V}$, $E^\circ_{cell} = +2.015\text{ V}$ (from Example 12.7), $z = 6$ e (from Example 12.2), $T = 298.15$ K, and $[Cu^{2+}] = 1.00\times10^{-2}$ M and rearrange to find:

$$E_{cell} = E^\circ_{cell} - \left(\frac{RT}{zF}\right)\ln Q = E^\circ_{cell} - \left(\frac{RT}{zF}\right)\ln\left[\frac{C^\circ[Al^{3+}]^2}{[Cu^{2+}]^3}\right]$$

$$\left(\frac{RT}{zF}\right)\ln\left[\frac{C^\circ[Al^{3+}]^2}{[Cu^{2+}]^3}\right] = E^\circ_{cell} - E_{cell}$$

$$\ln\left[\frac{C^\circ[Al^{3+}]^2}{[Cu^{2+}]^3}\right] = \frac{zF(E^\circ_{cell} - E_{cell})}{RT}$$

$$\frac{C^\circ[Al^{3+}]^2}{[Cu^{2+}]^3} = \exp\left[\frac{zF(E^\circ_{cell} - E_{cell})}{RT}\right]$$

$$[Al^{3+}] = \sqrt{\frac{[Cu^{2+}]^3}{C^\circ}\exp\left[\frac{zF(E^\circ_{cell} - E_{cell})}{RT}\right]}$$

$$= \left[\frac{(1.00\times10^{-2}\text{ M})^3}{(1\text{ M})}\exp\left[\frac{(6)(96,485\text{ C/mol})[(+2.015\text{ J/C}) - (1.98\text{ J/C})]}{(8.314\text{ J/mol - K})(298\text{ K})}\right]\right]^{1/2}$$

$$[Al^{3+}] = 5.97\times10^{-2}\text{ M}$$

Example 12.16: The Cell Potential for an Air/Aluminum Battery

An air/aluminum battery can be constructed using aluminum foil, crushed activated carbon, a concentrated aqueous solution of sodium chloride, a paper towel, a paperclip, and a pair of banana clips (Tamez and Yu, *J. Chem. Educ.* 84, 1936A, 2007). This battery operates based on the following redox reaction:

$$4\,Al(s) + 3\,O_2(g) + 6\,H_2O(l) \rightarrow 4\,Al(OH)_3(s) \qquad \text{Rxn [12.E7]}$$

with $z = 12\ e^-$ and a standard cell potential of $E^\circ_{cell} = +2.076\ V$. What is the cell potential (in V) when this battery is operated using air at room temperature?

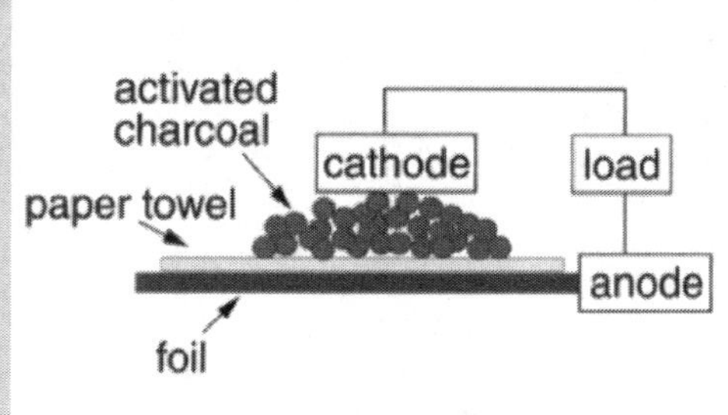

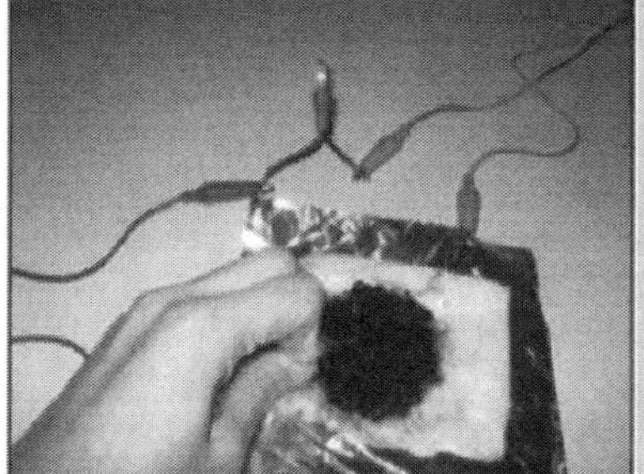

Solutions

Apply the Nernst equation with $P_{O_2} = 0.21$ bar, based on air as 21 mol-% oxygen:

$$Q = \frac{a_{Al(OH)_3(s)}^4}{a_{Al(s)}^4\, a_{O_2(g)}^3\, a_{H_2O(l)}^6} = \frac{(1)^4}{(1)^4\left(\frac{P_{O_2}}{P^\circ}\right)^3 (1)^6} = \left(\frac{P^\circ}{P_{O_2}}\right)^3$$

$$E_{cell} = E^\circ_{cell} - \left(\frac{RT}{zF}\right)\ln Q = E^\circ_{cell} - \left(\frac{RT}{zF}\right)\ln\left[\left(\frac{P^\circ}{P_{O_2}}\right)^3\right]$$

$$= (+2.076\ V) - \frac{(8.314\ J/mol-K)(298.15\ K)}{(12)(96,485\ C/mol)}\ln\left[\left[\frac{(1\ bar)}{(0.21\ bar)}\right]^3\right]$$

$$E_{cell} = +2.1\ V$$

This operating voltage is indistinguishable from the standard-state value.

It can be shown that the maximum possible electrical work that can be done by an electrochemical system, w_e, can be expressed either as the change in Gibbs free energy for the reaction or the negative of the product of the cell potential and the total charge transferred, $q_{e^-,total}$:

$$w_e = \Delta G_{rxn} = -E_{cell}\, q_{e^-,total} \qquad [12.8]$$

This relationship indicates that we can interpret the change in Gibbs free energy for a redox reaction physically as the maximum amount of work possible from the reaction. We can also interpret $q_{e^-,total}$ as the total charge stored in an electrochemical system. For car batteries we frequently express the total charge stored in terms of "cold cranking amps" (CCA), where 1 CCA is equal to 1 A-hr.

Example 12.17: Sizing a Car Battery

Consider a 12-V lead acid car battery (created by linking 6 2.02-V cells in series) rated 1000 CCA.

a) How much electrical work (in kJ) can be done by this battery?

b) What is the mass (in kg) of this battery?

Solutions

a) First, convert CCA to C:

$$q_{e^-,total} = 1000 \frac{\text{C}}{\text{s}} - \text{hr} \times \frac{60 \text{ s}}{1 \text{ min}} \times \frac{60 \text{ min}}{1 \text{ hr}} = 3.6 \times 10^6 \text{ C}$$

The maximum electrical work can then be determined as:

$$w_e = -E_{cell} q_{e^-,total} = -(12 \text{ J/C})(3.6 \times 10^6 \text{ C}) \times \frac{1 \text{ kJ}}{10^3 \text{ J}} = -4.3 \times 10^5 \text{ kJ}$$

b) A rough estimate for the mass of the battery is the mass of $PbSO_4$ when the battery has fully discharged. We can use the Faraday constant and stoichiometry to covert the total charge stored to its equivalent mass of $PbSO_4$:

$$3.6 \times 10^6 \text{ C} \times \frac{1 \text{ mol } e^-}{96{,}485 \text{ C}} \times \frac{2 \text{ mol PbSO}_4}{2 \text{ mol } e^-} \times \frac{303 \text{ g PbSO}_4}{1 \text{ mol PbSO}_4} \times \frac{1 \text{ kg}}{10^3 \text{ g}} \cong 11 \text{ kg}$$

This mass is equivalent to about 25 lb_m. The large molar mass of lead and lead compounds means that this material has a low energy density; however, the lead acid battery is very efficient, with greater than 90% of the stored charge available for discharge because of minimal side reactions.

12.7 Faraday's Laws

Observations by the father of electrochemistry, the British scientist Michael Faraday, in the 1830s offer a framework to generalize relationships between charge transferred and changes in mass. Faraday showed that the mass of a substance *i* produced or consumed at an electrode, Δm_i, is directly proportional to both the total charge passing through the electrode (*i.e.*, $\Delta m_i \alpha It$, where *t* is the time over which a current *I* flows) and the ratio of molar mass of the substance to the number of electrons transferred (*i.e.*, $\Delta m_i \alpha \frac{M_i}{z}$). We combine these observations to obtain Faraday's laws:

$$|\Delta m_i| = \frac{\upsilon_i M_i It}{zF} \qquad [12.9]$$

where υ_i is the stoichiometric coefficient for species i. We find Δm_i is positive if the substance is produced and negative if the substance is consumed. Faraday's laws can be applied to a range of applications, including determining the deposition of metals in an electroplating processes, evaluating the amount of metal purified in electrorefining, and characterizing the electrolysis of water.

Example 12.18: Electroplating Copper

How many hours must a 20.00 A current be applied to electroplate 1.00 kg of Cu(*s*) from a $CuSO_4$(*aq*) solution?

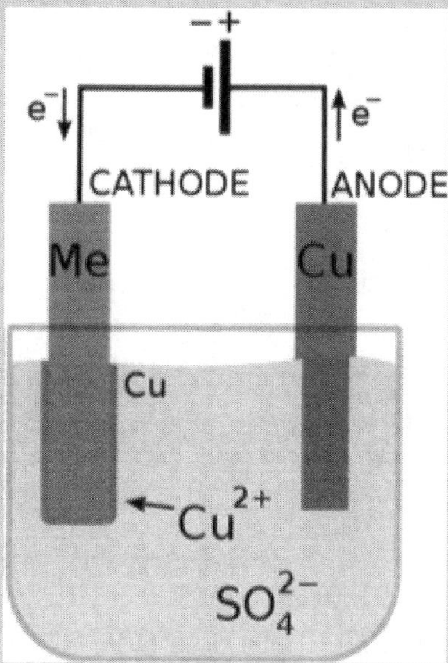

Solutions

Electroplating metallic copper from aqueous Cu^{2+}(*aq*) involves the reduction half-reaction $Cu^{2+} + 2\,e \rightarrow Cu$, such that $\upsilon_{Cu} = 1$ with $z = 2$. With a molar mass of 63.546 g/mol for Cu we obtain, after rearranging Faraday's laws:

$$t = \frac{zF\Delta m_{Cu}}{\upsilon_{Cu} M_{Cu} I} = \frac{(2)(96{,}485\text{ C})(1.00\times 10^3\text{ g})}{(1)(63.546\text{ g/mol})(20.00\text{ C/s})} = 1.52\times 10^5\text{ s}\times\frac{1\text{ min}}{60\text{ s}}\times\frac{1\text{ hr}}{60\text{ min}} = 42.2\text{ hr}$$

Example 12.19: The Electrolysis of Water

Consider the electrolysis of water, described by the following balanced redox reaction:

$H_2O(l) \rightarrow H_2(g) + \frac{1}{2} O_2(g)$ Rxn [11.1]

The following half-reactions (with associated standard reduction potentials, E^o, at 298 K) are <u>relevant</u> for an electrolyzer operated under <u>acidic</u> conditions:

$O_2 + 4\,H_3O^+ + 4\,e^- \rightarrow 6\,H_2O$ $E^\circ = +1.2288$ V Half-rxn [12.12]

$2\,H_3O^+ + 2\,e^- \rightarrow H_2 + 2\,H_2O$ $E^o = 0.0000$ V Half-rxn [12.13]

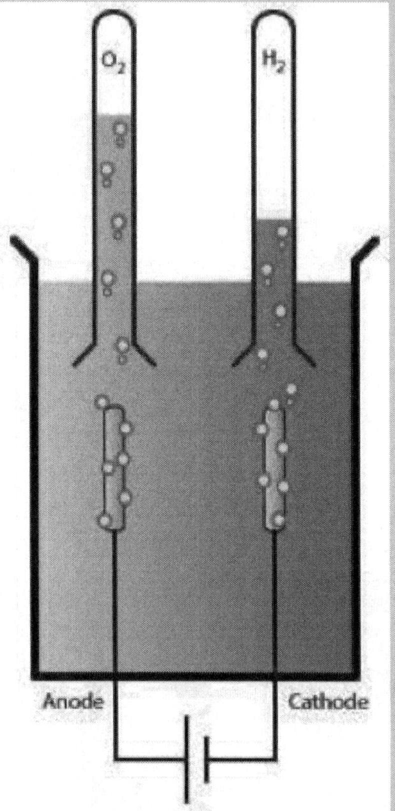

a) Identify the half-reaction occurring at the anode and the half-reaction occurring at the cathode.

b) Determine the number of electrons transferred per mole of reaction, z, for the balanced reaction.

c) What is the standard cell potential, E^o_{cell} (in V), at room temperature for the electrolyzer?

d) What is the thermodynamic efficiency, evaluated as $\eta = \frac{\Delta g^o_{rxn}}{\Delta h^o_{rxn}}$, for the electrolyzer operating at 298 K if $\Delta h^o_{rxn} = 285.8$ kJ/mol?

e) Predict the volume of $H_2(g)$ (in mL) generated at 298 K and 1 atm after 3.0 min with a 1.2 A current based on Faraday's laws and the ideal gas law as:

$$\Delta V_{H_2} = \frac{RT\Delta n_{H_2}}{P} = \frac{RT\Delta m_{H_2}}{PM_{H_2}} = \frac{RT}{PM_{H_2}}\left(\frac{\upsilon_{H_2}M_{H_2}It}{zF}\right) = \frac{\upsilon_{H_2}RTIt}{zFP} \quad [12.E1]$$

Solutions

a) In the overall redox reaction, water is both the reductant (the oxidation state of oxygen increases from -2 in water to 0 in molecular oxygen) and the oxidant (the oxidation state of hydrogen decreases from +1 in water to 0 in molecular hydrogen), such that the oxidation half-reaction that occurs at the anode is given as:

$$6\ H_2O \rightarrow O_2 + 4\ H_3O^+ + 4\ e^- \qquad \text{Half-rxn [12.E1]}$$

and the reduction half-reaction that occurs at the cathode is Half-rxn [12.13].

b) Combine the half-reactions to obtain the overall balanced redox reaction with no net change in the number of electrons by dividing the oxidation half-reaction by 2×, yielding:

Oxidation half-rxn: $\frac{1}{2} \times [6\ H_2O \rightarrow O_2 + 4\ H_3O^+ + 4\ e^-]$

+ Reduction half-rxn: $2\ H_3O^+ + 2\ e^- \rightarrow H_2 + 2\ H_2O$

$H_2O(l) \rightarrow H_2(g) + \frac{1}{2}\ O_2(g)$

There are $z = 2\ e^-$ transferred.

c) The standard cell potential is given as:

$$E^\circ_{cell} \equiv E^\circ_{cathode} - E^\circ_{anode} = (0.0000\ \text{V}) - (+1.2288\ \text{V}) = -1.2288\ \text{V}$$

The negative value for the standard cell potential means that the electrolysis of water is non-spontaneous under standard-state conditions, such that for this reaction to occur a voltage of at least +1.2288 V must be applied. In practice, the observed decomposition voltage, V_d, that must be applied exceeds +1.2288 V because of kinetic limitations.

d) First, determine the change in Gibbs free energy for the reaction at standard state based on the standard cell potential:

$$\Delta g^\circ_{rxn} = -zFE^\circ_{cell} = -(2)(96{,}485\ \text{C/mol})(-1.2288\ \text{J/C}) = 2.3712 \times 10^5\ \text{J/mol} \times \frac{1\ \text{kJ}}{10^3\ \text{J}}$$

$$\Delta g^\circ_{rxn} = 237.12\ \text{kJ/mol}$$

such that the efficiency is:

$$\eta = \frac{\Delta g^\circ_{rxn}}{\Delta h^\circ_{rxn}} = \frac{(237.12\ \text{kJ/mol})}{(285.8\ \text{kJ/mol})} = 0.8297 = 82.97\%$$

Note that this efficiency greatly exceeds the efficiency of heat engines ($\eta \sim 0.2\text{-}0.3$), as discussed in Chapter 11.

e) Apply the equation based on Faraday's laws and the ideal gas law, using the gas constant expressed in units of atm-L/mol-K (because pressure is expressed in units of atm and volumes in units of mL):

$$\Delta V_{H_2} = \frac{\upsilon_H RTIt}{zFP} = \frac{(1)(8.206 \times 10^{-2}\ \text{atm} - \text{L/mol} - \text{K})(298\ \text{K})(1.2\ \text{C/s})\left(3.0\ \text{min} \times \frac{60\ \text{s}}{1\ \text{min}}\right)}{(2)(96{,}485\ \text{C/mol})(1\ \text{atm})}$$

$$\Delta V_{H_2} = 2.8 \times 10^{-2}\ \text{L} \times \frac{10^3\ \text{mL}}{1\ \text{L}} = 28\ \text{mL}$$

Appendix A : Using the TI Equation Solver

Some mathematical descriptions for problems in science and engineering cannot be solved analytically but rather require a numerical approach. For example, cubic equations frequently cannot be factored for determining roots, and more complex problems involve forms for which we cannot solve explicitly for the unknown value for a variable. The TI-83 Plus (or better) calculator, however, has the capability to solve such problems numerically, as does MS Excel using the Solver tool/application. The guided example below illustrates step-by-step procedures for finding the roots of a quadratic equation using a TI-83 Plus calculator.

Guided example: Find the Roots for the Equation $x^2 - x - 2 = 0$

Press:	Displayed Result:
MATH Δ	MATH NUM CPX PRB 4↑$^3\sqrt{}$ 5: $^x\sqrt{}$ 6:fMin(7:fMax(8:nDeriv(9:fnInt(0:Solver...
ENTER	EQUATION SOLVER eqn:0=
Note: If you do not see eqn:0=, press Δ CLEAR to erase the existing equation	
X,T,Θ,*n* x^2 – X,T,Q,n – 2	EQUATION SOLVER eqn:0=X^2-X-2
ENTER	X^2-X-2=0 X=0 Bound={-1E99,1...
ALPHA SOLVE	X^2-X-2=0 ■ X=-1.000000000... Bound={-1E99,1... ■ left-rt=0
SOLVE is ALPHA function (in green) for ENTER	

5	X^2-X-2=0 X=5 Bound={-1E99,1… Left-rt=0
ALPHA SOLVE	X^2-X-2=0 ■ X=1.9999999999… Bound={-1E99,1… ■ left-rt=0
2nd QUIT	
QUIT is 2nd function (in yellow) for MODE	
We find values for the roots as x = -1 and x = 2.	

Appendix B : Table of Standard Reduction Potentials at 298.15 K

Reduction Half-Reaction	$E°$ (V)
$Li^+ + e^- \rightarrow Li$	-3.0401
$K^+ + e^- \rightarrow K$	-2.931
$Ca^{2+} + 2\ e^- \rightarrow Ca$	-2.868
$Na^+ + e^- \rightarrow Na$	-2.7144
$Mg^{2+} + 2\ e^- \rightarrow Mg$	-2.3568
$Al(OH)_3 + 3\ e^- \rightarrow Al + 3\ OH^-$	-2.31
$Al^{3+} + 3\ e^- \rightarrow Al$	-1.676
$OCl^- + H_2O + 2\ e^- \rightarrow Cl^- + 2\ OH^-$	-0.890
$2\ H_2O + 2\ e^- \rightarrow H_2 + 2\ OH^-$	-0.828
$Cd(OH)_2 + 2\ e^- \rightarrow Cd + 2\ OH^-$	-0.81
$Zn^{2+} + 2\ e^- \rightarrow Zn$	-0.7621
$Cr^{3+}\ 3\ e^- \rightarrow Cr$	-0.744
$Fe^{2+} + 2\ e^- \rightarrow Fe$	-0.440
$Cd^{2+} + e^- \rightarrow Cd$	-0.4022
$Tl^+ + e^- \rightarrow Tl$	-0.3358
$Co^{2+} + 2\ e^- \rightarrow Co$	-0.280
$PbSO_4 + 2\ H_3O^+ + 2\ e^- \rightarrow Pb + H_2SO_4 + 2\ H_2O$	-0.28
$Ni^{2+} + 2\ e^- \rightarrow Ni$	-0.257
$Sn^{2+} + 2\ e^- \rightarrow Sn$	-0.1410
$Pb^{2+} + 2\ e^- \rightarrow Pb$	-0.1266
Standard hydrogen (SHE): $2\ H_3O^+ + 2\ e^- \rightarrow H_2 + 2\ H_2O$	0.0000
$CO_2 + 6\ H_3O^+ + 6\ e^- \rightarrow CH_3OH + 7\ H_2O$	+0.02
$CO_2 + 12\ H_3O^+ + 12\ e^- \rightarrow CH_3CH_2OH + 15\ H_2O$	+0.079
$HgO + 2\ H_2O + 2\ e^- \rightarrow Hg + 2\ OH^-$	+0.0977

Sn^{4+} + 2 *e*– → Sn^{2+}	+0.1539
SO_4^{2}– + 4 H_3O^+ + 2 *e*– → 5 H_2O + H_2SO_3(*aq*)	+0.1576
Cu^{2+} + *e*– → Cu^+	+0.1607
S + 2 H_3O^+ + 2 *e*– → H_2S + 2H_2O	+0.1739
AgCl + *e*– → Ag + Cl–	+0.2221
Saturated Calomel (SCE): Hg_2Cl_2 → 2 Hg + 2 Cl–, sat'd KCl	+0.2412
Hg_2Cl_2 + 2 *e*– → 2Cl– + 2 Hg	+0.2680
Bi^{3+} + 3 *e*– → Bi	+0.286
Cu^{2+} + 2 *e*– → Cu	+0.3394
$Fe(CN)_6^{3-}$ + *e*– → $Fe(CN)_6^{4}$–	+0.3557
O_2 + 2 H_2O + 4 *e*– → 4 OH–	+0.401
NiO(OH) + H_2O + *e*– → $Ni(OH)_2$ + OH–	+0.49
Cu^+ + *e*– → Cu	+0.5180
I_2(*s*) + 2 *e*– → 2 I–	+0.5345
I_3– + 2 e– → 3 I–	+0.5354
2 $HgCl_2$ + 2 *e*– → Hg_2Cl_2 + 2 Cl–	+0.6011
I_2(*aq*) + 2 *e*– → 2 I–	+0.6195
O_2 + 2 H_3O^+ + 2 *e*– → H_2O_2 + 2 H_2O	+0.6237
Fe^{3+} + *e*– → Fe^{2+}	+0.769
2 MnO_2 + H_2O + 2 *e*– → Mn_2O_3 + 2 OH–	+0.78
Hg_2^{2+} + 2 *e*– → 2Hg	+0.7955
Ag^+ + *e*– → Ag	+0.7991
Hg^{2+} + 2 *e*– → Hg	+0.8519
2 Hg^{2+} + 2 *e*– → Hg_2^{2+}	+0.9083
NO_3– + 3 H_3O^+ + 2 *e*– → HNO_2 + 4 H_2O	+0.9275
NO_3– + 4 H_3O^+ + 3 *e*– → NO + 6 H_2O	+0.956
HNO_2 + H_3O^+ + *e*– → NO + 2 H_2O	+1.0362
Br_2 + 2 *e*– → 2 Br–	+1.0775
MnO_2 + 4 H_3O^+ + 2 *e*– → Mn^{2+} + 6 H_2O	+1.1406
Pt^{2+} + 2 *e*– → Pt	+1.18
2 IO_3– + 12 H_3O^+ + 10 *e*– → 18 H_2O + I_2	+1.2093
O_2 + 4 H_3O^+ + 4 e– → 6 H_2O	+1.2288
$Cr_2O_7^{2}$– + 14 H_3O^+ + 6 *e*– → 2 Cr^{3+} + 21 H_2O	+1.232

$O_3 + H_2O + e- \rightarrow O_2 + 2\ OH-$	+1.246
$Cl_2 + 2\ e- \rightarrow 2Cl-$	+1.3601
$MnO_4- + 8\ H_3O^+ + 5\ e- \rightarrow 12\ H_2O + Mn^{2+}$	+1.5119
$Au + e- \rightarrow Au-$	+1.692
$PbO_2 + H_2SO_4 + 2\ H_3O^+ + 2\ e- \rightarrow PbSO_4 + 4\ H_2O$	+1.74
$Ce^{4+} + e- \rightarrow Ce^{3+}$	+1.7432
$H_2O_2 + 2\ H_3O^+ + 2\ e- \rightarrow 4\ H2O$	+1.776
$F_2 + 2\ e- \rightarrow 2\ F-$	+2.866

Index

A

B

C

J

L

M

N

O

P

Q

R

S

T

U

V

W

Important Physical Constants

Avogadro's number	$N_A = 6.022\times10^{23}$ mol^{-1}
Gas constant	$R = 8.314$ J/mol-K = $8.206\ 10^{-2}$ atm-L/mol-K
	= $8.314\ 10^{-2}$ bar-L/mol-K
Charge of an electron	$q_{e^-} = -1.609\ 10^{-19}$ C
Rest mass of an electron	$m_{e^-} = 9.11\ 10^{-31}$ kg
Speed of light	$c = 3.00\ 10^{8}$ m/s
Planck's constant	$h = 6.626\ 10^{-34}$ J-s
Rydberg constant	$R_H = 1.097\ 10^{5}$ cm^{-1}
Dielectric permittivity of a vacuum	$\varepsilon_o = 8.85\times10^{-12}$ C^2/J-m
Boltzmann constant	$k_B = \dfrac{R}{N_A} = 1.38\times10^{-23}$ J/K
Faraday constant	$F = 96{,}485$ C/mol

Special Conversion Factors

1 amu = 1.6605×10^{-27} kg

1 yard = 3 ft = 36 in

2.54 cm = 1 in

1 Å = 10^{-10} m

1 lb_m = 0.4536 kg

1 lb_f = 4.448 N

1 gal = 3.7854 L

1 atm = $1.013\ 10^{5}$ Pa = 1.013 bar = 760 mm Hg (torr) = 33.9 ft H_2O = 14.7 psi

$$T(^{\circ}\mathrm{F}) = \frac{9}{5}T(^{\circ}\mathrm{C}) + 32$$

1 mol = 22.4 L at STP (1 bar & 0 °C)

1 eV = 1.602×10^{-19} J

1 D = 3.336×10^{-30} C-m

1 cal = 4.184 J

1 BTU = 1055 J

1 hp = 746 W

1 V = 1 J/C

1 A = 1 C/s

PERIODIC TABLE OF THE ELEMENTS

1 1A	2 2A	3 3B	4 4B	5 5B	6 6B	7 7B	8	9 8B	10	11 1B	12 2B	13 3A	14 4A	15 5A	16 6A	17 7A	18 8A
1 **H** 1.0079																	2 **He** 4.003
3 **Li** 6.941	4 **Be** 9.012											5 **B** 10.811	6 **C** 12.011	7 **N** 14.007	8 **O** 15.999	9 **F** 18.998	10 **Ne** 20.180
11 **Na** 22.990	12 Mg 24.305											13 **Al** 26.982	14 **Si** 28.086	15 **P** 30.974	16 **S** 32.066	17 **Cl** 35.453	18 **Ar** 39.948
19 **K** 39.098	20 **Ca** 40.08	21 **Sc** 44.956	22 **Ti** 47.88	23 **V** 50.942	24 **Cr** 51.996	25 **Mn** 54.938	26 **Fe** 55.847	27 **Co** 58.933	28 **Ni** 58.69	29 **Cu** 63.546	30 **Zn** 65.39	31 **Ga** 69.723	32 **Ge** 72.61	33 **As** 74.922	34 **Se** 78.96	35 **Br** 79.904	36 **Kr** 83.80
37 **Rb** 85.47	38 **Sr** 87.62	39 **Y** 88.906	40 **Zr** 91.224	41 **Nb** 92.906	42 **Mo** 95.95	43 **Tc** (98)	44 **Ru** 101.07	45 **Rh** 102.91	46 **Pd** 106.42	47 **Ag** 107.87	48 **Cd** 112.41	49 **In** 114.82	50 **Sn** 118.17	51 **Sb** 121.75	52 **Te** 127.60	53 **I** 126.90	54 **Xe** 131.29
55 **Cs** 132.90	56 **Ba** 137.33	57 **La** 138.91	72 **Hf** 178.49	73 **Ta** 180.95	74 **W** 183.85	75 **Re** 186.21	76 **Os** 190.2	77 **Ir** 192.22	78 **Pt** 195.08	79 **Au** 196.97	80 **Hg** 200.59	81 **Tl** 204.38	82 **Pb** 207.2	83 **Bi** 208.98	84 **Po** (209)	85 **At** (210)	86 **Rn** (222)
87 **Fr** (223)	88 **Ra** (226)	89 **Ac** 227.03	104 **Rf** (261)	105 **Db** (262)	106 **Sg** (263)	107 **Bh** (262)	108 **Hs** (265)	109 **Mt** (266)	110 (269)	111 (272)	112 (272)		114 (287)		116 (289)		118 (293)

Lanthanide Series	58 **Ce** 140.12	59 **Pr** 140.91	60 **Nd** 144.24	61 **Pm** (145)	62 **Sm** 150.36	63 **Eu** 151.96	64 **Gd** 157.25	65 **Tb** 158.92	66 **Dy** 162.50	67 **Ho** 164.93	68 **Er** 167.26	69 **Tm** 168.93	70 **Yb** 173.04	71 **Lu** 174.97
Actinide Series	90 **Th** 232.04	91 **Pa** 231	92 **U** 238.03	93 **Np** 237.05	94 **Pu** (244)	95 **Am** (243)	96 **Cm** (247)	97 **Bk** (247)	98 **Cf** (251)	99 **Es** (252)	100 **Fm** (257)	101 **Md** (258)	102 **No** (259)	103 **Lr** (260)